Modern Optics Simplified

Modern Optics Simplified

B. D. Guenther

OXFORD
UNIVERSITY PRESS

Great Clarendon Street, Oxford, OX2 6DP,
United Kingdom

Oxford University Press is a department of the University of Oxford.
It furthers the University's objective of excellence in research, scholarship,
and education by publishing worldwide. Oxford is a registered trade mark of
Oxford University Press in the UK and in certain other countries

First Edition published in 2020

Impression: 1

Published in the United States of America by Oxford University Press
198 Madison Avenue, New York, NY 10016, United States of America

British Library Cataloguing in Publication Data
Data available

Library of Congress Control Number: 2019945064

ISBN 978–0–19–884285–9 (hbk.)
ISBN 978–0–19–884286–6 (pbk.)

DOI: 10.1093/oso/9780198842859.001.0001

Printed and bound by
CPI Group (UK) Ltd, Croydon, CR0 4YY

Contents

Introduction

All the studies of light from the antiquity until the middle of nineteenth century were based on incoherent light sources such as the Sun, candlelight, sodium lamp, or light bulb. In 1950s a new coherent source of light was invented, first in the microwave region and then in the optical region. This new kind of light source, the laser, is one of the greatest inventions of the second part of the twentieth century. It has helped to revolutionize many branches of science and technology, ranging from biotechnology and precision measurements to communication and remote sensing. A Nobel Prize was awarded in 1964 for the invention of the laser and since then some 10 additional prizes have been awarded involving this technology.

The physical process behind conventional light sources is spontaneous emission and the source operates in thermal equilibrium. Initially, the majority of atoms and molecules are in their ground state. When energy is supplied to the atoms or molecules, some of them go to the excited states and then radiate via spontaneous emission. The spontaneous emission process is due to the ubiquitous vacuum fluctuations[1] and each atom radiates independently of each other. The resulting light is a white light sent in all directions and is incoherent. On the other hand, the dominant emission process in a laser is stimulated emission. By a clever design, the radiated photons by the atoms or molecules are able to stimulate other atoms to radiate with the same frequency and same direction. The resulting radiation is coherent, i.e., monochromatic, and highly directional.

In this text we want to look at optical science using the fact that the encounter of coherent light is probable rather than improbable. We will view light as an electromagnetic wave described completely by Maxwell's equations. This is a set of vector equations but we will do our best to limit our discussion to scalar, one-dimensional waves. We only will use vectors where we must to characterize the properties of a light wave such as the fact that solutions of Maxwell's equations are solutions of a wave equation concealed within Maxwell's equations. The amplitude of the electric and magnetic waves that make up a light wave are orthogonal to the direction of propagation and to each other. We will introduce the method used to handle the vector nature of the wave's displacement vectors using the theory of polarization.

We must also use vectors to discuss reflection and refraction since the magnitude of the wave that undergoes reflection depends on the orientation of the electric field displacement. We are able to prove that interference requires that the polarization of the waves participating in interference must be parallel. This allows us to limit discussions to scalar waves.

Interference and coherence are treated together. These are the keys to our modern view of optics based on Fourier theory. Because Fourier theory is the basis for our modern description of diffraction and imaging, we spend a little time describing the theory but we limit our discussion to very simple functions such as a rectangle.

---

[1] This is the temporary change in the energy at a point in space.

There are some topics discussed in the book that can be skipped because of their advanced nature. They have been indicated by teal boxes setting off the paragraphs containing the advanced material. The instructor can decide, as the class progresses, whether to include the material.

Interference is only observed if the interfering waves are coherent and if the polarization of the waves is parallel. The presence of coherence allows a number of useful experimental techniques to be used, such as the quality of optical components using Newton's rings or a Michelson interferometer; use of a Michelson interferometer and Fourier transforms makes it possible to evaluate the spectrum of a source. By using a source with a known spectral distribution, it is possible to create an optical coherence tomographic three-dimensional image of biological material, such as the retina or the cornea of the human eye. An interferometer based on Young's two-slit experiment can be used to measure the size of stellar objects. Finally, interference generated by dielectric layers can be used to reduce the reflections from the surface of optical components for complex camera lenses or high index eyeglasses.

A number of different theories are based on wave theory but most of them assume a scalar wave; i.e., the amplitude of the wave is a scalar function. Light, however, is an electromagnetic wave with the wave displacement described by a magnetic and an electric vector. We can describe the vector wave in terms of only the electric field vector. In this book we outline the math used to describe the vector nature of the wave in terms of its polarization and discuss optical devices used to experimentally manipulate the polarization vector. The polarization of light is affected by chemical compounds possessing optical activity; the chiral property and a brief introduction to that property is given.

In many optics textbooks, emphasis is placed on the thin lens equation, and simple optical systems involve often no more than one element. This is of little use to the student since single optical elements are dominated by aberrations and in the lab a combination of at least two lenses would be preferred to eliminate spherical aberration. We have introduced the student to the math needed to handle multiple optical components and selected a sign convention based on that used in a Cartesian coordinate system that does not result in utter confusion when multiple elements are treated. Actual optical design is complicated and hard to do well. For that reason, aberrations and their examples are only described and experimentally obtained examples shown.

Triggered by the development of lasers, a new method of communications based on the use of optical fibers was developed in the 1970s. When a propagating light wave encounters a barrier between a high index of refraction material and a material with low index of refraction, we can observe total reflection. If a light wave is confined to the high index dielectric medium by the total reflection, the light can propagate long distances without attenuation and can be made to carry large quantities of information. Beginning about 2010 the ability to conduct communications without conversion from the optical regime made it possible to stream movies and sporting events, causing a rapid expansion of optical fiber technology. We explore the description of fiber optical systems based on a geometrical model called the zig-zag theory.

Because it is easy to construct coherent light sources it is easy to see the interference of multiple waves scattered from complicated structures. This has led to the invention of the previously mentioned optical coherence tomography microscope of retinal imaging. This is an extension of the concept of interference into a much more complicated theory which is only mentioned in passing in this text. We have found it possible to derive simple mathematical expressions that make it possible to calculate diffraction from simple geometric constructs. The theory allows us to build a theory that can be used to predict limits to our imaging capability. The theory makes it possible to compare the capabilities of a variety of imaging systems, such as cell phone cameras. It also allows us to develop way of processing imagery that permits, for example, the extraction of the image of a planet orbiting a distant star.

1 Waves

1.1 Wave Parameters

The basic character of light is explained using the theory of wave motion. Mathematically, a wave is the solution of a second-order, partial differential equation called the wave equation. The solution is viewed as an oscillation that transfers energy from one point to another without any mass flow. In one dimension, the equation of a scalar wave propagating without change in the x-direction has the form

$$\left(\frac{\partial^2 y}{\partial x^2}\right) = \frac{1}{c^2} \cdot \frac{\partial^2 y}{\partial t^2}, \tag{1.1}$$

where c is the velocity of the wave. This differential equation can be used to describe a number of physical situations: a vibrating guitar string, sound waves, a vibrating drum head, elastic waves in solids such as seismic waves in the earth, ocean waves, electric signals in a cable, and electromagnetic waves—viz. a light wave that is of interest to us.

The idea that sound was a wave came from observing water waves and was introduced around 240 BC by the Greek philosopher Chrysippus. Pythagoras, the ancient Greek philosopher and mathematician, studied the vibration of strings around 500 BC. The connection between the production of sound by vibrating bells or strings and sound waves in air was made by Galileo. Gassendi, a contemporary of Galileo, argued for a particle view of the propagation of sound.

Daniel Bernoulli, a Swiss mathematician, tried to develop an understanding of how a violin string made sound. In the 1720s, he developed a way of describing a string as it vibrated by imagining the string as composed of tiny masses, connected by springs. Applying Isaac Newton's laws of motion for the individual masses allowed Bernoulli to predict that the simplest shape of a vibrating violin string, fixed at each end, was a single sine curve. Some decades later, mathematician Jean Le Rond d'Alembert generalized the string problem into a mathematical expression of the wave equation. He found that the acceleration of any segment of the string was proportional to the tension acting on it. The waves created by different tensions of the string produce different notes. With this theoretical development, the design of string instruments became possible.

The success of the treatment of sound waves suggested to scientists of the early seventeenth century that they should view the universe as a large mechanical device and to apply a mechanical model to all observables, even light. Everything was made of matter and motion. Matter was made up of atoms that were indivisible. Corpuscles were larger particles with properties that controlled their function and they acted as vehicles, carrying stuff through

Modern Optics Simplified. B. D. Guenther. © B. D. Guenther 2020.
Published in 2020 by Oxford University Press. DOI: 10.1093/oso/9780198842859.001.0001

THE ELECTROMAGNETIC SPECTRUM

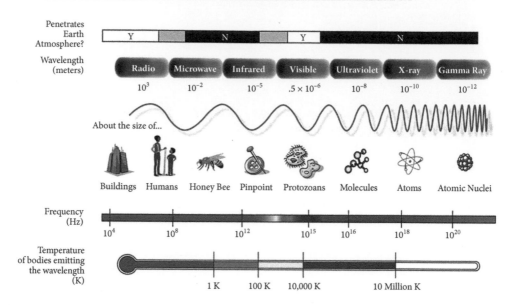

Figure 1.1 Frequency regions of the electromagnetic spectrum. The theories discussed in the text cover the frequencies from 10^{12} to 10^{16} Hz.

By NASA [Public domain], via Wikimedia Commons.

space. René Descartes described light as a pressure wave transmitting at an infinite speed through some type of elastic medium.

Christiaan Huygens in 1672 developed a mathematical wave theory of light. He retained a mechanical view of light by requiring the wave to travel in a *luminiferous ether*. Isaac Newton latched on to the particle (corpuscle) theory of Gassendi to build his particle theory of light. He rejected the concept of wave propagation because light seemed to travel in a straight line. The weakness of his theory was that it explained refraction by assuming that light traveled faster in a dense material, opposite to the predictions of Huygens's theory. Newton's theory was dominant throughout the eighteeenth century.

In 1800, Thomas Young demonstrated through diffraction experiments using two slits that light behaved as a wave. In 1817 Augustin-Jean Fresnel worked out a wave theory of light that helped cast Newton's theory in poor light. It was not until propagation velocity measurements made by Léon Foucault in 1850 demonstrated that the propagation velocity in a dense medium was less than in a vacuum. With this experimental evidence, Newton's particle theory was abandoned.

The need to explain light resulted in the development of a theory of electromagnetic waves by James Clerk Maxwell in 1862. Electromagnetic waves have frequencies that extend over many orders of magnitude (see Figure 1.1).

We will develop the parameters used to identify a wave in this chapter. We will discuss Maxwell's equations that explain the behavior of the electromagnetic spectrum of Figure 1.1 in the next chapter.

The wave must be a solution of Eq. (1.1); i.e., it is a function of two variables: a spatial variable and a temporal variable:

$$y \equiv f(x, t)$$

To make the problem as simple as possible, we will assume that a wave propagates in our medium (currently undefined) without change. If the wave propagates between x_1 and x_2, then the wave function evaluated at (x_1, t_1) must have the same value as at (x_2, t_2). The velocity of propagation along the x-direction in our medium is equal to c and the velocity allows us to calculate the new position of the wave from the old position using the equation

$$x_2 = x_1 + c(t_2 - t_1)$$

One way to satisfy the requirement of propagation without change is to assume the wave function has the form[1]

$$y = f(ct - x). \tag{1.2}$$

It is easy to show this functional form meets our requirement

$$f(x_1, t_1) = f(ct_1 - x_1)$$
$$f(x_2, t_2) = f(ct_2 - x_2) = f(ct_2 - x_1 - c[t_2 - t_1])$$
$$= f(ct_1 - x_1)$$

To our requirement of propagation without change, we add a second requirement that each point on the wave oscillate transversely, i.e., perpendicular to the direction of propagation, with simple harmonic motion. A good physical example of such a wave would be a guitar string vibrating at its fundamental frequency—for example, the fifth guitar string (A) vibrating at 110 Hz. The equations of motion of a point on the wave vibrating with simple harmonic motion is given by

$$m\frac{d^2x}{dt^2} + sx = 0. \tag{1.3}$$

The point only moves transversely to the wave motion. When the sign of the x and t components of the wave phase differ, the wave travels in the positive x-direction. The point on the wave, having been displaced, experiences a linear restoring force that defines s and is given by

$$F = -sx = ma = m\frac{d^2x}{dt^2}.$$

The constant s is called the spring constant.

It would appear that such a simple, second-order, differential equation would have limited applicability, but such is not the case. The force acting upon a mass can be written in terms of the potential energy function, V,

$$F = -\frac{dV}{dx}.$$

The potential function is used because it is a scalar and thus easier to manipulate than a vector. We see that our force $F = -sx$ is due to a potential energy function $V(x)$ that is

[1] The expression $y = f(ct - x)$ is shorthand notation to denote a function that contains x and t only in the combination $(ct - x)$; i.e., the function can contain combinations of the form $2(ct - x)$, $(t \pm x/c)$, $(x - ct)$, $(ct - x)^2$, $\sin(ct - x)$, etc., but not expressions such as $(2ct - x)$ or $(ct^2 - x^2)$.

proportional to x^2. If we have a more complicated potential function, we can expand the potential function, about the equilibrium point, in a Taylor series

$$V = V_0 + \frac{1}{2}\left(\frac{d^2V}{dx^2}\right)x^2 + \frac{1}{6}\left(\frac{d^3V}{dx^3}\right)x^3 + \ldots.$$

There is no term involving x, because $V(x)$ is a minimum at the equilibrium position and the derivative is zero. We see our model is applicable when the Taylor expansion beyond the first non-zero term is not needed, i.e., when the displacements, x, of the oscillations about the equilibrium position are small.

The solution of Equation Eq. (1.3) is

$$x = A\cos\left(\underbrace{\frac{\sqrt{\frac{s}{m}}t + \delta}{phase}}\right) \tag{1.4}$$

where δ is set by the initial conditions. If we assume that the amplitude of displacement is A at $t = 0$, then $\delta = 0$.

We can use this harmonic motion to define several wave parameters:

- The *period*, T, is the time required to complete one oscillation. The value of x at time t and $(t + T)$ must be equal; thus, the two phases differ by 2π:

$$\left[\sqrt{\frac{s}{m}}\,(t + T) + \delta\right] = \left[\sqrt{\frac{s}{m}}t + \delta\right] + 2\pi$$
$$T = 2\pi\sqrt{\frac{m}{s}}. \tag{1.5}$$

- The *frequency* of oscillation, i.e., the number of times x has the same value in a unit of time, is the reciprocal of the period

$$\nu = \frac{1}{T} = \frac{1}{2\pi}\sqrt{\frac{s}{m}} \tag{1.6}$$

We use the angular frequency to keep from continually writing 2π in our equations:

$$\omega = 2\pi\nu.$$

Using our assumptions of the properties of the wave's amplitude, we can write the function of $(ct - x)$, which will reduce to harmonic motion at $x = 0$, as

$$y = f(ct - x) = Y\cos\left[\frac{\omega}{c}(ct - x)\right].$$

This is called a *harmonic wave*.

We will now add another parameter that we will use to characterize a wave; it is called the *propagation constant* or the *wave number*:

$$k = \frac{\omega}{c}. \tag{1.7}$$

This is the spatial frequency of the wave and is measured in the number of waves per unit distance as compaired to the temporal frequency, ω, which is equal to the number of waves per unit time. The generalized harmonic wave can then be written

$$y = Y \cos (\omega t - kx) \tag{1.8}$$

The values of x, with t fixed, for which the phase $(\omega t - kx)$ changes by 2π is the *spatial period* and is called the *wavelength*, λ. Let $x_2 = x_1 + \lambda$, so that

$$\omega t - kx_2 = \omega t - kx_1 - k\lambda = \omega t - kx_1 - 2\pi;$$

Thus,

$$k = \frac{2\pi}{\lambda}. \tag{1.9}$$

The final parameter we would like to define for our wave is its propagation velocity. To determine the speed of the wave in space, a point on the wave is selected and the time it takes to go some distance is measured. This is equivalent to asking how fast a given value of phase propagates in space. Assume that in the time $\Delta t = (t_2 - t_1)$ the disturbance y_1 travels a distance $\Delta x = (x_2 - x_1)$. Since the disturbance at the two points is the same, i.e., y_1, then the phases must be equal:

$$\omega t - kx = \omega (t + \Delta t) - k (x + \Delta x),$$

$$\frac{\Delta x}{\Delta t} = \frac{\omega}{k}.$$

In the limit as $\Delta t \to 0$, we obtain the *phase velocity*

$$c \equiv \frac{dx}{dt} = \frac{\omega}{k};$$

since $k = \omega/c = 2\pi\nu/c$, we also have the relationship between wavelength, frequency, and propagation velocity, $c = \nu\lambda$.[2]

Dimensional Representation

We now have defined parameters that can be used to characterize an arbitrary wave but only in one dimension. We know that light travels in a three-dimensional space. What changes to the parameters will be necessary to describe the wave in three dimensions? The wave equation, generalized to three dimensions, becomes

$$\frac{\partial^2 f(\mathbf{r}, t)}{\partial x^2} + \frac{\partial^2 f(\mathbf{r}, t)}{\partial y^2} + \frac{\partial^2 f(\mathbf{r}, t)}{\partial z^2} = \frac{1}{c^2} \frac{\partial^2 f(\mathbf{r}, t)}{\partial t^2}. \tag{1.10}$$

---

[2] Another way to determine the phase velocity is to use a result from partial differential calculus

$$\left(\frac{\partial x}{\partial t}\right)_y = -\frac{\left(\frac{\partial y}{\partial t}\right)_x}{\left(\frac{\partial y}{\partial x}\right)_t} = \frac{\omega}{k}.$$

This relationship is useful even without knowledge of is origin.

Or using vector calculus notation, we can write Eq. (1.10) in terms of the del (nabla, ∇) operator:

$$\nabla^2 f(\mathbf{r}, t) = \frac{1}{c^2} \frac{\partial^2 f(\mathbf{r}, t)}{\partial t^2} \tag{1.11}$$

In Cartesian coordinates

$$(\nabla \cdot \nabla)f = \nabla^2 f = \Delta f = \frac{\partial^2 f}{\partial x^2} + \frac{\partial^2 f}{\partial y^2} + \frac{\partial^2 f}{\partial z^2}$$

is called the Laplace operator and is the divergence of a function's gradient. A generalized harmonic wave solution of Eq. (1.11) is

$$f(\mathbf{r}, t) = E(\mathbf{r}) \cos [\omega t - \mathcal{S}(\mathbf{r})]$$

where we have replaced kx in Eq. (1.8) by a surface described by $\mathcal{S}(\mathbf{r})$. This quality defines a surface over which $\mathcal{S}(\mathbf{r}) = $ constant at a fixed time. That surface is call the *wavefront*. If the surface defined by $\mathcal{S}(\mathbf{r})$ is a plane with a unit vector, $\hat{\mathbf{n}}$, normal to its surface (in three dimensions, the unit vector normal to a point in a plane is the definition of the plane), then

$$\mathbf{r} \cdot \hat{\mathbf{n}} = s$$

defines the plane located a distance $\mathbf{r}$ from the origin.

We will limit most of our discussions in this text to waves with a plane wave front (Figure 1.2). If $\mathcal{S}(\mathbf{r})$ describes a plane wave then it must be given by

$$\mathcal{S}(\mathbf{r}) = k(\mathbf{n} \cdot \mathbf{r})$$

We define the wave vector as

$$\mathbf{k} = k\mathbf{n}$$

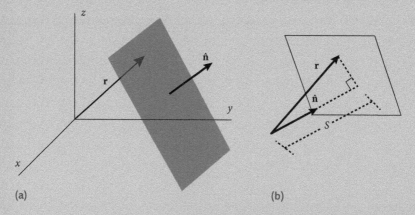

(a)   (b)

Figure 1.2 (a) A plane wave. Its normal is the unit vector **n** that points in the direction of propagation. A surface of constant phase is the shaded plane passing through the point defined by the vector **r**. (b) The projection of **r** on the plane's normal defines the distance s from the origin.

and the plane wave solutions of the wave equation in 3D become

$$f(\mathbf{r}, t) = E(\mathbf{r}) \cos(\omega t - \mathbf{k} \cdot \mathbf{r}) \tag{1.12}$$

The manipulation of sines and cosines places a demand on us to remember a variety of trigonometric identities. We can remove that unreasonable demand by using complex notation. The generalized solution of the wave equation can be expressed in complex notation, using Euler's theorem

$$f(\mathbf{r}, t) = E(r) [\cos(\omega t - \mathbf{k} \cdot \mathbf{r} + \delta) \pm i \sin(\omega t - \mathbf{k} \cdot \mathbf{r} + \delta)]$$
$$= E(r) e^{\pm i(\omega t - \mathbf{k} \cdot \mathbf{r})}. \tag{1.13}$$

The amplitude $\mathbf{E}(\mathbf{r})$ can be complex and contain an arbitral phase term.

If we substitute $f(\mathbf{r}, t)$ into the wave equation Eq. (1.11), we obtain

$$\frac{\partial^2 f(\mathbf{r}, t)}{\partial t^2} = -\omega^2 \mathbf{E}(\mathbf{r}) e^{i\omega t}$$

$$\nabla^2 f(\mathbf{r}, t) = e^{i\omega t} \nabla^2 \mathbf{E}(\mathbf{r})$$

Using the relationship $\omega/c = k$, the wave equation becomes

$$(\nabla^2 + k^2) \mathbf{E}(\mathbf{r}) = 0 \tag{1.14}$$

If we are interested in the spatial properties of the wave but not the temporal, we need only seek solutions of this equation, which is called the *Helmholtz equation*.

Use of the complex notation will simplify our calculations involving waves. The only requirement that must be met then is to remember to retain only the real part of the solution when we complete our analysis. If our analysis yields Eq. (1.13), then the solution we will retain is

$$\boldsymbol{Re}\{f(\mathbf{r}, t)\} = \mathbf{E}(\mathbf{r}) \cos(\omega t - \mathbf{k} \cdot \mathbf{r})$$

This result follows from

$$\boldsymbol{Re}\{f(\mathbf{r}, t)\} = 1/2 \left[f(\mathbf{r}, t) + f^*(\mathbf{r}, t) \right]$$
$$= \mathbf{E}/2 \left[\{\cos(\omega t - \mathbf{k} \cdot \mathbf{r}) + i \sin(\omega t - \mathbf{k} \cdot \mathbf{r})\} \right.$$
$$\left. + \{\cos(\omega t - \mathbf{k} \cdot \mathbf{r}) - i \sin(\omega t - \mathbf{k} \cdot \mathbf{r})\} \right]$$

1.2 Fourier Theory

A question probably has arisen in the reader's mind as to how a general theory of optics can be developed using only a single cosine representation of a plane wave. *Jean Baptiste Joseph Baron de Fourier (1768–1830)* developed a technique to make it possible to solve heat flow problems. This theory makes it possible to utilize plane harmonic waves to construct a description of any general waveform. Fourier's first paper on the subject was rejected because Lagrange did not believe the series would converge. In the eighteenth century, mathematicians did not consider it possible that a finite function could contain an infinite series of terms. The lack of understanding about convergence of infinite series is highlighted by one of Zeno's paradoxes:

> Achilles was to race a tortoise and the tortoise was given a lead. The argument was made that Achilles could never catch the tortoise because he first had to reach the point where the tortoise started but by then the tortoise had moved ahead. No matter how fast Achilles ran, the tortoise added a finite distance to the separation that allowed him to remain ahead of Achilles.

The unstated assumption was that the sum of an infinite series couldn't be finite. The solution to this paradox was found in developing a way to determine whether a series converged. In this particular problem if Achilles cuts the distance to the tortoise in half in each interval of time, he catches the tortoise. The distance Achilles traverses is given by

$$\sum_{n=1}^{\infty} \frac{1}{2^n} = 1.$$

Our current understanding of convergence has eliminated Zeno's paradox for mathematicians and any uncertainty of the correctness of Fourier theory.

1.2.1 Fourier theory for periodic functions

Fourier theory states that any periodic functions, $f(t) = f(t + T)$, where T is the period, can be described a sum of sinusoidal functions called the Fourier series,

$$\int_{-\frac{\pi}{\omega}}^{\frac{\pi}{\omega}} f(t)dt = \int_{-\frac{\pi}{\omega}}^{\frac{\pi}{\omega}} \frac{a_0}{2}dt + \sum_{\ell=1}^{\infty} \int_{-\frac{\pi}{\omega}}^{\frac{\pi}{\omega}} a_\ell \cos \ell\omega t \, dt + \sum_{\ell=1}^{\infty} \int_{-\frac{\pi}{\omega}}^{\frac{\pi}{\omega}} b_\ell \sin \ell\omega t \, dt.$$

where

$$a_n = \frac{\omega}{\pi} \int_{-\frac{\pi}{\omega}}^{\frac{\pi}{\omega}} f(t) \cos \ n\omega t \, dt. \tag{1.15}$$

$$b_n = \frac{\omega}{\pi} \int_{-\frac{\pi}{\omega}}^{\frac{\pi}{\omega}} f(t) \sin \ n\omega t \, dt. \tag{1.16}$$

There is a set of criteria that $f(t)$ must meet for the expansion to be correct.[3] Functions that do not meet these criteria can be constructed, but experimentally the conditions are almost always met. To understand these equations let us explore the description of a periodic set of rectangular pulses:

$$f(t) = \begin{cases} 1 & -T/q < t < T/q \\ 0 & T/q \leq t \leq -T/q \end{cases} \tag{1.17}$$

where $T = 2\pi/\omega$ is the period of the wave. The constant q in Eq. (1.17) allows us to vary the width of the positive-going part of the rectangular wave relative to the period. This parameter can be varied to increase the duration of the period.

By properly selecting the origin we can simplify our calculations of the Fourier series by forcing the function to be even or odd. An even(odd) function can be described by a single cosine(sine) series made up of harmonics of the fundamental frequency of the periodic function. In Figure 1.3 we have set the origin so that our square wave is an even function and so the Fourier series will involve only cosine harmonics.

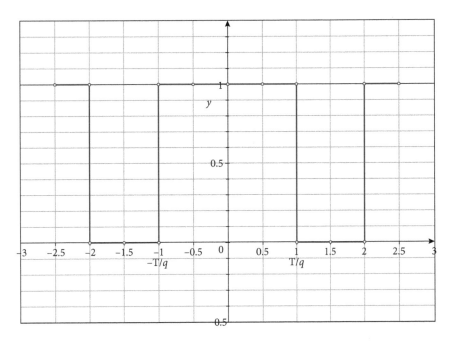

Figure 1.3 Generalized square wave where q is a constant. The duration of the positive portion of the square wave is inversely related to the size of q. The use of q allows us to vary the period of the function.

[3] Dirichlet conditions form a sufficient condition for the description of a periodic function by its Fourier series. The requirements are that f must:

- be single valued,
- have a finite number of maxima and minima in any finite interval,
- have a finite number of finite discontinuities but they cannot be infinite, and
- lead to a finite frequency spectrum.

To clarify the statement about even and odd functions, suppose f is an even function

$$f(t) = f(-t),$$

then $f(t)$ will be represented by a series of cosines. This occurs because the integral over one period, about zero (from $-\pi/\omega$ to π/ω), of an even function is nonzero. The integral of an odd function over the same interval is zero. Using this fact we can determine that Eq. (1.16) will be zero whenever $f(t)$ is an even function because the sine is an odd function and the product of an odd function and an even function is an odd function.

If $f(t)$ is an odd function

$$f(t) = -f(-t),$$

then it can be represented by a series of sine terms because the product of $f(t)$ and the sine is even. If $f(t)$ is neither odd nor even (for example, $f(t) = e^t$), then both the sine and cosine series are required.

The calculation is a little simpler if we use complex notation for the Fourier series. Using the identities

$$\cos \ell \omega t = \frac{1}{2} \left[e^{i\ell\omega t} + e^{-i\ell\omega t} \right]$$

$$\sin \ell \omega t = \frac{-i}{2} \left[e^{i\ell\omega t} - e^{-i\ell\omega t} \right]$$

the Fourier series is then

$$f(t) = \frac{a_0}{2} + \frac{1}{2} \sum_{\ell=1}^{\infty} (a_\ell - ib_\ell) e^{i\ell\omega t} + \frac{1}{2} \sum_{\ell=1}^{\infty} (a_\ell - ib_\ell) e^{-i\ell\omega t} \tag{1.18}$$

$$= \sum_{\ell=1}^{\ell=\infty} \alpha_\ell e^{i\ell\omega t} \tag{1.19}$$

$$\alpha_\ell = \frac{\omega}{2\pi} \int_{-\frac{\pi}{\omega}}^{\frac{\pi}{\omega}} f(t) e^{-i\ell\omega t} dt. \tag{1.20}$$

The coefficients of the series can be obtained by evaluating the integral Eq. (1.20). For the square wave $f(t)=1$ in Eq. (1.17) over the interval $(-T/2 = -\pi/\omega) \leq t \leq (\pi/\omega = T/2)$

$$\alpha_\ell = -\frac{1}{2\pi\ell i} \left[e^{-i\ell\frac{2\pi}{q}} - e^{i\ell\frac{2\pi}{q}} \right] = \frac{2}{q} \cdot \frac{\sin(2\pi\ell/q)}{2\pi\ell/q}. \tag{1.21}$$

The function in Eq. (1.21) is encountered so often it is given its own name:

$$\operatorname{sinc} x = \frac{\sin x}{x}.$$

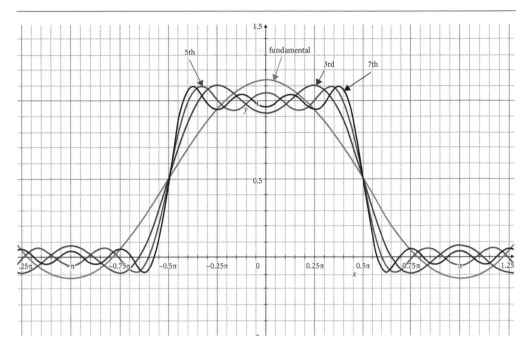

Figure 1.4 The Fourier series approximation of a square wave with the series terminated after the fundamental, third, fifth, and seventh harmonic.

By increasing q we can let the square wave period approach infinity. To start, let $q = 4$, giving us exactly a square wave. The area enclosed by the square wave is

$$\alpha_0 = \frac{1}{T} \int_{-\frac{T}{q}}^{\frac{T}{q}} dt = \frac{1}{T}\left(\frac{T}{q} + \frac{T}{q}\right) = \frac{2}{q} = \frac{1}{2}. \tag{1.22}$$

We can combine the positive and negative exponents of Eq. (1.19) in order to express the expansion in terms of cosine functions. The Fourier series becomes

$$f(t) = \frac{1}{2} + 2\left[\frac{\cos \omega t}{\pi} - \frac{\cos 3\omega t}{3\pi} + \frac{\cos 5\omega t}{5\pi} \cdots\right].$$

We have a sum of odd harmonics of the fundamental frequency of the square wave. As we add terms we get a curve approaching a square wave in Figure 1.4.

Another way to represent the function is in terms of its power spectrum, i.e. a plot of the coefficients of the series, Eq. (1.21).

The spectrum shown in Figure 1.5 is made up of a set of discrete frequencies. The width of the bars representing each frequency is exaggerated to make it easy to see. As we increase q,

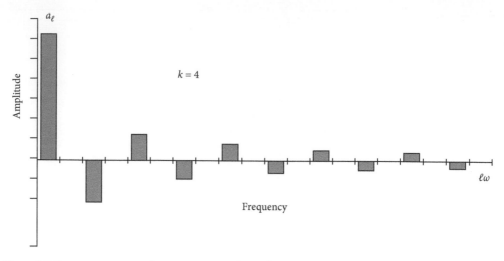

Figure 1.5 Frequency spectrum of a square wave, i.e., the coefficients of the Fourier series of the square wave where $q = 4$.

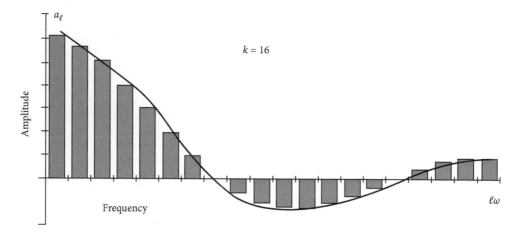

Figure 1.6 Frequency spectrum of a square wave of width $T/8$. The line traced over the histogram aids the eye in seeing the approach of the discrete spectrum to a continuous curve shown in Figure 1.7.

the period grows and the spectrum becomes a finer and finer set of discrete frequencies with an envelope indicated by a smooth curve shown in Figure 1.6.

1.2.2 **Non-periodic function**

If we let $q \to \infty$, the period T goes to infinity, creating a non-periodic square pulse. This leads us to the definition of the Fourier transform, defined for a non-periodic function, $g(t)$, as

$$\mathbf{F}\{g(t)\} \equiv G(\omega) \equiv \int_{-\infty}^{\infty} g(\tau)\, e^{-i\omega\tau}\, d\tau. \tag{1.23}$$

or in terms of sines and cosines

$$G(\omega) = \int_{-\infty}^{\infty} g(\tau) \cos \omega\tau \; d\tau - i \int_{-\infty}^{\infty} g(\tau) \sin \omega\tau \; d\tau$$

Let us turn our periodic square wave (1.17) into a single pulse by letting q become very large; the function then becomes

$$f(\tau) = \begin{cases} 1 & \tau \leq \tau_0/2 \\ 0 & \text{all other } \tau \end{cases}. \tag{1.24}$$

Now the spectrum given by $G(\omega)$ is a continuous distribution of frequencies, shown in Figure 1.7, rather than the discrete distribution of frequencies that characterize a periodic function.

1.2.3 Spatial Fourier Transforms

We have written relationships using time and frequency but we could replace time with a space variable, say x. The transform or conjugate variable must have reciprocal units; thus, when a space variable is used, the conjugate units are "distance" and its reciprocal, 1/"distance". The conjugate variable to the space variable is called *spatial frequency* and in optics is the propagation constant $\mathbf{k}$ (remember $k = 1/\lambda$). We can use the spatial frequencies for a set of plane waves propagating over a spectrum of angles to construct a propagating wave front of any shape. This means a simple theory constructed around a harmonic plane wave can form the basis for the description of any propagating light wave. This mathematical formulation serves as the foundation of modern optics. Later we will establish a connection between $\mathbf{k}$ and physical parameters that we can measure in optical experiments. For now we will use time and frequency in our discussions.

Uncertainty Relationship

There is an important relationship, called the Gabor limit,[4] between two conjugate dimensions: $\omega \Longleftrightarrow t$ or $x \Longleftrightarrow k_x$. The uncertainty arises because the conjugate variables are related via the Fourier transform. Let's look at a temporal signal. Mathematically, we would know the values of $f(t)$ at some time t_0 accurately if it is compact, i.e. if $f(t_0)$ is zero beyond some limit, as in Eq. (1.23) where $t_0 = 0$ beyond $t = \pm\tau_0/2$. We know $f(t)$ with an uncertainty of τ_0. Stated simply, the temporal spread is small. The smaller the temporal spread is, the more frequencies are required by Fourier theory to describe the temporal signal. Thus, the more accurate the temporal signal is, smaller Δt, the lower the frequency resolution, larger $\Delta\omega$, we can simultaneously state:

$$\Delta t \Delta \omega \geq 1/2$$

Another way to state this fact is through the use of the scaling property of a Fourier transform:

$$\mathcal{F}\{g(at)\} = \frac{1}{a}G\left(\frac{1}{\omega}\right).$$

Using Eq. (1.23) in Eq. (1.22) we can calculate the Fourier transform of the pulse with width τ_0:

$$G(\omega) = \int_{-\tau_0/2}^{\tau_0/2} \cos \omega\tau \, d\tau = \frac{1}{\omega}[\sin \omega\tau]_{-\tau_0/2}^{\tau_0/2} = \tau_0 \frac{\sin \omega\tau_0/2}{\omega\tau_0/2}.$$

Here again appears the sinc function

$$\text{sinc}\,(x) = \frac{\sin x}{x} \tag{1.25}$$

shown in Figure 1.7.

The minima of Eq. (1.25) occur when $\omega\tau_0/2 = n\pi$. The spread in frequencies is thus determined by the reciprocal of the spatial extent of the pulse

$$\frac{\omega\tau_0}{2} = n\pi$$
$$\omega = \frac{2\pi n}{\tau_0}.$$

Later we will develop a mathematical description using space and spatial frequencies to describe diffraction of light and generate a similar result for space and spatial frequency. In anticipation of that result Figure 1.8 is an optical representation of the spatial distribution of Eq. (1.23), i.e., as slit.

An optical representation of the transform of Figure 1.8 is shown in Figure 1.9 and matches the sinc of Eq. (1.25) shown in Figure 1.7. In optics this is called diffraction and we will spend a little more time on the theory later.

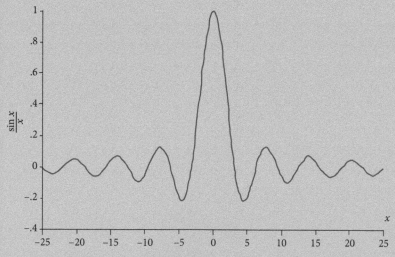

Figure 1.7 A plot of Eq. (1.25) where $x = \omega\tau_0/2$.

[4] It is also called the Heisenberg uncertainty principle.

Figure 1.8 Spatial distribution of a pulse similar to the temporal pulse constructed by letting k in the function shown in Figure 1.2 go to infinity.

Figure 1.9 Fourier transform of Figure 1.7 created by an optical wave.

The width of the central maximum in intensity is inversely related to the width of the slit, a result equivalent to that obtained for the temporal variables.

Windowing of Functions

As an aside, let us review an example of another method of handling nonperiodic functions that is very useful. Let us assume we wish to describe the function

$$g(t) = a + bt,$$

where a and b are constants with respect to time. This function is definitely not a pulse or a periodic function, but we can still utilize Fourier theory to produce a frequency description of this linear function. To make the discussion easy we will select $a = 0$ and $b = 1$, yielding an odd function with a slope of 1, shown in Figure 1.10.

We are only interested in describing this function over the interval

$$-T/2 \leq x \leq T/2,$$

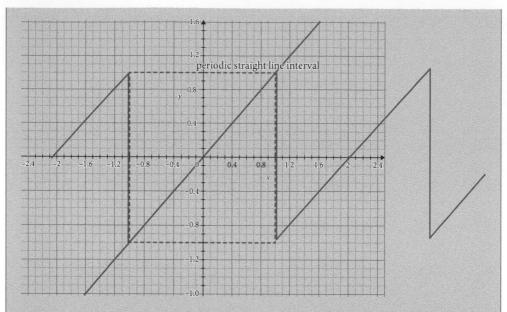

Figure 1.10 The dark line is the function we would like to describe with Fourier theory. We are only interested in the description over the region of interest shown by the window created by a dotted line that marks off the interval $-1 \leq t \leq 1$. We are assuming that the portion of the straigh-line function $g(t)$ is part of a periodic sawtooth function. Outside the region of interest this assumption is incorrect but not important to us.

shown by the dotted lines in Figure 1.10. You can think of this as a temporal window and data will only be taken within this window. We assume that outside the window, the function can be treated as a periodic sawtooth function, shown as an overlay in Figure 1.10:[5]

$$f_P(x) = Ax \qquad -T/2 < x < T/2.$$

Because the selected $g(t)$ is an odd function, the Fourier series contains only sine functions

$$f(t) = \sum_{n=1}^{\infty} b_n \sin \frac{2n\pi t}{T},$$

where

$$b_n = \frac{2}{T} \int_{-T/2}^{T/2} f(t) \sin \frac{2n\pi x}{T} dx = \frac{2A}{T} \int_{-T/2}^{T/2} x \sin \frac{2n\pi x}{T} dx.$$

Rather than using the complex notation we integrate this equation by parts

$$u = x \qquad dv = \sin\left(\frac{2n\pi x}{T}\right) dx$$

$$du = dx \quad v = -\frac{\cos\left(\frac{2n\pi x}{T}\right)}{x} \cdot \left(\frac{T}{2n\pi}\right)^2$$

$$\int u \, dv = uv - \int v \, du.$$

The integral for b_n is then

$$b_n = \frac{2A}{T}\left[-\frac{x\cos{(2n\pi x/T)}}{(2n\pi/T)^2}\right]_{-T/2}^{T/2} + \frac{2A}{T}\int_{-T/2}^{T/2}\frac{\cos\left(\frac{2n\pi x}{T}\right)}{(2n\pi/T)^2}\,dx.$$

The second term is zero because the $\sin(n\pi) = 0$:

$$f(x) = \sum_{n=1}^{\infty}\frac{2A(-1)^{n+1}}{n}\sin{nx}$$

Let A = 1

$$= 2\left[\sin x - \frac{\sin 2x}{2} + \frac{\sin 3x}{3} - \frac{\sin 4x}{4} + \frac{\sin 5x}{5} - \cdots\right].$$

(1.26)

In Eq. (1.26) we display five terms of the series. In Figure 1.11 we plot the result in the window using 20 terms.

The Gibbs phenomenon seen at the ends of the straight line shown in Figure 1.11 is of great interest to mathematicians. It always occurs at a jump discontinuity but does

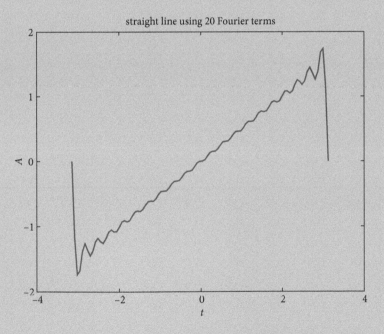

straight line using 20 Fourier terms

Figure 1.11 Fourier series approximation of the function $g(t) = t$ using 20 terms of the series. Forty terms give a better realization of a straight line. The wiggles at the window edges are called Gibbs phenomenon. They do not die out as more terms are added to the series approximation.

not represent our desired function. It will not disappear by using more terms in the expansion. There are signal-processing techniques that have been developed to reduce but not eliminate what is often called ringing.

[5] As an aside, the sawtooth waveform is an important function in signal processing and commonly used in analog music synthesizers.

1.3 Problem Set 1

(1) Use the complex notation to find a representation for $\cos(A + B)$.

(2) Use Euler's equation to find the value of

$$(\cos\theta + i\sin\theta)^n.$$

(3) Prove the Linearity Theorem of Fourier transforms:

$$\mathcal{F}\{ag(x) + bh(x)\} = a\mathcal{F}\{g(x)\} + b\mathcal{F}\{h(x)\}$$
$$= aG(k) + bH(k),$$

where

$$\mathcal{F}\{g(x)\} = \int_{-\infty}^{\infty} g(x)e^{-ikx}\, dx = G(k).$$

(4)* Find the Fourier series for the function $f(x) = x^2$ over the range $a \le x \le a$.

(5) Show that

$$\int_{-\frac{\pi}{\omega}}^{\frac{\pi}{\omega}} \cos(m - n)\omega t\, dt = 0,$$

unless $m = n$.

(6)* Find the Fourier series for (periodic extension of)

$$f(t) = \begin{cases} 1 & 0 \le t < 2 \\ -1 & 2 \le t < 4. \end{cases}$$

(7) Assume that the Fourier transform of $f(t)$ is $F(\omega)$. What is the Fourier transform of $f(t + t') + f(t - t')$?

(8) Assume that the Fourier transform of $f(t)$ is $F(\omega)$. What is the Fourier transform of $f(t)\sin(\omega'/2)t$?

(9) Find the direction of travel of the two waves

$$s(x, t) = A\sin(kx + \omega t) \qquad s(x, t) = A\cos(\omega t - kx).$$

(10) If s_1 and s_2 are solutions of the wave equation

$$\frac{\partial^2 s}{\partial x^2} = \frac{1}{v^2}\frac{\partial^2 s}{\partial t^2},$$

prove the superposition principle, i.e., $a\, s_1 + b\, s_2$ is also a solution, where a and b are both constants.

2 Electromagnetic Theory

Before 1860, the nature of light was not known but a lot of different opinions were voiced. Some are not understandable, like Aristotle's claim (384–322 BC) that it is a state of transparency. Ptolemy (100–70) said it was a form of energy that fits with our current view but he thought the light originated in the eye and traveled to the object, like Superman's X-ray vision. Alhazen (965–1040) dispelled the view of Ptolemy using experiments to show that vision was the result of light bouncing off of objects and traveling to the eye. In 1678 Christiaan Huygens (1629–95) suggested that light was a wave similar to sound traveling in an elastic solid, but for light it traveled in a "luminiferous ether." In competition with this view was that held by Isaac Newton (1643–1727) that light was made up of weightless particles. A hundred years after Newton's death, wave theory won out when Thomas Young (1773–1829) demonstrated that light could produce interference fringes.

In 1864 James Clerk Maxwell[1] (1831–79) published a paper presenting four equations that established the theoretical foundation of light by connecting the theory of electromagnetism with the propagation of light. Before Maxwell the mathematical description of electrostatics and magnetostatics was well established, as was the connection between currents and magnetic fields. Michael Faraday used the concept of fields to connect electric fields not only from currents but also from changing magnetic fields. Maxwell developed a mechanical model that added to Faraday's law of induction a statement that changing electric fields were a source of magnetic fields. The combined fields were found to propagate as a wave moving at the velocity of light.

Electromagnetic waves fall into the wavelength classes shown in Table 2.1. Visible light waves are electromagnetic waves that have frequencies between 7.5×10^{14} Hz with a wavelength of 400 nm and frequency 4.3×10^{14} Hz at a wavelength of 700 nm, as shown in Figure 1.1.

An important property of the theory of an electromagnetic field is known as *scale invariance*. This property states that solutions of a set of equations we will introduce in a moment, Maxwell's equations [1], have no natural length scale. Solutions for centimeter scale structure at microwave frequencies are the same as for nanometer scale structures at optical frequencies. For this reason, the theory we will develop is applicable to frequency regions well outside the visible spectrum. However, there are assumptions we will use in our discussion that do impact the applicability of the theory to particular frequency regions. We should particularly tread carefully in using assumptions about the *constitutive relations* that we will discuss in the next section. As an example, we will assume that available detectors cannot record the oscillations of a simple light wave and we must record the time average of a wave. This means we should apply

---

[1] The family name was Clerk but James' dad inherited a large estate from the Maxwell family, so he added "Maxwell" to their last name. No one ever hyphenates the two surnames but rather treats Maxwell as the family name.

Modern Optics Simplified. B. D. Guenther. © B. D. Guenther 2020.
Published in 2020 by Oxford University Press. DOI: 10.1093/oso/9780198842859.001.0001

Table 2.1 Electromagnetic wavelength regions

Type	Wavelength, λ	source
Gamma rays	$10^{-14} - 10^{-10}$ m	Nuclei
X-rays	$10^{-12} - 10^{-8}$ m	Intershell transitions and Bremsstrahlung
Ultraviolet	$6 \times 10^{-10} - 4 \times 10^{-7}$ m	Sunlight
Visible	$4 \times 10^{-7} - 7 \times 10^{-7}$ m	Sunlight and atomic transitions
Infrared	$7 \times 10^{-7} - 10^{-3}$ m	Molecules and hot objects
Microwaves	$10^{-4} - 0.3$ m	Solid state and vacuum tubes
Radio	$0.1 - 10$ m	Currents in wires and vacuum

caution using our results when the frequency of the electromagnetic wave is less than about 300 GHz where we can detect the amplitude of a wave. This frequency limitation also applies to our assumption about the response of the magnetic properties of matter. Similar caution should be applied in treating waves with wavelengths shorter than about 100 nm.

The theory will describe the propagation of a classical electromagnetic wave in terms of the fields, **E**, **D**, **B**, and **H**. Some explanation is needed to justify having a theory involving four fields instead of two. The fields labeled **D** and **B** are defined as the internal fields, in a material induced by the applied fields **E** and **H** acting on the atoms and molecules making up the material:

$$\mathbf{D} = \varepsilon_0 \mathbf{E} + \mathbf{P}$$
$$\mathbf{B} = \mu_0 \mathbf{H} + \mathbf{M},$$

(2.1)

where **P** is the polarization field associated with bound charges in a material and **M** is the magnetization field associated with bound currents in the material. This formulation emphasizes that the internal fields of a material are due not only to the applied field but also due to fields created by microscopic sources within the material [2].[2] A simple example is shown in Figure 2.1.

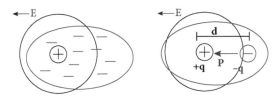

Figure 2.1 On the left is the classical model of an atom or molecule with a cloud of negative charge surround a positive point charge. When an electric field is applied, the charges shift is opposite directions, creating an electric dipole, as shown on the right.

[2] The addition of the applied and induced fields is given by the Ewald–Oseen extinction theorem. The applied wave, traveling at the speed of light, is replaced in a distance called the *extinction length* by the induced waves traveling at the material's light velocity. The induced field has two components: one cancels out the applied field and the second has a velocity that yields the index of refraction.

The induced dipole created when the positive and negative charges of an atom or molecule are separated by the applied field (we will ignore the added complication of polar molecules like water that have permanent as well as induced dipole moments). The result of separating the positive and negative charges is an electric dipole moment

$$\mathbf{p} = \sum_i q_i \mathbf{r}_i - \sum_j q_j \mathbf{r}_j,$$

where the sum over i is over the positive charges multiplied by their displacement, $\mathbf{r}_i$, relative to the center of mass and the sum over j is over the negative charges multiplied by their displacement, $\mathbf{r}_j$. To take this contribution due to "bound charge" into account, we sum over all the electric dipole moments, $\mathbf{p}$, in a microscopically large volume, with respect to the elemental charge,[3] and macroscopically small volume, ΔV, with respect to the illuminating wavelength.[4] The resulting vector, $\mathbf{P}$, is called the polarization vector.

$$\mathbf{P} = \frac{1}{\Delta V} \sum_i \mathbf{p}_i$$

The constant ε_0 is called the vacuum permittivity and μ_0 is the vacuum permeability:

$$\mu_o = 4\pi \times 10^{-7} \qquad \text{weber/ampere} - \text{meter}$$
$$\varepsilon_0 = 1/\mu_0 c^2 = 8.854187817 \times 10^{-12} \qquad \text{farad/meter}$$

The dipole in Figure 2.1 can be modeled by a harmonic oscillator that can be used to derive an equation predicting the temporal response of ε, σ, and μ. A qualitative understanding of the processes can be developed using classical mechanics, but to obtain a quantitative predictive capability, quantum mechanics must be used.

Maxwell originally introduced his equation through the use of 20 component equations. Oliver Heaviside rewrote the equations in a much more compact format using vector calculus. Before we examine the current standard set of Maxwell's equations for electromagnetic waves, we need to provide several definitions from vector calculus that allow the compact presentation generated by Heaviside.

- A *gradient* is a vector describing the slope of a function:

$$\nabla f(x, y, z) = \frac{\partial f}{\partial x}\hat{\mathbf{i}} + \frac{\partial f}{\partial y}\hat{\mathbf{j}} + \frac{\partial f}{\partial z}\hat{\mathbf{k}}. \tag{2.2}$$

- *Divergence* is a scalar that provides a measure of the flow of the vector $\mathbf{E}$ in or out of a volume:

$$\text{div } \mathbf{E} = \nabla \bullet \mathbf{E} = \frac{\partial E_x}{\partial x} + \frac{\partial E_y}{\partial y} + \frac{\partial E_z}{\partial z}. \tag{2.3}$$

[3] Atoms are spaced on the order of ½ a nanometer, so we must use a volume bigger than the cube of this number.
[4] The wavelength of green light is about 500 nm, so our volume must be smaller than the cube of this number.

- *Curl* is a vector field that wraps around **E**

$$\text{curl } \mathbf{E} = \nabla \times \mathbf{E} = \begin{vmatrix} \hat{\mathbf{i}} & \hat{\mathbf{j}} & \hat{\mathbf{k}} \\ \partial/\partial x & \partial/\partial y & \partial/\partial z \\ E_x & E_y & E_z \end{vmatrix}$$

$$= \left(\frac{\partial E_z}{\partial y} - \frac{\partial E_y}{\partial z} \right) \hat{\mathbf{i}} + \left(\frac{\partial E_x}{\partial y} - \frac{\partial E_z}{\partial z} \right) \hat{\mathbf{j}} + \left(\frac{\partial E_y}{\partial y} - \frac{\partial E_x}{\partial z} \right) \hat{\mathbf{k}}, \tag{2.4}$$

where $\hat{i}, \hat{j}$, and $\hat{k}$ are unit vectors in the x, y, and z directions, respectively. The direction of the curl is the axis of rotation and the magnitude is the magnitude of the rotation of the vector field.

2.1 Maxwell's Equations

Using the vector calculus notation, we can write a set of equations called Maxwell's equations:

$$\nabla \cdot \mathbf{D} = \rho \qquad \nabla \cdot \mathbf{B} = 0$$
$$\nabla \times \mathbf{E} + \frac{\partial \mathbf{B}}{\partial t} = 0 \quad \nabla \times \mathbf{H} = \mathbf{J} + \frac{\partial \mathbf{D}}{\partial t} \tag{2.5}$$

All of the equations are physical laws that were experimentally generated. Maxwell's contribution was to gather them together and add the term $\partial \mathbf{D}/\partial t$ to the fourth equation. Let's discuss the physics behind the various equations that make up Maxwell's equations:

1. *Gauss' law*
$$\nabla \cdot \mathbf{D} = \rho \tag{2.6}$$

This is called the divergence of the displacement **D**. Maxwell viewed the fields graphically using a model with lines of force developed by Faraday. At that time in the development of physics, mechanical models were expected to yield the "theory of everything." Maxwell viewed the displacement as a set of flow lines similar to those used to model a flowing liquid.[5] Mathematically, the divergence simply adds up the lines of flux crossing a closed boundary surrounding a charge. The divergence is a scalar function that describes how **D** expands or contracts across a boundary around the charge density ρ. The charge density ρ is measured in coulombs/(meter)^3 and the flux of **D** is given by the charge density contained in the volume around the point in space we are evaluating. We imagine **D** as consisting of a set of field lines that converge onto a negative charge (a sink) and diverge from a positive charge (a source).

2. *Gauss's law for magnetism*
$$\nabla \cdot \mathbf{B} = 0 \tag{2.7}$$

---

[5] You have seen an image of the field lines associated with a magnetic field in grade school when you laid a piece of paper over a bar magnet and sprinkled iron filings over the paper, as is reproduced in Figure 2.2.

We always look for symmetries in physics and here we come up short because the equation for magnetism does not look like the electric field equation, Eq. (2.6). The lack of symmetry in the equations is due to the fact that we have never been able to discover a magnetic monopole in nature. For the magnetic field, the field lines form closed loops because the divergence is zero; i.e., there is no net flow in or out of the volume around a point in space. You can visualize the field lines associated with a bar magnet by using iron fillings, as shown in Figure 2.2.

These first two equations, Eqs. (2.6) and (2.7), have no temporal dependence but retain their validity under both static and dynamic conditions. We will spend most of our time discussing the properties of the other two equations, which describe how light propagates.

3. *Faraday's law*

$$\nabla \times \mathbf{E} + \frac{\partial \mathbf{B}}{\partial t} = 0 \tag{2.8}$$

A magnetic field that is changing in time produces an **E** field circulating around the changing **B** field—a result of the "$\nabla \times$."

4. *Ampere's law*

$$\nabla \times \mathbf{H} = \mathbf{J} + \frac{\partial \mathbf{D}}{\partial t} \tag{2.9}$$

A flowing current given by the current density, **J**, will produce a magnetic field circulating about the current. This only works for steady-state currents. Maxwell's major contribution was the addition of the time derivative of **D** to predict that a time-varying electric field will also produce a magnetic field circulating about the electric field.

These last two equations explain how light propagates through space in what is called the far field. A temporally changing magnetic field produces a time-varying electric field. The

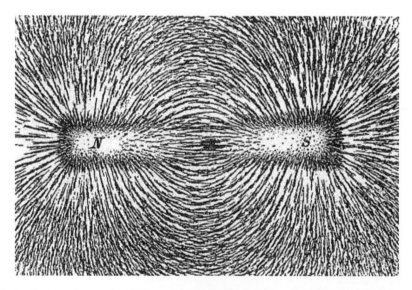

Figure 2.2 The direction of magnetic *field lines* represented by the alignment of iron filings sprinkled on paper placed above a bar magnet.
Image by Newton Henry Black. Newton Henry Black and Harvey N. Davis (1913), *Practical Physics*, The MacMillan Co., New York, USA, p. 242, Fig. 200, Public Domain, *https://commons.wikimedia.org/w/index.php?curid=73846*

time-varying electric field in turn produces a time-varying magnetic field and the process is repeated as light travels through space. There is no need for currents or charges to obtain this effect and, in fact, the propagating field, called the far field, is independent of the currents and charges that originally created it. Fields produced by currents and charges appear in the near field[6] and fall off in value as $1/r^n$, $n > 1$. We will ignore the near field effects because we seldom work as close as a wavelength to the source at optical wavelengths. New technology was developed in the early 1980s called near field scanning optical microscopy (NSOM) that is based on the use of near field waves and the reader is encouraged to read about that technology [3].

One would expect that because these two equations explain the propagation of a electromagnetic wave, then they must somehow contain the wave equation. To discover the wave equation within Eqs. (2.8) and (2.9), we will make some assumptions to simplify Maxwell's equations to make the task easier.

As stated earlier, the theory is not limited to visible optical frequencies but the order of magnitude of the frequency allows us to make several simplifying approximations about the dynamic response of atoms and molecules in a propagation medium. The dynamic response of atoms and molecules are considered through what are called the *constitutive relations*:

$$\mathbf{D} = f(\mathbf{E})$$
$$\mathbf{J} = g(\mathbf{E})$$
$$\mathbf{B} = h(\mathbf{H})$$

These relationships are normally determined experimentally and are described theoretically using the model shown in Figure 2.1. In general the field pairs $\mathbf{D}$ and $\mathbf{E}$ and $\mathbf{B}$ and $\mathbf{H}$ may point in different directions, making the constitutive relations, in general, tensors. Here we will assume that the constitutive relations are simple scalar relationships

$$\mathbf{D} = \varepsilon\mathbf{E} \qquad \varepsilon = \text{dielectric constant (relative permittivity)}$$
$$\mathbf{J} = \sigma\mathbf{E} \qquad \sigma = \text{conductivity (Ohm's law)}$$
$$\mathbf{B} = \mu\mathbf{H} \qquad \mu = \text{permeability}$$

where the constants ε, σ, and μ contain the description of the material.

2.1.1 Free Space Medium

To reduce the complexity of our discussions further, we will assume that the materials our electromagnetic waves are propagating in have the following properties:

1. *Uniform*: ε and μ have the same value at all points and are thus independent of space coordinates.

2. *Isotropic*: ε and μ do not depend upon the direction of propagation so can be treated as scalar quanities.

3. *Nonconducting*: $\sigma = 0$; thus, $\mathbf{J} = 0$.

---

[6] In quantum theory, the near field is a mixture of real and virtual photons, whereas the far field only has real photons.

4. *No free charge*: $\rho = 0$.

5. *Nondispersive*: ε and μ are not functions of the frequency; i.e., they have no time dependence. In general, this is not true but it allows a simplified first look.

With these assumptions about the propagation medium for the electromagnetic wave, Maxwell's equations can be simplified to read as follows:

$$\nabla \cdot \mathbf{E} = 0 \qquad \text{(a)} \qquad \nabla \cdot \mathbf{B} = 0 \qquad \text{(d)}$$

$$\nabla \times \mathbf{E} = -\frac{\partial \mathbf{B}}{\partial t} \qquad \text{(b)} \qquad \nabla \times \mathbf{H} = \frac{\partial \mathbf{D}}{\partial t} \qquad \text{(e)} \qquad (2.10)$$

$$\varepsilon \mathbf{E} = \mathbf{D} \qquad \text{(c)} \qquad \mathbf{B} = \mu \mathbf{H} \qquad \text{(f)}$$

Wave Equation

To find the wave equation within Eq. (2.10), Maxwell's equations must be rearranged to display explicitly the time and coordinate dependence. We could work out the following derivations using only one dimension and time but by simply accepting a few identity relations from vector calculus we can do a three-dimensional derivation. Using Eqs. (2.10c) and (2.10f) along with properties 1 and 5, we can rewrite Eq. (2.10e) as

$$\frac{1}{\mu} \nabla \times \mathbf{B} = \varepsilon \frac{\partial \mathbf{E}}{\partial t}.$$

The curl of Eq. (2.10b) is taken and the magnetic field dependence is eliminated by using the rewritten Eq. (2.10e):

$$\nabla \times (\nabla \times \mathbf{E}) = \nabla \times \left(-\frac{\partial \mathbf{B}}{\partial t}\right) = -\frac{\partial}{\partial t}(\nabla \times \mathbf{B}) = -\frac{\partial}{\partial t}\left(\varepsilon \mu \frac{\partial \mathbf{E}}{\partial t}\right).$$

We have not made any special mathematical moves except for the interchange of space and time derivatives $\nabla \iff \partial/\partial t$. That move was made possible by our assumptions 1 and 5. The assumption that ε and μ are independent of time allows the constants ε and μ to be removed from the derivative with respect to time and the equation to be rewritten as

$$\nabla \times (\nabla \times \mathbf{E}) = -\varepsilon \mu \frac{\partial^2 \mathbf{E}}{\partial t^2}.$$

We now need to borrow an identity from vector calculus:

$$\nabla \times (\nabla \times \mathbf{A}) = \nabla(\nabla \cdot \mathbf{A}) - \nabla^2 \mathbf{A}.$$

We can then write

$$\nabla(\nabla \cdot \mathbf{E}) - \nabla^2 \mathbf{E} = -\varepsilon \mu \frac{\partial^2 \mathbf{E}}{\partial t^2}.$$

Because free space is free of charge (property 4), $\nabla \cdot \mathbf{E} = 0$. This means physically that the electric field will propagate without change. Also we have used the fact that ε is uniform and isotropic. The result is the wave equation for the electric field

$$\nabla^2 \mathbf{E} = \mu\varepsilon \frac{\partial^2 \mathbf{E}}{\partial t^2}. \tag{2.11}$$

We can use the same procedure to obtain

$$\nabla^2 \mathbf{B} = \mu\varepsilon \frac{\partial^2 \mathbf{B}}{\partial t^2}. \tag{2.12}$$

We have now found hidden in Maxwell's equations the wave equations, with a wave's velocity given by

$$v = \frac{1}{\sqrt{\mu\varepsilon}}. \tag{2.13}$$

The connection of the velocity of light with the electric and magnetic properties of a material was one of the most important results of Maxwell's theory. In a vacuum

$$\mu_0\varepsilon_0 = \left(4\pi \times 30^{-7}\right)\left(8.8542 \times 0^{-12}\right) = 1.113 \times 30^{-17} \frac{s^2}{m^2}$$
$$\frac{1}{\sqrt{\mu_0\varepsilon_0}} = 2.998 \times 30^8 \frac{m}{s} = c. \tag{2.14}$$

In a material the velocity of light is less than c. We can characterize a material by defining the *index of refraction*, the ratio of the speed of light in a vacuum to its speed in a medium:

$$n = \frac{c}{v} = \sqrt{\frac{\varepsilon\mu}{\varepsilon_0\mu_0}}. \tag{2.15}$$

Because we have selected the order of magnitude of a optical wave's frequency to be around 10^{14} we find that the magnetic permeability of all but magnetic materials to be nearly equal to the vacuum value [4]. This means that the magnetic field of the electromagnetic wave does not strongly interact with the atoms and molecules of most materials at optical frequencies.

The data in Table 2.2 demonstrate that if magnetic materials are not considered, then $\mu/\mu_0 \approx 1$ so that the velocity of propagation of an electromagnetic wave is determined only by interactions with the electric field:

$$n = \sqrt{\frac{\varepsilon}{\varepsilon_0}} \tag{2.16}$$

Modern calculations about electrodynamics are carried out by a technique known as finite-difference, time-domain method (FDTD), developed by Kane S. Yee. The technique utilizes only the two equations Eqs. (2.10b) and (2.10e). The solution duplicates the propagation method we just described:*The changing electric field generates a changing magnetic field that in turn generates a changing electric field.* In addition, there are housekeeping details that must be included, like the additions of sources to handle the power, material properties to handle dispersion, etc., and a technique for preventing reflections from a finite boundary. The result of the calculation is a movie of the propagation of light and makes it possible to observe transient effects. We will not incorporate this technique into our discussion but rather generate analytic equations that allow us to generate a visual understanding of the electromagnetic wave.

Table 2.2 Representative Magnetic Permeability

Material	μ/μ_o	Class
Silver	0.99998	Diamagnetic
Copper	0.99999	Diamagnetic
Water	0.99999	Diamagnetic
Air	1.00000036	Paramagnetic
Aluminum	1.000021	Paramagnetic
Iron	5000	Ferromagnetic
Nickel	600	Ferromagnetic

2.2 Characteristics of the Electromagnetic Wave

2.2.1 Transverse wave

The electric field, plane wave solution of the wave equation, is given by

$$\mathbf{E} = \mathbf{E}_0 e^{i(\omega t - \mathbf{k}\cdot\mathbf{r} + \phi)}$$

If we use the plane wave to calculate the divergence of the electric field, we get for the x-component of the wave

$$\frac{\partial E_x}{\partial x} = \frac{\partial}{\partial x}\left[E_{0x}e^{i(\omega t - \mathbf{k}\cdot\mathbf{r} + \phi)}\right] = iE_{0x}e^{i(\omega t - \mathbf{k}\cdot\mathbf{r} + \phi)}\frac{\partial}{\partial x}\left(\omega t - k_x x - k_y y - k_z z + \phi\right)$$

$$\frac{\partial E_x}{\partial x} = -ik_x E_x. \tag{2.17}$$

Carrying out the same calculation for the y and z component of the field yields the vector equation

$$\nabla \cdot \mathbf{E} = -i\mathbf{k} \cdot \mathbf{E} = 0. \tag{2.18}$$

From vector calculus, if the dot product of two vectors, $\mathbf{E}$ and $\mathbf{k}$, is zero, then the vectors $\mathbf{E}$ and $\mathbf{k}$ must be perpendicular to each other.[7] This means for our plane wave propagating in a medium with the characteristics we have specified, the electromagnetic wave is a transverse wave with the electric field perpendicular to the direction of propagation. The same calculation can be done for the magnetic field, concluding that the magnetic field of the wave also must be perpendicular to the direction of propagation. The electromagnetic wave is therefore transverse; i.e., it has its amplitudes perpendicular to the direction of propagation. Maxwell's equations will also provide us information about the relative orientation of $\mathbf{E}$ and $\mathbf{B}$.

---

[7] The dot product is usually called the inner product but a more descriptive name is projection product since it is the projection of one vector onto the other. Numerically, it is the product of the length of each vector times the cosine of the angle between the two vectors.

2.2.2 Interdependence of E and B

The electric and magnetic fields are not independent of each other. To give us some ammunition to evaluate the interdependence of **E** and **B**, let us calculate several derivatives of the plane wave. We will need

$$\frac{\partial \mathbf{B}}{\partial t} = \frac{\partial}{\partial t}\mathbf{B}_0 e^{i(\omega t - \mathbf{k}\cdot\mathbf{r}+\phi)} = i\mathbf{B}\frac{\partial}{\partial t}(\omega t - \mathbf{k}\cdot\mathbf{r}+\phi)$$

$$\frac{\partial \mathbf{B}}{\partial t} = i\omega\mathbf{B} \tag{2.19}$$

and in a similar manner

$$\frac{\partial \mathbf{E}}{\partial t} = i\omega\mathbf{E}. \tag{2.20}$$

An expression for the curl of a plane wave of **E** is

$$\nabla \times \mathbf{E} = \left(\frac{\partial E_z}{\partial y} - \frac{\partial E_y}{\partial z}\right)\hat{\mathbf{i}} + \left(\frac{\partial E_x}{\partial z} - \frac{\partial E_z}{\partial x}\right)\hat{\mathbf{j}} + \left(\frac{\partial E_y}{\partial x} - \frac{\partial E_x}{\partial y}\right)\hat{\mathbf{k}}.$$

The terms making up the x-component of the curl are

$$\frac{\partial E_z}{\partial y} = E_{0z}\frac{\partial}{\partial y}e^{i(\omega t - \mathbf{k}\cdot\mathbf{r}+\phi)} = -ik_y E_z,$$

$$\frac{\partial E_z}{\partial y} = -ik_z E_y.$$

By the evaluation of each component, we find that the curl of **E** for a plane wave is

$$\nabla \times \mathbf{E} = -i\mathbf{k} \times \mathbf{E}. \tag{2.21}$$

A similar derivation leads to the curl of **B** for a plane wave

$$\nabla \times \mathbf{B} = -i\mathbf{k} \times \mathbf{B}. \tag{2.22}$$

With these vector operations on a plane wave defined, we can evaluate Eq. (2.10b) for a plane wave. The left side of

$$\nabla \times \mathbf{E} = -\frac{\partial \mathbf{B}}{\partial t}$$

is replaced with Eq. (2.21) and the right side by Eq. (2.19), resulting in an equation connecting the electric and magnetic fields:

$$-i\mathbf{k} \times \mathbf{E} = -i\omega\mathbf{B}.$$

Using the relationship between ω and **k** given by Eq. (1.7) and the relationship for the wave velocity in terms of the electromagnetic properties of the material, Eq. (2.13), we can write

$$\frac{\sqrt{\mu\varepsilon}}{k}\mathbf{k} \times \mathbf{E} = \mathbf{B}. \tag{2.23}$$

From the definition of the cross product, we see that the electric and magnetic fields are perpendicular to each other, in phase, and form a right-handed coordinate system with the propagation direction, **k**; see Figure 2.3.

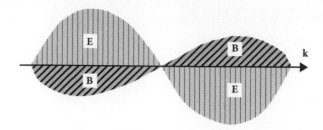

Figure 2.3 Graphical representation of an electromagnetic plane wave. Note **E** and **B** are perpendicular to each other and individually perpendicular to the propagation vector, **k**, in phase, and form a right-handed coordinate system as required by Eqs. (2.22) and (2.23).

If we are only interested in the magnitude of the two fields we can use Eq. (2.15) to write

$$n|\mathbf{E}| = c|\mathbf{B}|. \tag{2.24}$$

The ratio of E to H is called the impedance, z_0, of the medium. In a vacuum

$$z_0 = \sqrt{\mu_0/\varepsilon_0} = \mu_0 c \equiv 376.730\ldots\,\Omega. \tag{2.25}$$

Because it is the product of two defined constants, Eq. (2.25) is exactly defined.

2.2.3 Energy density and flow

The energy density (in joules/m^3) associated with an electromagnetic wave is given by

$$U = \frac{(\mathbf{D} \cdot \mathbf{E} + \mathbf{B} \cdot \mathbf{H})}{2}. \tag{2.26}$$

We can simplify Eq. (2.26) by using the simple constitutive relations $\mathbf{D} = \varepsilon\mathbf{E}$ and $\mathbf{B} = \mu\mathbf{H}$, if they apply to the propagation medium,

$$U = \frac{1}{2}\left(\varepsilon E^2 + \frac{B^2}{\mu}\right) = \frac{1}{2}\left(\varepsilon + \frac{1}{\mu c^2}\right)E^2. \tag{2.27}$$

In a vacuum we have $\varepsilon_0 = 1/\mu_0 c^2$ and the relationship reduces to

$$U = \varepsilon_0 E^2 = \frac{B^2}{\mu_0}. \tag{2.28}$$

2.2.4 Poynting vector

The wave carries the energy U from point to point as the wave propagates. The flow of the field energy is described by the *Poynting vector*

$$\mathbf{S} = \mathbf{E} \times \mathbf{H}, \tag{2.29}$$

with units of joules/(m^2•sec) and a direction parallel to the propagation vector, **k**. The magnetic field **H** is given by

$$H = \frac{B}{\mu} = \frac{\sqrt{\mu\varepsilon}}{\mu k} k \times E.$$

If we substitute the plane wave,

$$E = E_0 \cos(\omega t - k \cdot r + \phi),$$

we get an equation for the energy flowing with the wave

$$S = \frac{\sqrt{\mu\varepsilon}}{\mu k} E_0 \times (k \times E_0) \cos^2(\omega t - k \cdot r + \phi).$$

The vector identity used earlier makes it possible to simplify:

$$A \times (B \times C) = B(A \cdot C) - C(A \cdot B)$$

$$k \perp E_0 \Rightarrow k \cdot E_0 = 0.$$

Remember that the dot product is a scalar, $E \cdot k = \|E\| \cdot \|k\| \cos\theta$, where θ is the angle between the two vectors. The Poynting vector becomes

$$S = \frac{n}{\mu c} |E_0|^2 \frac{k}{k} \cos^2(\omega t - k \cdot r + \phi).$$

2.2.5 Temporal average of energy flow

At light frequencies ($\approx 10^{15}$ Hz), we do not detect S but rather detect a temporal average of S with the average taken over a time, ΔT, determined by the response time of the detector used. The time average of any function is given by

$$\langle f(t) \rangle = \frac{1}{\Delta T} \int_t^{t+\Delta T} f(\tau) \, d\tau.$$

The time average of S is called the *flux density* and has units of watts/m^2.[8] We will call this quantity the *intensity* of the light wave,

$$I = |\langle S \rangle| = \frac{1}{T} \int A \cos^2(\omega t - k \cdot r + \phi) \, dt, \tag{2.30}$$

where we have simplified our notation by creating a vector, **A,** containing all of the time-independent qualities and moving **A** outside the integral

$$A = \frac{n}{\mu\varepsilon} |E_0|^2 \frac{k}{k}.$$

---

[8] The integral can be a moving average, allowing the Poynting vector to have a time dependence that is slow with respect to ΔT.

The time average of the Poynting vector is then

$$\langle \mathbf{S} \rangle = \frac{\mathbf{A}}{\omega T} \int\limits_{t_0 \omega}^{(t_0+T)\omega} \cos^2 (\omega t - \mathbf{k} \cdot \mathbf{r} + \phi) \, d(\omega t).$$ (2.31)

Using the trig identity

$$\cos^2\theta = \frac{1}{2}(1 + \cos 2\theta),$$

the integral result becomes

$$\langle \mathbf{S} \rangle = \frac{\mathbf{A}}{2} + \frac{\mathbf{A}}{4\omega T}[\sin 2(\omega t_0 + \omega T - \mathbf{k} \cdot \mathbf{r} + \phi) - \sin 2(\omega t_0 - \mathbf{k} \cdot \mathbf{r} + \phi)]$$

$$\leq \frac{\mathbf{A}}{2} + \frac{\mathbf{A}}{2\omega T}.$$ (2.32)

The largest value the term in brackets can assume is 2.[9] The period T is the response time of the detector to the light wave. Normally, it is much longer than the period of light oscillations so that $\omega T \gg 1$ and we can neglect the second term of Eq. (2.32).

As an example, suppose our high-speed optical detection system has a 10-GHz bandwidth, yielding a response time of $T = 10^{-11}$s (the reciprocal of the bandwidth). Green light has a frequency of $\nu = 6 \times 10^{14}$Hz or $\omega \approx 4 \times 10^{15}$rad/s.

With these values $\omega T = 4 \times 10^4$ and the neglected term would be no larger than 10^{-3} of the first term. Therefore, in optics the assumption that $\omega T \gg 1$ is reasonable and allows the average Poynting vector to be written as

$$\langle \mathbf{S} \rangle = \frac{\mathbf{A}}{2} = \frac{n}{2\mu c}|\mathbf{E}_0|^2 \frac{\mathbf{k}}{k}.$$ (2.33)

The intensity flows in the direction of the unit vector $\mathbf{k}/k$, in the direction of wave travel, parallel to the wavevector. Its magnitude depends on the square of the amplitude of the wave with a proportionality constant that depends on the electric and magnetic properties of the medium it is traveling in.

The energy crossing a unit area, A, in time Δt is contained in a volume $A(\nu\Delta t)$ (in a vacuum $\nu = c$), as shown in Figure 2.4. To find the magnitude of this energy, we must multiply this

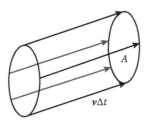

Figure 2.4 The energy or momentum of a wave crossing a unit area A in a time Δt.

---

[9] It is simply the sum of two sine functions that each never exceed 1.

volume by the average energy density, $\langle U \rangle$. Thus, we expect the magnitude of the energy flow to be given by

$$|\langle S \rangle| = \frac{\text{energy}}{A\Delta t} \propto \frac{Av\Delta t \langle U \rangle}{A\Delta t} = v\langle U \rangle. \tag{2.34}$$

We may use the definitions of the wave velocity and index of refraction

$$v = \frac{1}{\sqrt{\mu\varepsilon}} \quad \text{and} \quad n = \frac{c}{v}$$

to rewrite Eq. (2.34) as

$$|\langle S \rangle| = \frac{\varepsilon v E_0^2}{2} = v\langle U \rangle. \tag{2.35}$$

This is, as we expect, the energy of the light wave flowing with the wave, through space, at the speed of the wave in the medium. This is a general property of all waves, not just light waves.

We have only expressed a light wave in terms of mathematical variables. To get an idea of what the expression for an actual wave looks like using measurable qualities, let us assume we have the following light wave traveling in the positive z-direction with its electric field making an angle of $45°$ with respect to the x-axis (this is the polarization of the wave and will be discussed later) with an amplitude of 42.42 V/m. The frequency of the wave is 6×10^{14}Hz. At $45°$ the x and y components of the electric field will be equal and

$$|\,\mathbf{E}\,| = \sqrt{E_{0x}^2 + E_{0y}^2} = \sqrt{2}E_{0x} = 42.42\text{V/m}$$

$$E_{0x} = 30\text{V/m}$$

The electric component of the light wave would be

$$\mathbf{E} = \hat{\mathbf{i}} \cdot 30 \cos\left[2\pi\left(6\times10^{14}\right)\left(t - \frac{Z}{3\times10^8}\right)\right] + \hat{\mathbf{j}} \cdot 30\cos\left[2\pi\left(6\times10^{14}\right)\left(t - \frac{Z}{3\times10^8}\right)\right]$$

and the magnetic component would be

$$\mathbf{B} = -\hat{\mathbf{i}} \cdot 10^{-7} \cos\left[2\pi\left(6\times10^{14}\right)\left(t - \frac{Z}{3\times10^8}\right)\right] + \hat{\mathbf{j}} \cdot 10^{-7}\cos\left[2\pi\left(6\times10^{14}\right)\left(t - \frac{Z}{3\times10^8}\right)\right].$$

The flux density would be

$$I = \frac{c\varepsilon_o E^2}{2} = \frac{\left(3\times10^8\right)\left(8.85\times10^{-12}\right)(42.42)^2}{2} = 2.4\text{W/m}^2.$$

2.3 Quantization

Classical electromagnetic theory is successful in explaining all of the experimental observations to be discussed in this book. There are, however, experiments that cannot be explained by classical wave theory, especially those conducted at short wavelengths or very low light levels.

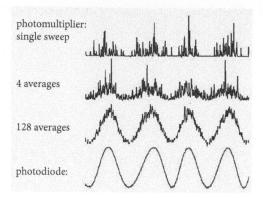

Figure 2.5 Detector current from an imaging detector of interference fringes as the intensity of the light increases from top to bottom. The classical interference pattern emerges when many single-photon traces are averaged [5]. Reproduced from Dimitrova, T. L., and A. Weis, The wave–particle duality of light: A demonstration experiment. *Am. J. Phys.* (2007) **76**(2): 137–42, with the permission of the American Association of Physics Teachers.

For example, at low light levels, the energy flow just introduced in Eq. (2.35) appears to be made up of discrete energy packets that are each consumed to excite an electron making up a photocurrent such as seen in Figure 2.5. The minimum packet size energy is given by

$$\hbar\omega = \frac{\langle U \rangle}{N}$$

$$\hbar = h/2\pi = 6.63 \times 10^{-34} \text{J} \times \text{s}/2\pi = 1.05 \times 10^{-34} \text{J} \times \text{s/rad}$$

where ω is the frequency of the light wave, N is the number of energy packets, and h is Planck's constant. The sensitivity of a detector, called the *quantum efficiency*, is the percent of energy packets that produce charge carriers.[10] The energy packets are called photons and are considered a fundamental particle. We can observe the energy packets by measuring the current produced by the incident light in a photoconductive or photoemissive detector.

In Figure 2.5 we display the current in an imaging detector (vertical axis) as a function of position on the detector (horizontal axis). The light distribution on the detector is a set of optical fringes giving the sinusoidal variation across the horizontal axis (we will discuss how to make these fringes later). At the lowest light intensity, the top three traces, a photoemissive detector called a photomultiplier was used. You see bursts of electrons in the detector circuit. These bursts of electrons are due in part to photons but we are looking at electrons and all of the electron spikes are not generated by photons. Some are due to various noise sources. These noise sources are:

1. *Johnson noise*: Electrons contributing to the observed current are generated by thermal energy.

---

[10] Units are electrons/photon or amps/watt and can range from less than 10% for photographic film to 90% for some CCDs.

2. *Shot noise*: This noise is associated with current fluctuations during the flow of a macroscopic current; thus, the quantization of the light energy and the discrete electric charges generate the observed noise.

3. *Photon noise*: Not all of the photons arriving at the detector produce charge carriers. This can lead to a larger effective shot noise as can also the zero-point fluctuations of the quantized electromagnetic field.

This subject is quite complicated and we will limit our discussion to pointing out that shot noise is not a distinctive feature of the quantization of the optical field but can be explained using semi-classical theory. The observed behavior has been successfully treated by Mandel and Wolf [5] by quantizing the electron detector system and treating the light wave classically. This theoretical treatment shows that it is not only photons but also electrons (the only quantum system in the semi-classical treatment) that are responsible for the effects shown in Figure 2.5.

We can overcome these various noise sources by using signal averaging to reduce the spikey behavior in the detector current. The observed intensity grows with the number of observations, N, while the noise grows as $\sqrt{N}$. The resulting observed current approaches a smooth curve as we show in the second and third traces. At high light intensities a less sensitive detector must be used to observe the electric current. At the higher light intensities, the observed current is a smoothly varying function, as one would expect from classical theory; see the bottom curve in Figure 2.5.

The name "photon" is also given in a rather loose fashion to the virtual exchange particle of the electromagnetic force between two charged particles (viz. electrons). The virtual particle is an important element in quantum electrodynamics (QED), but it cannot be detected experimentally. Some of the confusion in understanding the concept of a photon is due to the two definitions, but the basic properties do not agree with the historical view of a particle with mass.

The characteristics of a photon are the following:

- To exist, photons must travel at the speed of light within the medium they are passing through. The wave they are associated with is not a probability wave but an electromagnetic wave.

- A photon is a massless boson with spin 1 and electric charge 0. The lack of mass is key to QED. Experiments have set the mass to $< 10^{-23}$ of an electron mass.

- A photon has no antiparticle.

- All other quantum numbers of the photon (such as lepton number, baryon number, and flavor) are zero.

- Being bosons, photons don't obey Pauli exclusion; they like to crowd together.

- Photons are quanta of the electromagnetic field. Some typical energies are

$$U(\nu) = hc/\lambda \Rightarrow \begin{cases} 450\text{nm} & 2.76\text{eV} & 4.4 \times 10^{-19}\text{J} \\ 700\text{nm} & 1.8\text{eV} & 2.8 \times 10^{-19}\text{J} \end{cases}.$$

- The photons are not localized. They do have a cross section for interaction but that is a function of the interacting particle.

- Photons (virtual) are interacting particles of electromagnetism; they interact with charged particles.

While these characteristics lead to confusion when compared to particles with mass, there are additional sources of confusion. For example, when an electron absorbs a photon, the photon is completely destroyed and when an electron emits a photon, the photon is created instantaneously out of the vacuum. Photons aren't things. They pop in and out of being in a rapid way like glitter. What we call "particles" are really just knots or bundles of energy fields. Since the theory utilizing photons makes such accurate predictions, it has been suggested that attempts to discuss what is really happening are not particularly productive [6].

Many famous experiments once regarded as evidence for photons[11] were later explained qualitatively or semi-quantitatively based on classical electromagnetic (EM) theory. This has led to the formation of a minority group of physicists who have avoided the concept of photons altogether and prefer the idea that the utility of classical EM wave theory is not exhausted [7, 8]. We will not participate in the debate of this topic but simply limit our treatment of light using wave theory, and if we need to talk about interaction with matter we will treat the properties of matter using quantum theory; this is often called semi-classical theory.

We need to get a feel for what we are neglecting by ignoring photons. Let us look at the scale of two common optical events encountered daily.

In quantum theory we look for changes in the photon number by ± 1. How many of the photons would we encounter in a typical optical event? Assume we have a 1-mW laser pointer operating at a wavelength of 633 nm. The energy of a single photon is

$$U = \frac{hc}{\lambda} = \frac{\left(6.63 \times 10^{-34} \text{J} \cdot \text{s}\right) \cdot \left(3 \times 10^{8} \text{m/s}\right)}{6.33 \times 10^{-7} \text{m}} .$$
$$= 3.14 \times 10^{-19} \text{J}$$

The power of the laser is (energy/time) in watts. The 1-mW laser introduces 10^{-3} J/s. The number of photons generated by the laser pointer is then

$$N = \frac{1 \times 10^{-3} \text{J/s}}{3.14 \times 10^{-19} \text{J/photon}} = 3.18 \times 10^{15} \text{photon/s}.$$

The large number of photons makes it unlikely that a quantum event will impact most measurements at light levels normally encountered in our daily experience.

For another estimate of how many photons are found in everyday activities let us turn to photochemistry and ask how many photons we need to generate a photoreaction of a mole of something. We will optimistically assume we have a quantum efficiency of 100% and calculate the energy to produce a mole of photons to trigger the reaction. We assume the wavelength of the light we will use is 500 nm:

---

[11] For example, blackbody radiation, photoelectric effect, and Compton scattering.

$$U = \frac{hc}{\lambda} = \frac{\left(6.63 \times 10^{-34}\right) \cdot \left(3 \times 10^{8}\right)}{5 \times 10^{-7}} = 3.9 \times 10^{-19} \text{J}$$
$$\text{energy/mole} = \left(3.9 \times 10^{-19}\right) \cdot \left(6.03 \times 10^{23}\right) = 2.3 \times 10^{5} \text{J}$$

To get a feel for how much energy this is, compare it to burning a mole of propane in a Bunsen burner where we get 2.02×10^{6}J.

If the experiment involves shorter wavelengths, we see in Table 2.3 that fewer and fewer photons are involved in the experiment and quantum effects are more likely to be observed.

Before leaving QED, let's look at how the theory is formulated. Rather than using a quantized version of the classical fields, the vector and scalar potentials are quantized to create the QED description

$$\mathbf{E} = -\nabla\phi - \frac{\partial \mathbf{A}}{\partial t}$$
$$\mathbf{B} = \nabla \times \mathbf{A}.$$

where $\mathbf{A}$ is the vector potential and ϕ is the scalar potential. In classical theory you have the option to use either fields or potentials to describe the electromagnetic wave. The potentials are viewed as an optional way to solve Maxwell's equations but contain no unique physical significance and the preference is to use fields. In quantum theory, the vector potential becomes physically significant. It leads to the Aharonov–Bohm effect. For this reason, the potential representation is required in quantum theory.

When the perturbative approach of quantum field theory was applied in first order it was successful but additional accuracy was desired. When additional perturbation orders were added, divergences appeared. To handle these divergences a renormalization approach was taken. In this approach, the divergent terms were made to occur within terms that involved experimentally determined constants. By adding in the experimentally determined terms the infinities were made to disappear. The renormalization approach was given a theoretical basis by Richard Feynman, Julian Schwinger, and Sin-itiro Tomonaga, who shared the Nobel Prize for their effort. The resulting theory was able to predict the outcome of all optical experiments with an accuracy never before experienced.[12] In particular, the theoretical calculation of the anomalous magnetic dipole moment of the electron was combined with the experimentally

Table 2.3 Number of Photons Needed to Carry One Joule of Energy

Region	Wavelength meters	No. of photons
γ-rays	10^{-12}	5×10^{12}
Visible	5×10^{-7}	3×10^{18}
Microwave	10^{-2}	5×10^{22}

---

[12] The anomalous magnetic moment of the electron and the Lamb shift in the spectrum of hydrogen are calculated with a high degree of agreement with high-precision experiments.

measured spin g-factor[13] to give the fine structure constant[14] a precision of better than a part in a billion (10^9). Some people are not happy with a theory that depends on the subtraction of infinities from infinities to get a finite answer. Also, QED does not explain why or how but does provide correct, highly accurate answers. The accuracy of the theory has led everyone to use the theory despite its shortcomings. An excellent elementary introduction to QED has been written by Richard Feynman [9].

Momentum

An argument for the need to use a photon model for light is that the photon has momentum. We can easily prove that a classical electromagnetic wave also has momentum, as Johannes Kepler recognized in 1619. The origin of momentum, associated with a classical electromagnetic wave, is easier to understand than is the source of momentum associated with an abstract wave, Eq. (1.9). The electric field of the electromagnetic wave acts on a charged particle in a material with a force

$$\mathbf{F}_E = q\mathbf{E}. \tag{2.36}$$

This force accelerates the charged particle to a velocity v in a direction transverse to the direction of light propagation and parallel to the electric field. The moving charges interact with the magnetic field of the electromagnetic wave with a force, parallel to the propagation vector, of

$$\mathbf{F}_H = q\,(\mathbf{v} \times \mathbf{B})\,. \tag{2.37}$$

The combined action of these two forces (called the Lorentz force) creates a radiation pressure.

While the actual derivation will not be carried out here, it is possible by the comparison of the sum of these two forces, $(\mathbf{F}_E + \mathbf{F}_H)$, with results that can be derived from Maxwell's theory [10], to postulate a momentum density associated with the electromagnetic wave, given by

$$\mathbf{g} = \left(1/\mu_0 c^2\right) (\mathbf{E} \times \mathbf{B}) - \frac{\mathbf{S}}{c^2}. \tag{2.38}$$

A dimensional analysis can be used to verify that Eq. (2.38) is a momentum density; the units of $\mathbf{g}$ are

$$\frac{\dfrac{J}{m^2 \cdot s}}{\left(\dfrac{m}{s}\right)^2} = \frac{\dfrac{kgm \cdot m}{s}}{m^3}.$$

The pressure on a surface, of area A, is defined as

$$P = \frac{\mathbf{F} \cdot \hat{\mathbf{n}}}{A} = \frac{\dfrac{\Delta \mathbf{p}}{\Delta t} \cdot \hat{\mathbf{n}}}{A}.$$

---

[13] Classical theory predicts that the g-factor $= 2$ but the observed value differs by a few percent.

[14] This is a fundamental physical constant, $\alpha = e^2 Z_0/2h$.

We assume the light is totally absorbed; i.e., the momentum change is equal to the total momentum contained in the light wave, $\Delta\mathbf{p} = \mathbf{p}$. The total momentum in the light wave is given by the momentum density, $\mathbf{g}$, multiplying by a unit volume, V, $\mathbf{p} = \mathbf{g}V$:

$$P = \frac{\frac{V\mathbf{g}}{\Delta t} \cdot \hat{\mathbf{n}}}{A}.$$

We will choose a volume $c\Delta t$ long by A in surface area (see Figure 2.4.), enabling the pressure to be expressed as

$$P = \frac{\left(\frac{\mathbf{g} \cdot \hat{\mathbf{n}}}{\Delta t}\right) c \cdot \Delta t \cdot A}{A} = \frac{\mathbf{S} \cdot \hat{\mathbf{n}}}{c} = \frac{I}{c}. \tag{2.39}$$

At the Earth's surface and normal to it, sunlight has a flux density of 1.36×10^3 J/(m²·s).[15] We will make the incorrect assumption that the flux density of sunlight is equal to the Poynting vector, making it possible, through the use of Eq. (2.32), to estimate the pressure of sunlight to be

$$P = 4.53 \times 10^{-6} \frac{\text{Newtons}}{m^2}.$$

As a point of reference, atmospheric pressure is about 10^5 N/m². For our example, a wave described earlier has an energy of

$$I = \frac{c\varepsilon_o E^2}{2} = 2.4 \text{W/m}^2$$

$$P = \frac{I}{c} = \frac{\varepsilon_o E^2}{2} = \frac{\left(8.85 \times 10^{-12}\right)\left(42.42\right)^2}{2} = 8 \times 10^{-9} \text{N/m}^2.$$

[15] This is the solar constant, an experimentally determined value averaged over time. It has varied less than 0.2% over the past 400 years.

Substituting Eq. (2.28) into Eq. (2.34), we discover that the radiation pressure is equal to the energy density of the incident radiation

$$P = \langle U \rangle.$$

Combining Eq. (2.28) with Eq. (2.33) suggests that it is proper to associate momentum with the ratio of wave energy to velocity. This is consistent with relativistic principles. In the theory of relativity, the energy is given by

$$U = mc^2,$$

which implies a mass of U/c^2 and a momentum of U/c. The idea is also consistent with quantum theory where $U = h\nu$ so that

$$p = \frac{h}{\lambda} = \frac{h\nu}{c} = \frac{U}{c}.$$

An interesting question is how many photons does it take to exert a pressure of 1 N/m²? We need to pick a wavelength of the light so lets use a HeNe laser with $\lambda = 633$ nm. We will assume that the surface has an area $A = 1$ m² and is totally reflecting so that the momentum change of a single photon is

$$\Delta p = 2\,(h/\lambda)$$

and the pressure is given by the number of photons times the momentum of a single photon:

$$P = F/A = N\left(\frac{2h}{\lambda}\right)$$

$$N = \frac{663 \times 10^{-9}}{2\left(6.63 \times 10^{-34}\right)} = 5 \times 10^{26} \text{ photons}$$

Optical Tweezer

In 1970, Arthur Ashkin of Bell Labs reported on observing the forces of light on micrometer-sized particles [11]. This was not the first time light pressure was observed but was the first time that forces sufficient to actually move microscopic particles were observed. The invention of the laser made it possible for Ashkin to demonstrate acceleration, deceleration, and trapping of micrometer-sized neutral particles. The initial objective was to create traps to hold neutral atoms, but applications in the biological sciences were developed by the end of the 1980s. To see the forces on a micrometer-sized object, consider the scattering of light by a dielectric sphere, in Figure 2.6. As a light ray is refracted by the dielectric sphere, the direction of the ray changes and thus there is a momentum change. Newton's third law states there will be an equal and opposite momentum change on the dielectric sphere.

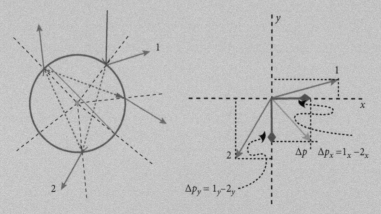

Figure 2.6 Optically generated scattering force. Trajectory of a light ray through a dielectric sphere. The momentum change felt by the sphere is opposite to the momentum change due to refraction of light beams 1 and 2, shown as Δp.

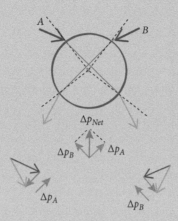

Figure 2.7 Trapping of a dielectric sphere in a focused laser beam. The momentum change felt by the sphere is opposite to the path of the light paths created by a focused laser beam. The vector addition of the momentum change keeps the sphere centered in the beam. If we did the analysis in three dimensions, we would find the momentum change also keeps the sphere just behind the beam focus.

If the sphere is uniformly illuminated, then the total force on the sphere is equal and opposite to the change in momentum due to the refraction of the light beam. We only consider the transmission of one very thin beam. We are able to use the geometrical calculation of force shown in Figures 2.6 and 2.7, because we assume the particle is much larger than the wavelength.[16] A second approximation called the Rayleigh regime or dipole approximation assumes the particle is smaller than the wavelength of light and can be treated as a dipole with homogeneous fields within the particle. The *size parameter* that determines the regime is

$$ka = \frac{2\pi n_1}{\lambda_0} a \quad \begin{array}{l} \text{Geometric} \quad ka \ll 1 \\ \text{Rayleigh} \quad \frac{n_2}{n_1} ka \ll 1 \end{array},$$

where the index is measured in the medium supporting the particle of diameter a and index n_2.

Instead of using a uniform beam of light, if we focus the beam, a trap of the sphere can be created; see Figure 2.7. This is called an optical tweezer. The tweezer can be used not only to manipulate microparticles but also to measure the forces generated by biological motors. Biological motors are ubiquitous in biology and used for locomotion and mechanical motions within a cell.

---

[16] A Nd:YAG laser with a wavelength of 1056 nm is normally used because of the low absorption in the biological samples.

2.4 **Polarization**

The displacement of a transverse electromagnetic wave shown in Figure 2.3 is defined as a vector quantity in Maxwell's equations. Before light was identified as an electromagnetic wave, a mechanical view built on sound wave theory dominated the scientific community.[17] Sound in the atmosphere was a longitudinal wave and the concept of polarization did not make sense. Robert Hooke proposed in 1665 that the vibrations of a light wave could be transverse. Michael Faraday showed that the polarization of light passing through a transparent dielectric, along a magnetic field, would be rotated; this is now called Faraday rotation. This was the first evidence that light was an electromagnetic wave and led James Clerk Maxwell to study the relationship between electromagnetic radiation and light.

We have just examined Maxwell's equations and determined from Figure 2.3. that we must specify not only the frequency, phase, and direction of the wave but also the magnitude and direction of the wave displacement, which by convention is the electric field. The direction of the electric vector is called the *direction of polarization*, and the plane containing the direction of polarization and the propagation vector is called the *plane of polarization*. This quantity has the same name as the field quantity introduced in Eq. (2.1). Because the two terms describe completely different physical phenomena, there should be no danger of confusion.

The experimental demonstration of polarization occurred before we had a mathematical representation of electromagnetic waves. In 1808, *Etienne Louis Malus* (1775–1812) accidentally observed, while watching a sunset reflected from the windows across the street from his apartment, that the two images of the sun produced by a piece of calcite would alternately dim and brighten as the calcite was rotated.[18] He coined the word *polarization* for the observed property of light, relating the orientation of the preferred direction of the light to the Earth's poles. Malus defined the plane of polarization of the reflected light as the plane of incidence. This was an arbitrary definition made with no knowledge of electromagnetic theory; in fact, Malus' view of light was based upon Newton's particle theory. The definition of the plane of polarization used by Malus is orthogonal to the current definition of polarization in terms of the electric field. Many authors use the term *plane of vibration* for the plane containing the electric field to differentiate the modern definition from the one made by Malus.

As an interesting aside, Malus undertook a study of birefringence (double refraction), the property of the calcite crystal that led to his discovery of polarization. His motivation was a money prize offered for an experimental and theoretical explanation of the effect. Everyone hoped his solution would support the current belief in Newton's particle model of light. He won the prize in 1810 by demonstrating the opposite, that Huygens's wave theory explained all observations.

The present concept of polarization had to await the suggestion by Young, in 1817, that light propagated as a transverse wave. Augustin-Jean Fresnel claimed to have made the association of light and transverse waves prior to Young, but even if he did not, he did construct a formal theory of the concept in 1824. The final step in understanding the polarization properties of light came with the introduction of Maxwell's theory.

---

[17] Newton wrote in *The Principia*: "I wish we could derive the rest of phenomena of nature by the same kind of reasoning from mechanical principles; for I am induced by many reasons to suspect that they may all depend upon certain forces."

[18] The origin of the light variation will be covered after we introduce reflection and refraction in the next chapter.

2.4.1 **Polarization ellipse**

The first step in discussing polarization is to define a coordinate system. Assume that a plane wave is propagating horizontally from left to right in the positive z-direction and the electric field, determining the direction of polarization, is oriented in the x, y-plane. We pick the y-axis as vertical, floor to ceiling, and the x-axis as in the horizontal direction, perpendicular to our z-axis. The electric field direction will hop over time from position to position in the x, y-plane, as shown in Figure 2.8.[19] In Figure 2.8. we are viewing the light coming toward us from the source located somewhere in the negative z-direction.

In complex notation, the plane wave is written in terms of the x and y components of $\mathbf{E}_0$:

$$\tilde{\mathbf{E}} = E_{0x}e^{i(\omega t - kz + \phi_1)}\hat{\mathbf{i}} + E_{0y}e^{i(\omega t - kz + \phi_2)}\hat{\mathbf{j}}. \tag{2.40}$$

When these unit vectors are added together, the result will be a vector whose tip moves along a closed curve called a *Lissajous figure*. The geometrical construction shown in Figure 2.9. can be used to visualize the generation of the Lissajous figure. The harmonic motion, along the x-axis, is found by projecting a vector rotating around a circle of diameter E_{ox} onto the x-axis. The

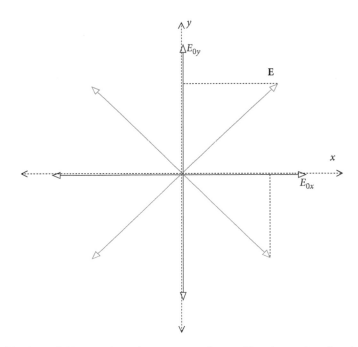

Figure 2.8 Possible electric field vectors located in a transverse plane positioned at $z = 0$, as viewed toward the source located in the negative z-direction. The x,y-coordinate system is oriented with respect to the lab frame with x horizontal and y vertical. In unpolarized light the electric field vectors will randomly hop from orientation to orientation, like the vectors shown here. Because the hopping occurs at near the light frequency we cannot detect any preferred orientation.

---

[19] In astronomy this lab coordinate system is replaced by the equatorial coordinate system. The proper coordinate system is at the discretion of the experimenter.

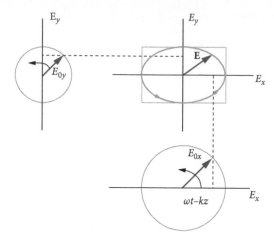

Figure 2.9 Geometrical construction showing how the Lissajous figure is constructed from harmonic motion along the x and y coordinate axes. Projecting a vector rotating around a circle onto the axis creates the harmonic motion along each coordinated axis. By projecting the vectors on perpendicular x and y-axes we guarantee that the phase difference between the two harmonic motions will be 90°.

harmonic motion, along the y-axis, is generated the same way using a circle of diameter E_{oy}. The resulting x and y components are added to obtain **E**:

$$\frac{E_x}{E_{0x}} = \cos(\omega t - kz + \phi_1) = \cos(\omega t - kz)\cos\phi_1 - \sin(\omega t - kz)\sin\phi_1,$$

$$\frac{E_y}{E_{0y}} = \cos(\omega t - kz)\cos\phi_2 - \sin(\omega t - kz)\sin\phi_2.$$

We use these equations to generate a mathematical description of the Lissajous figure in Figure 2.9. We eliminate the dependence on $(\omega t - kz)$. First, multiply the equations by $\sin\phi_2$ and $\sin\phi_1$, respectively and then subtract the resulting equations. Second, multiply the two equations by $\cos\phi_2$ and $\cos\phi_1$, respectively, and then subtract the new equations:

$$\frac{E_x}{E_{0x}}\sin\phi_2 - \frac{E_y}{E_{0y}}\sin\phi_1 = \cos(\omega t - kz)[\cos\phi_1\sin\phi_2 - \sin\phi_1\cos\phi_2],$$

$$\frac{E_x}{E_{0x}}\cos\phi_2 - \frac{E_y}{E_{0y}}\cos\phi_1 = \sin(\omega t - kz)[\cos\phi_1\sin\phi_2 - \sin\phi_1\cos\phi_2].$$

(2.41)

The terms in brackets can be simplified by using the trig identity

$$\sin\delta = \sin(\phi_2 - \phi_1) = \cos\phi_1\sin\phi_2 - \sin\phi_1\cos\phi_2.$$

A similar trig identity is

$$\cos\delta = \cos(\phi_2 - \phi_1) = \cos\phi_1\cos\phi_2 + \sin\phi_1\sin\phi_2.$$

Using these identities and adding the square of equations Eq. (2.41) we get

$$\left(\frac{E_x}{E_{0x}}\right)^2 + \left(\frac{E_y}{E_{0y}}\right)^2 - \left(\frac{2E_xE_y}{E_{0x}E_{0y}}\right)\cos\delta = \sin^2\delta.$$

(2.42)

2.4.2 **Circular polarization**

In Figure 2.9, the two harmonic oscillators both have the same frequency, $(\omega t - kz)$, but differ in phase by

$$\delta = \phi_2 - \phi_1 = -\frac{\pi}{2}.$$

The minus sign arises because the x-component reaches its maximum 90° before the y component. The tip of the electric field, **E**, in Figure 2.9. traces out an ellipse, with its axes aligned with the coordinate axes. To determine the direction of the rotation of the vector assume that $\phi_1 = 0$, $\phi_2 = -\pi/2$, and $z = 0$, so that

$$\frac{E_x}{E_{0x}} = \cos \omega t, \quad \frac{E_y}{E_{0y}} = \sin \omega t,$$

$$\mathbf{E} = \left(\frac{E_x}{E_{0x}}\right)\hat{\mathbf{i}} + \left(\frac{E_y}{E_{0y}}\right)\hat{\mathbf{j}}.$$

The normalized vector **E** can easily be evaluated at a number of values of ωt to discover the direction of rotation. Table 2.4 shows the approach used to determine the direction of motion of the electric field vector as the light evolves in time by calculating the value of the electric vector as ωt increases.

The rotation of the vector **E** in Figure 2.9. is seen to be in a counterclockwise direction, moving from the positive x-direction, to the positive y-direction and finally to the negative x-direction. If the motion of the electric vector is moving in a counterclockwise direction as we face the source, then the polarization is *right-handed*.[20] If the electric vector moves around the ellipse in a clockwise direction, as we face the source, the polarization is defined as *left-handed* and the light is said to be left-circularly polarized. A Java Applet is available for the reader to explore the graphical richness of Lissajous figures beyond a display of polarization.[21]

Table 2.4 Rotating E-Field Vecto

ωt	**E**
0	$\hat{\mathbf{i}}$
$\frac{\pi}{4}$	$\frac{1}{\sqrt{2}}(\hat{\mathbf{i}} + \hat{\mathbf{j}})$
$\frac{\pi}{2}$	$\hat{\mathbf{j}}$
$\frac{3\pi}{4}$	$\frac{1}{\sqrt{2}}(-\hat{\mathbf{i}} + \hat{\mathbf{j}})$
π	$-\hat{\mathbf{i}}$

[20] It is recommended that the reader commit only one of these definitions to memory.
[21] Available at https://ngsir.netfirms.com/englishhtm/Lissajous.htm.

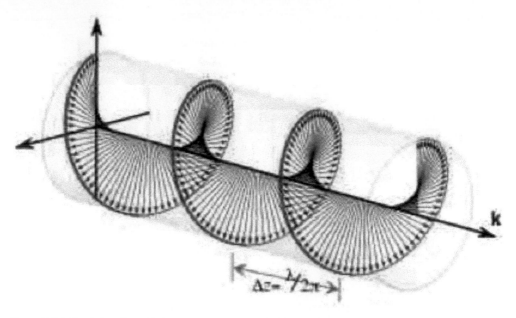

Figure 2.10 Circularly polarized light with the source on the left and the receiver on the right. Looking back at the source the polarization moves in a counter-clockwise direction in the *x,y*-plane at the origin and is considered right-handed circularly polarized light. Never consider the polarization by looking at the receiver or you will constantly confuse yourself.

By Dave3457 - Own work, Public Domain, *https://commons.wikimedia.org/w/index.php?curid=9861553*

If we compare the definition for optical polarization to the one used in particle physics, where light would be said to have a negative helicity if it rotated in a clockwise direction we find agreement. If we look at the source, the electric vector seems to follow the threads of a left-handed screw agreeing with the nomenclature that left-handed quantities are negative. To summarize, the description from optics of *left-circularly polarized* light maps directly onto the description of negative helicity used in particle physics.

The association of right-circularly polarized light with "right-handedness" in optics came about by looking at the path of the electric vector in space at a fixed time; see Figure 2.10.

2.4.3 Linear polarization

Now consider when $\delta = 0$ or π, the ellipse defined in Eq. (2.40) collapses into a straight line with slope E_{oy}/E_{ox}. The equation of the straight line is

$$\frac{E_x}{E_{0x}} = \pm \frac{E_y}{E_{0y}}.$$

At a fixed point in space, the x and y components oscillate in phase (or 180° out of phase) according to the equation

$$\mathbf{E} = \left(E_{0x}\hat{\mathbf{i}} \pm E_{0y}\hat{\mathbf{j}}\right) \cos\left(\omega t - \phi\right).$$

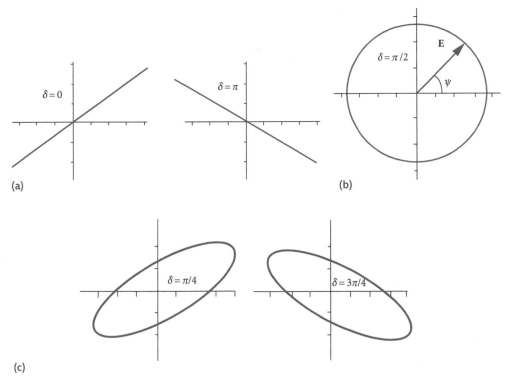

Figure 2.11 Lissajous figures for elliptical polarized light: (a) linear polarization, (b) circular polarization, and (c) elliptical polarization. The amplitudes E_{ox} and E_{oy} of curves in (a) and (b) were set equal. The elliptical curves in (c) were calculated with $E_{ox} = 0.75$ and $E_{oy} = 0.25$.

The electric vector undergoes simple harmonic motion along the line defined by E_{0x} and E_{0y}. At a fixed time, the electric field varies sinusoidally along the propagation path (the z-axis) according to the equation

$$\mathbf{E} = \left(E_{0x}\hat{\mathbf{i}} \pm E_{0y}\hat{\mathbf{j}}\right)\cos\left(\phi - kz\right).$$

This light said to be *linearly polarized*.

Figure 2.11 displays the Lissajous figures for five phase differences. This is the time average of the rotation of the electric field vector as the optical wave propagates toward the viewer.

2.4.4 Stokes parameters

We have seen how to visualize the polarization of an electromagnetic wave. Mathematically, you can demonstrate that four independent measurements are needed to characterize polarization. There is no unique set of measurements but a set of easily remembered measurements were defined by George Gabriel Stokes in 1852 and have become the standard.

The Stokes parameters of a light wave are measurable quantities, i.e. time-averaged intensity measurements defined as:

$s_0 \Rightarrow$ Total flux density

$s_1 \Rightarrow$ Difference between flux density transmitted by a linear polarizer oriented parallel to the x-axis and one oriented parallel to the y-axis. The x- and y-axes are usually selected to be parallel to the horizontal and vertical directions in the laboratory.

$s_2 \Rightarrow$ Difference between flux density transmitted by a linear polarizer oriented at 45° to the x-axis and one oriented at 135°.

$s_3 \Rightarrow$ Difference between flux density transmitted by a right-circular polarizer and that by a left-circular polarizer.

Only three of the Stokes parameters are independent for totally polarized light waves:

$$s_0^2 = s_1^2 + s_2^2 + s_3^2. \tag{2.43}$$

As you see, we need to find a means for measuring the components of the time-averaged electric field in the x, y-plane. The Stokes's parameters can be used to describe the *degree of polarization* defined as

$$V = \frac{1}{s_0}\sqrt{s_1^2 + s_2^2 + s_3^2}, \tag{2.44}$$

This set of parameters can completely characterize any physically realizable electromagnetic wave. Completely polarized light would have $V = 1$.

H. *Mueller*[12] pointed out that the Stokes parameters could be thought of as elements of a column matrix or a 4-vector:

$$\begin{pmatrix} s_0 \\ s_1 \\ s_2 \\ s_3 \end{pmatrix}.$$

We will generate polarized light using reflection in the next chapter, which will allow us to construct an experimental Stokes vector. In Chapter 5 we will discuss devices that allow us to measure the Stokes parameters.

Jones Vector

There is one other representation of polarized light, complementary to the Stokes parameters, developed by *R. Clark Jones* in 1941 and called the *Jones vector*. It is superior to the Stokes vector for discussing coherent light in that it handles light of a known phase and amplitude with a reduced number of parameters. It is inferior to the Stokes vector in that, unlike the Stokes representation, which is experimentally determined, the Jones representation is theoretically generated and cannot handle unpolarized or partially polarized light. It can only describe light with a well-defined phase and frequency.

Assuming the coordinate system is such that the electromagnetic wave is propagating along the z-axis, it was shown earlier that any polarization could be decomposed into two orthogonal **E** vectors, say for this discussion, parallel to the x- and y-directions. The Jones vector is defined as a two-row, column matrix consisting of complex components in the x- and y-directions:

$$\mathbf{E} = \begin{bmatrix} E_{0x}e^{i(\omega t - \mathbf{k}\cdot\mathbf{r} + \phi_1)} \\ E_{0y}e^{i(\omega t - \mathbf{k}\cdot\mathbf{r} + \phi_2)} \end{bmatrix}. \tag{2.45}$$

If absolute phase is not an issue, then we may normalize the vector by dividing by that number (real or complex) that simplifies the components but keeps the sum of the square of the components equal to 1. For example, assume that $E_{0x} = E_{0y}$, then

$$\mathbf{E} = E_{0x}e^{i(\omega t - \mathbf{k}\cdot\mathbf{r} + \phi_1)} \begin{bmatrix} 1 \\ e^{i\delta} \end{bmatrix}$$

The normalized vector would be the terms contained within the bracket, each divided by $1/\sqrt{2}$.

The general form of the Jones vector is

$$\mathbf{E} = \begin{bmatrix} \mathbf{A} \\ \mathbf{B} \end{bmatrix}, \quad \mathbf{E}^* = \begin{bmatrix} \mathbf{A}^* & \mathbf{B}^* \end{bmatrix}.$$

Some examples of Jones vectors[22] (on the left) and Stokes vectors (on the right) are shown in Table 2.5.

Table 2.5 Jones and Stokes Vectors

Horizontal polarization

$$\begin{bmatrix} 1 \\ 0 \end{bmatrix} \quad \begin{bmatrix} 1 \\ 1 \\ 0 \\ 0 \end{bmatrix}$$

Vertical polarization

$$\begin{bmatrix} 0 \\ 1 \end{bmatrix} \quad \begin{bmatrix} 1 \\ -1 \\ 0 \\ 0 \end{bmatrix}$$

(*Continued*)

---

[22] Throughout this book the phase of the light wave traveling along the z-direction is given by $\phi = (\omega t - kz)$, so that the Jones vector of $i = e^{i\pi/2}$ corresponds to an advancement of $\pi/2$. In other books the phase is $\phi = (kz - \omega t)$ and everything is reversed. The Jones vector component of $i = e^{i\pi/2}$ corresponds to a retardation of $\pi/2$.

Table 2.5 Continued

+45° polarization

$$\frac{1}{\sqrt{2}}\begin{bmatrix} 1 \\ 1 \end{bmatrix} \qquad \begin{bmatrix} 1 \\ 0 \\ 1 \\ 0 \end{bmatrix}$$

−45° polarization

$$\frac{1}{\sqrt{2}}\begin{bmatrix} 1 \\ -1 \end{bmatrix} \qquad \begin{bmatrix} 1 \\ 0 \\ -1 \\ 0 \end{bmatrix}$$

Right circular polarization

$$\frac{1}{\sqrt{2}}\begin{bmatrix} 1 \\ i \end{bmatrix} \qquad \begin{bmatrix} 1 \\ 0 \\ 0 \\ 1 \end{bmatrix}$$

Left circular polarization

$$\frac{1}{\sqrt{2}}\begin{bmatrix} 1 \\ -i \end{bmatrix} \qquad \begin{bmatrix} 1 \\ 0 \\ 0 \\ -1 \end{bmatrix}$$

We have the matrix math to handle polarization but how do we measure and control the various quantities, for example, in a Stokes vector? We need an optical device that treats each component of polarization differently. We will introduce some of the devices used to control polarization in a later chapter.

2.5 **Problem Set 2**

(1) What is the refractive index of a material if the speed of light in the material is 50% of the vacuum speed?

(2) Light is traveling in glass (n = 1.5). If the amplitude of the electric field of the light is 100 V/m (volts/meter), what is the amplitude of the magnetic field? What is the magnitude of the Poynting vector?

(3) A 60-W monochromatic point source is radiating equally in all directions in a vacuum. What is the electric field amplitude 2 m from the source?

(4) The flux density at the Earth's surface due to sunlight is $I = 1.36 \times 10^3$ J/(m²•s). Calculate the magnitude of the electric and magnetic fields at the Earth's surface by assuming that the average Poynting vector is equal to this flux density.

(5) In a vacuum a slab of GaAs is illuminated by a wavelength of 1033 nm. The index of refraction of GaAs at that wavelength is n = 3.49. What is the wave number and the effective wavelength of this illumination within GaAs?

(6) What is the flux density of light needed to keep a glass sphere of 10^{-8} g and 2×10^{-5} m in diameter floating in midair?

(7) Describe the polarization of a wave with the Jones vector

$$\begin{pmatrix} -i \\ 2 \end{pmatrix}.$$

Write the Jones vector that is orthogonal to this vector and describe its polarization.

(8) What is the polarization of the waves

$$\mathbf{E} = E_0 \left[\hat{\mathbf{i}} \cos(\omega t - kz) + \hat{\mathbf{j}} \cos\left(\omega t - kz + \frac{5\pi}{4}\right) \right],$$
$$\mathbf{E} = E_0 \left[\hat{\mathbf{i}} \cos(\omega t + kz) + \hat{\mathbf{j}} \cos\left(\omega t + kz - \frac{\pi}{4}\right) \right],$$
$$\mathbf{E} = E_0 \left[\hat{\mathbf{i}} \cos(\omega t - kz) - \hat{\mathbf{j}} \cos\left(\omega t - kz + \frac{\pi}{6}\right) \right]?$$

(9) Write an expression, in MKS units, for a plane electromagnetic wave, with a wavelength of 500 nm and an intensity of 53.2 W/m², propagating in the z-direction. Assume that the wave is linearly polarized at an angle of 45° to the x-axis.

(10) If a 1-kW laser beam is focused to a spot with an area of 10^{-9} m², what is the amplitude of the electric field at the focus?

(11)* Given the Stokes vector

$$\begin{pmatrix} 1 \\ 0 \\ -\dfrac{3}{5} \\ \dfrac{4}{5} \end{pmatrix},$$

(a) calculate the degree of polarization, (b) determine the orthogonal vector, and (c) draw the polarization ellipse.

(12) The first demonstrated solar sail vehicle, called the IKAROS, was launched in 2010, and communications were received from the vehicle until 2015. The sail was highly reflective. Why did they select reflective rather than absorptive? What is the difference in momentum change between totally absorbed light and totally reflected light?

REFERENCES

1. Maxwell, J. C., A dynamical theory of the electromagnetic field. *Phil. Trans. Ro. Soc. London* **155**: 459–512 (1865).

2. Mansuripur, M., The Ewald-Oseen extinction theorem, in *Classical Optics and its Applications*, pp. 168–82. Cambridge, UK: Cambridge University Press, 2009.

3. Novotny, L., The history of near-field optics, in E. Wolf (Ed.), *Progress in Optics*, p. 137–84. Amsterdam: Elsevier, 2007.

4. Landau, L. D., E. M. Liftshitz, and L. P. Pitaevskii, *Electrodynamics of Continuous Media*, 2nd edn. Oxford: Pergamon Press, 1984.

5. Mandel, L., and E. Wolf, *Optical Coherence and Quantum Optics*. Cambridge, UK: Cambridge University Press, 1995.

6. Mermin, N. D., Could Feynman have said this? *Physics Today* **57**(5): 10 (2004).

7. Kidd, R., J. Ardini, and A. Anton, Evolution of the modern photon. *Am. J. Phys.* **57** (1989).

8. Kuhlmann, M., What is real? *Scientific American* **309**(2): 40–7 (2013).

9. Feynman, R. P., *QED: The Strange Theory of Light and Matter*. Princeton Science Library. Princeton, NJ: Princeton University Press, 2006.

10. Wangsness, R. K., *Electromagnetic Fields*, 2nd edn. New York: Wiley, 1986.

11. Ashkin, A., Acceleration and trapping of particles by radiation pressure. *Phys Rev Letts.* **24**(4): 156–9 (1970).

12. Mueller, H., The foundation of optics. *J. Opt. Soc. Am.* *38*: 661 (1948).

3 Reflection and Refraction

3.1 Introduction

In Chapter 2, we treated the propagation of light in a uniform medium using Maxwell's equation and discovered the properties of the propagating light wave. In this chapter we wish to explore what happens to the propagation of a light wave when the electrical properties of the medium change in a discontinuous way. In most solids, the atoms are arranged in a periodic fashion with lattice constants on the order of ½ nanometer. This dimension is three orders of magnitude smaller than the wavelength of green light of about 500 nm. Thus, for a given direction of propagation, the light field sees a homogeneous medium and the atoms of the medium see a uniform electric field. This scale difference allows us to treat the propagation medium as uniform and isotropic, as we did in Chapter 2. The atoms and molecules that make up the medium respond in a collective fashion to the incoming light wave, as we described using Figure 2.1. Variation of the index of refraction over space can be used to control the optical properties[1] but involves complex theories we are not prepared to study. We will restrict our discussion to the collective behavior of naturally occurring materials.

We will find that the wave will experience reflection at the boundary between two media with different electromagnetic properties. The light transmitted across the boundary will undergo a change in propagation direction. The direction change is called *refraction*. We will rely only on the wave properties of light to obtain the fundamental laws of reflection and refraction. We will use boundary conditions developed in classical electromagnetic theory for Maxwell's equations to obtain equations that yield the amount of light transmitted or reflected.

Once equations for the reflected and transmitted amplitudes are obtained, we will consider light incident normal to the boundary to simplify the equations for the amplitudes of reflected and transmitted waves. With this simplification, it will become obvious that the fractions of the wave reflected and transmitted at the boundary between two media depend on the relative propagation velocities of the wave in the two media.

We will find that there is an angle, called Brewster's angle, for which light reflected from a boundary will be linearly polarized. There is also a set of conditions for which all light incident on a boundary will be reflected. A few of the properties of this reflected wave, called a totally reflected wave, will be discussed.

---

[1] The field of optics involves the study of photonic crystals, where the index of refraction varies periodically in two and three dimensions.

Modern Optics Simplified. B. D. Guenther. © B. D. Guenther 2020.
Published in 2020 by Oxford University Press. DOI: 10.1093/oso/9780198842859.001.0001

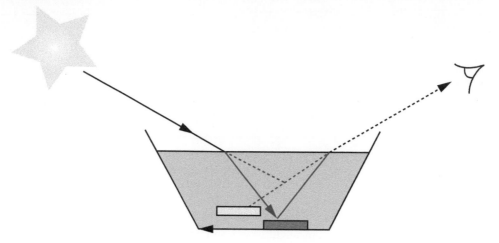

Figure 3.1 A coin hidden by the side of a bowl appears when covered by water due to the refraction of light at the water surface. The lines drawn represent the normal to the light's wave front and we call them rays. The bending of the ray demonstrates why the coin becomes visible.

Use of reflection existed before written history, as is evidenced by the discovery of a mirror in Turkey from the period around 8000 years ago [1]. Some of the earliest written comments about reflection can be found in Exodus 38:8 and Job 37:18. *Euclid*, in about 300 BC, discussed the focus of a spherical mirror in his book *Catoptrics*.*Cleomedes* (AD 50) discussed refraction of light at an air–water interface. He described an experiment whereby a coin at the bottom of a bowl, and hidden by the bowl's sides, could be made visible by pouring water in the bowl; see Figure 3.1.

Claudius Ptolemy of Alexandria (AD 139) made tables of the angles of incidence and refraction; see Figure 3.2. His work is one of the few examples of experiment during that time. The approach to experiment and theory was quite different then. Ptolemy theorized that the ratio of incident and refracted angles should be a constant for each surface. Experimentally it was not, so he modeled the corrections needed to the experiment to make it fit the theory.

Actual discovery of the law of refraction was delayed because the sine function had not yet been discovered. *Johannes Kepler* (1571–1630) gave a broad outline of the correct theory of the telescope and discussed total internal reflection without knowledge of the law of refraction. He used an empirical expression, $\theta_i = n\theta_r$, where $n = \dfrac{3}{2}$.

The law of refraction was discovered, evidently through experimentation, by *Willebrord Snellius* (Willebrord Snel van Royen, 1580–1626), a professor of mathematics at Leiden. He never published but Christiaan Huygens and Isaak Voss claimed to have examined Snell's manuscript. Ibn Sahl is credited with the first discovery in 984, about the same time that tables of sines and cosines were first published, but he failed to get credit, probably because his paper was misfiled in two libraries. A comparison of the measurements of Ptolemy and the prediction of Snell's law are found in Figure 3.2.

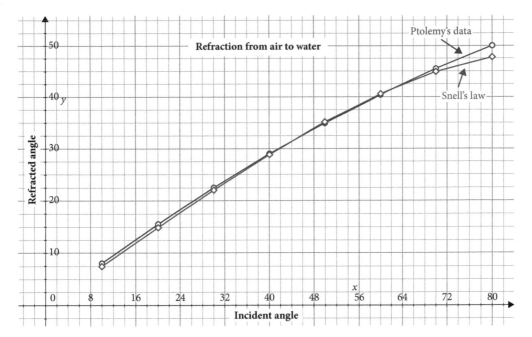

Figure 3.2 Refraction measurements at the air–water interface reported by Ptolemy and compared to Snell's law data with the index for water $n = 1.33$.

3.2 **Laws of Reflection and Refraction**

The three-dimensional problem of a wave propagating across a boundary, where a discontinuous change of propagation velocity occurs, can be studied by using the coordinate system shown in Figure 3.3. We will perform the derivation by using the boundary condition that the phase of the wave must vary smoothly across the boundary.

The wave velocity in the medium in the upper half plane of Figure 3.3 is v_i. For a light wave, this propagation velocity can be indicated by characterizing the medium with the index of refraction, $n_i = c/v_i$, Eq. (2.15).

The incident wave is assumed to be of the form $f(\omega t - \mathbf{k} \cdot \mathbf{r})$, as we introduced in Eq. (1.12). A transmitted wave is expected but the boundary conditions cannot be met without the addition of a reflected wave. To learn something about the geometry of reflective and refractive waves, we require that the phases of the three wave functions, incident, reflected, and transmitted, be the same on the boundary between the two half planes (at $z = 0$),

$$\omega_i t - \mathbf{k}_i \cdot \mathbf{r}\big|_{z=0} = \omega_r t - \mathbf{k}_r \cdot \mathbf{r}\big|_{z=0} = \omega_t t - \mathbf{k}_t \cdot \mathbf{r}\big|_{z=0}. \tag{3.1}$$

Since the equality of Eq. (3.1) must hold for all time, independent of the spatial coordinates, the frequency of the wave across the boundary does not change: $\omega_i = \omega_t = \omega$.

The frequency across the discontinuity is required, by the boundary conditions, to be a constant but the velocity does change. This means that the wavelength must change, as shown in Figure 3.4.

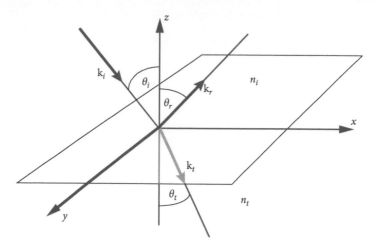

Figure 3.3 Coordinate diagram for light reflection and transmission across a boundary. The index of refraction in the upper half plane is n_i and in the lower half plane is n_t

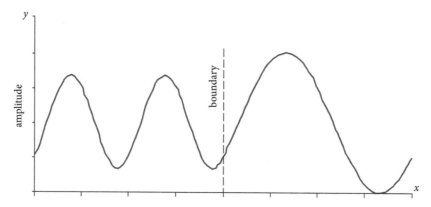

Figure 3.4 Two strings of unequal mass/unit length joined at $x = 0$. The wave velocity of the string on the left is 20 m/s and that on the right is 10 m/s. The wave is incident onto the junction from the left and has an amplitude of 3 cm and a wavelength of 1 m. The plot shows the resultant waves: on the right of $x = 0$ the transmitted wave, and on the left of $x = 0$ the sum of the incident and reflected waves.

There are two opposite propagating waves to the left of the origin. From the principle of superposition, the displacement of the wave at any time is a sum of these two waves: the incident and the reflected waves. The addition of the incident and the reflected wave leads to a standing wave.

At $z = 0$ we have the components of the various terms given using the following notation to identify the components:

$$\mathbf{a} \Rightarrow \left(|\mathbf{a}|_{mag}\right)\left(x_i, y_j, z_k\right)$$

$$\mathbf{r} \Rightarrow (x, y, 0), \quad \mathbf{k}_i \Rightarrow \left(\frac{n_i\omega}{c}\right)\left(k_x^i, k_y^i, k_z^i\right),$$

$$\mathbf{k}_r \Rightarrow \left(\frac{n_i\omega}{c}\right)\left(k_x^r, k_y^r, k_z^r\right), \quad \mathbf{k}_t \Rightarrow \left(\frac{n_t\omega}{c}\right)\left(k_x^t, k_y^t, k_z^t\right).$$

where (k_x^i, k_y^i, k_z^i) and (k_x^t, k_y^t, k_z^t) are the direction cosines of the incident and transmitted waves, i.e., for the incident wave in a rectangular coordinate system:

$$\mathbf{k}_i = \left(\frac{n_i\omega}{c}\right)\left[k_x^i\hat{\mathbf{i}} + k_y^i\hat{\mathbf{j}} + k_z^i\hat{\mathbf{k}}\right].$$

The spatial part of the phase ($\mathbf{k}\cdot\mathbf{r}$) must meet the equality specified by Eq. (3.1), independent of the temporal part; therefore, since the boundary has been positioned at $z = 0$, we only have x and y components in the phase

$$\frac{n_i\omega}{c}\left(k_x^i x + k_y^i y\right) = \frac{n_i\omega}{c}\left(k_x^r x + k_y^r y\right) = \frac{n_t\omega}{c}\left(k_x^t x + k_y^t y\right).$$

This relationship must hold independently for all values of x and of y:

$$n_i k_x^i = n_i k_x^r = n_t k_x^t, \quad n_i k_y^i = n_i k_y^r = n_t k_y^t. \tag{3.2}$$

The relationships in Eq. (3.2) require that both the transmitted and reflected waves lie in the same plane as the incident wave. We call that plane the *plane of incidence* and define it as the plane containing the incident wave vector $\mathbf{k}_i$ and the normal to the boundary. In Figure 3.3 the boundary is taken to be the $z = 0$, x, y-plane; the normal to the plane is the unit vector $\hat{\mathbf{k}}$ parallel to the z-axis.[2] The plane of incidence is arranged to lie in the x, z-plane, as shown in Figure 3.3. The direction cosines associated with the propagation vectors are thus

$$k_x^i = \sin\theta_i \quad k_y^i = 0 \quad k_z^i = \cos\theta_i$$
$$k_x^r = \sin\theta_r \quad k_y^r = 0 \quad k_z^r = \cos\theta_r$$
$$k_x^t = \sin\theta_t \quad k_y^t = 0 \quad k_z^t = \cos\theta_t.$$

Equation (3.2) can now be written

$$n_i \sin\theta_i = n_i \sin\theta_r = n_t \sin\theta_t.$$

From this expression we have in the first medium,

$$\sin\theta_i = \sin\theta_r \tag{3.3}$$

If the coordinate system is selected so that $\mathbf{k}_i$ is propagating in the positive direction and $\cos\theta_i \geq 0$, then it is apparent from Figure 3.3 that $\mathbf{k}_r$ is in the negative direction and $\cos\theta_r \leq 0$, resulting in $\theta_r = \pi - \theta_i$. The statement that the reflected wave is in the same plane as the incident wave and Eq. (3.3), together, form the *Law of Reflection*.

Returning to Eq. (3.2), the second relationship is

$$n_i \sin\theta_i = n_t \sin\theta_t, \tag{3.4}$$

which is the *Law of Refraction* or *Snell's Law*.

By requiring phase continuity across the boundary, we have obtained the laws of reflection and refraction. These relations hold for any solution of the wave equation and are not dependent

---

[2] Please note that $\hat{\mathbf{k}}$ by tradition is the unit vector along the z-direction and that $\mathbf{k}$ is customarily defined as the wave vector. The "hat" ($\hat{}$) on the unit vector and the context of the discussion should prevent confusion.

upon the electromagnetic properties of light waves. A more fundamental approach to obtain the laws of reflection and refraction is the use of Fermat's principle.

3.3 **Fermat's Principle**

Fermat's principle states that *light travels the path that takes the least time.*

Historically, geometrical optics developed from Fermat's principle and not from the wave theory. Fermat was inspired by the book *Catoptrica*,[3] written by *Hero of Alexandria* (AD *c.*100), to extend the concept of least time from reflection to refraction. Hero was an Egyptian or Greek who proved the law of reflection using geometry from the premise that the rays of light take the shortest path between any two points.

To discuss Fermat's principle, all path lengths must be converted to equivalent path lengths in a vacuum so that we can compare the time of travel over each path. The equivalent path length is called the *optical path length*. The optical path length between points A and B, in a medium with an index of refraction, n, is defined as the distance a wave in a vacuum would travel during the time it took light to travel from A to B in the actual medium. If the distance between A and B is r and the velocity of propagation in the medium is v then the time to travel from A to B is $\tau = r/v$. The distance light would travel in a vacuum in the time τ, i.e., the optical path length, is given by

$$\ell = c\tau = \frac{cr}{v} = nr \tag{3.5}$$

This is the definition of the optical path length. We will later use this concept to aid in the calculation of phase differences between two beams of light that start from the same point but travel different paths before they again meet.

To handle paths that may not be linear in a homogeneous medium with index n, we use a path integral to calculate the distance between points 1 and 2:

$$S = \int_1^2 ds$$

The wave travels this distance in a time S/v. In a vacuum, during this same time, light would travel

$$\frac{Sc}{v} = nS = n[S(\mathbf{r}_1) - S(\mathbf{r}_2)]$$

If light is traveling over a path, C, then the optical path length is the path integral

$$\Delta = \oint_C nds,$$

where n is the index of refraction that can vary over the path, C, and ds is an incremental path length. In Figure 3.5, we see that the optical path length is the separation between the wavefronts at positions $\mathbf{r}_1$ and $\mathbf{r}_2$ along the optical path.

In modern mathematical terms, Fermat's principle is a statement of the determination of the extrema of a function; the optical path length that we just defined will be the quantity whose

---

[3] We will encounter this word in various forms in titles of books that concern the optics of reflection.

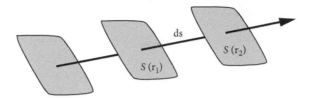

Figure 3.5 The integral along the optical ray between the wave fronts at $\mathbf{r}_1$ and at $\mathbf{r}_2$ is the optical path length.

extreme will be found. Fermat's principle states that *the optical path length of an actual wave path between any two points "1" and "2" is a minimum; i.e., any curve we choose which joins these points and lies in a "neighborhood" of the proper path has the same optical path length. The time required to traverse any of the paths in this neighborhood is the same.* (A physical interpretation of the "neighborhood" of the optical path will be given once we have discussed Fresnel diffraction.) Mathematically this path is found through the calculus of variations.[4] The mathematical statement of Fermat's principle is

$$\delta\Delta = \delta \int_{P_1}^{P_2} n\,ds = 0. \tag{3.6}$$

where the symbol "δ" indicates that the variation has been taken.

We can rewrite Fermat's principle in terms of the time taken to traverse the optical path:

$$\delta\Delta = \delta \int_{P_1}^{P_2} n\,ds = c\delta \int_{t_1}^{t_2} dt = 0. \tag{3.7}$$

From Eq. (3.7) we see that minimizing the optical path length is equivalent to minimizing the propagation time. In the following derivations of the fundamental laws of reflection and refraction we will apply Fermat's principle by minimizing the propagation time. From these examples, we will learn that Fermat's principle requires the time taken to traverse an optical path to remain constant, to first order, for incremental changes in the path length. Mathematically, this means that the first derivative of the propagation time with respect to optical path length must equal zero.

The first question that should arise when told of Fermat's principle is "*How does the light know what path to take?*" The answer lies in the fact that these are not rays but waves and we need to treat the light as a wave, not as a geometrical path. What we draw as a ray is simply the trajectory of the wavefront normal, $\mathbf{k}$. There is a wave that traverses each possible path. When the various waves reach their destination, they interfere.[5] The phase differences for those waves that travel over the "wrong" paths are such that they destructively interfere while those waves traveling over the "correct" paths constructively interfere. We can prevent light from traversing all possible paths by the use of obstructions; when we insert obstructions, we observe light in regions that geometrical optics say should be dark. This behavior is called diffraction. This same minimization principle is used in support of classical mechanics and quantum theory. There it is called Hamilton's principle.

---

[4] See Chapter 19 of the book by Feynman [2] for an excellent introduction into this subject.

[5] We will discuss interference at length in the next chapter.

3.3.1 Applications of Fermat's principle

The principle of Fermat makes some of the laws of optics self-evident. For example, the *principle of reciprocity* says that if light travels from point P_1 to P_2 over a path, then it will travel over the same path in going from P_2 to P_1. Since light travels at the same speed in both directions, the path of least time must be the same between P_1 and P_2, no matter which direction the light travels.

The laws of reflection and refraction can be quickly derived using Fermat's principle.

3.3.1.1 Law of reflection

Consider the optical path taken by light originating at point P_1, a distance a above the mirror M, and reflected by the mirror to point P_2, a distance c from point P_1 and a distance b above the mirror, as shown in Figure 3.6. There is some sort of obstruction between P_1 and P_2, not shown in Figure 3.6, which prevents the light from traveling between the two points without reflecting off the mirror. Our variation calculation will result in O moving from place to place by allowing x to vary.

The time for light to travel from P_1 to P_2 over the path $P_1 O P_2$ of Figure 3.6 is

$$t = \frac{\overline{P_1 O} + \overline{OP_2}}{v} = \frac{1}{v}\left[\sqrt{b^2 + x^2} + \sqrt{a^2 + (c - x)^2}\right].$$

For this time to be stationary the first derivative of the time with respect to the path length must be zero, where a, b, and c are constants since P_1 and P_2 are fixed in space

$$\frac{dt}{dx} = \frac{1}{v}\left[\frac{x}{\sqrt{b^2 + x^2}} - \frac{c - x}{\sqrt{a^2 + (c - x)^2}}\right] = 0,$$

$$\frac{x}{\sqrt{b^2 + x^2}} = \frac{c - x}{\sqrt{a^2 + (c - x)^2}},$$

$$\sin i = \sin r.$$

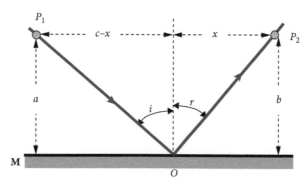

Figure 3.6 Geometry for use of Fermat's principle to derive the law of reflection.

Thus, the angle of incidence equals the angle of reflection. The second derivative, d^2t/dx^2, is greater than zero so the function is a minimum, as Hero stated 1800 years ago.

3.3.1.2 Law of refraction

To obtain the law of refraction using Fermat's principle, consider the path of a ray of light $P_1 O P_2$ in Figure 3.7. All of the paths under consideration start at P_1, a distance a above the interface, and end at P_2, a distance b below the interface, so that a, b, and c are constants. In Figure 3.7, as in Figure 3.6, we allow O to vary by letting x vary. The time to travel from P_1 to P_2 is

$$t = \frac{\overline{P_1 O}}{v_1} + \frac{\overline{OP_2}}{v_2} = \frac{1}{v_1}\sqrt{a^2 + (c-x)^2} + \frac{1}{v_2}\sqrt{b^2 + x^2},$$

where v_1 is the velocity of propagation of light in the upper medium of Figure 3.7 and v_2 is the velocity of propagation of the lower medium. The first derivative of the time with respect to the path length is

$$\frac{dt}{dx} = \frac{-(c-x)}{v_1\sqrt{a^2 + (c-x)^2}} + \frac{x}{v_2\sqrt{b^2 + x^2}} = 0,$$

$$\frac{\sin t}{v_2} = \frac{\sin i}{v_1},$$

$$\frac{\sin i}{\sin t} = \frac{v_1}{v_2} = \frac{n_2}{n_1}.$$

This is Snell's Law.

We have derived the basic properties of propagating light waves using Fermat's principle. We know the path the light will take but we do not know how much energy will travel along each of the paths. We will have to turn to Maxwell's equations and the boundary conditions to determine the quantity of light reflected and transmitted.

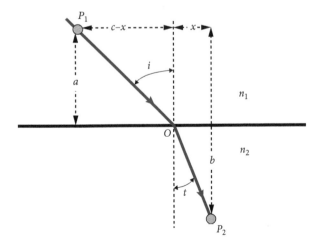

Figure 3.7 Geometric arrangement for using Fermat's principle to derive the law of refraction.

Fresnel's Formula

We must use Maxwell's equations and the boundary conditions associated with these equations to learn about the amplitudes of the reflected and transmitted waves. The waves that meet at the boundary have a vector amplitude. Calculation of the amplitude is important not only in physics and engineering but also in computer graphics (see Figure 3.8).

The geometry to be used in this discussion is shown in Figure 3.9. Two media are separated by an interface, the x, y-plane at $z = 0$, whose normal, $\hat{\mathbf{n}} = \hat{\mathbf{k}}$, is the unit vector along the z-direction. The incident wave is labeled with an "i," the reflected wave by an "r," and the transmitted wave by a "t." The incident wave's propagation vector, $\mathbf{k}_i$, which we assume lies in the x, z-plane, and the normal to the interface establish the plane of incidence that here is the surface of the page.

The electric field vectors for each of the three waves have been decomposed into two components, one in the plane of incidence, labeled P, and one normal to the plane and the drawing surface labeled N. This is an extension of the technique, discussed in Chapter 2, of using orthogonal vectors to describe the polarization of a light wave.[6] The upper half plane has a velocity of propagation v_i and an index of n_i and the lower half plane has a velocity of propagation v_t and an index of n_t.

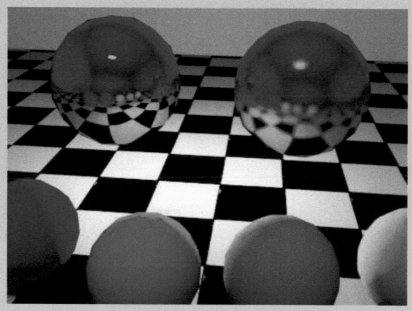

Figure 3.8 Computer-generated graphics using Fresnel equations.
By AlHart (Own work) [Public domain], via Wikimedia Commons.

[6] According to custom, the two polarizations are labeled: π for parallel to the plane of incidence and σ for perpendicular to the plane of incidence. The Greek letter σ denotes perpendicular because "s" is the first letter of the German word *senkrecht* that translates to vertical or perpendicular. We will use N (normal) and P (parallel) in this book in place of the Greek letters.

The actual vectors are written below. Note that the normal E field is parallel to the y-axis, pointing out of the page:

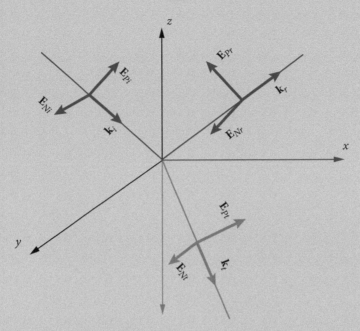

Figure 3.9 Orientation of the electric field and wave vectors in the coordinate system we selected for the discussion of reflection and refraction. The plane of incidence is the x,z-plane.

Incident wave

$$\mathbf{k}_i = k_i \left(\hat{\mathbf{i}} \sin\theta_i - \hat{\mathbf{k}} \cos\theta_i \right), \tag{3.8}$$

$$\mathbf{E}_i = E_{Pi} \left(\hat{\mathbf{i}} \cos\theta_i - \hat{\mathbf{k}} \sin\theta_i \right) + E_{Ni}\hat{\mathbf{j}}, \tag{3.9}$$

Reflected wave

$$\mathbf{k}_r = k_i \left(\hat{\mathbf{i}} \sin\theta_i - \hat{\mathbf{k}} \cos\theta_i \right), \tag{3.10}$$

$$\mathbf{E}_r = E_{Pr} \left(-\hat{\mathbf{i}} \cos\theta_i + \hat{\mathbf{k}} \sin\theta_i \right) + E_{Nr}\hat{\mathbf{j}}, \tag{3.11}$$

Transmitted wave

$$\mathbf{k}_t = k_i \left(\hat{\mathbf{i}} \sin\theta_t - \hat{\mathbf{k}} \cos\theta_t \right), \tag{3.12}$$

$$\mathbf{E}_t = E_{Pt} \left(\hat{\mathbf{i}} \cos\theta_t + \hat{\mathbf{k}} \sin\theta_t \right) + E_{Nt}\hat{\mathbf{j}}. \tag{3.13}$$

The boundary conditions associated with Maxwell's equations are the following:

- The normal components of **D** are continuous across the boundary if there are no surface charges.

- The normal components of **B** are continuous across the boundary.
- The tangential components of **E** are continuous across the boundary.
- The tangential components of **H** are continuous across the boundary if there are no surface currents.

Of the four boundary conditions of Maxwell's equations only two are needed to obtain the relationships between the incident, reflected, and transmitted waves in our special medium free of charges and currents; the conditions utilized are that the tangential components of **E** and **H** are continuous across the boundary. The boundary conditions place independent requirements on the polarizations parallel to and normal to the plane of incidence and generate two pair of equations that are treated separately. We can express the two boundary conditions using vector notation:

(1) From the Maxwell's equation containing $\nabla \times \mathbf{E}$, we have that the tangential component of **E** is continuous. This boundary condition is written

$$(\mathbf{E}_i + \mathbf{E}_r + \mathbf{E}_t) \times \hat{\mathbf{n}} = 0. \tag{3.14}$$

Each one of the vector products is of the form

$$\mathbf{E} \times \hat{\mathbf{n}} = E_y \hat{\mathbf{i}} - E_x \hat{\mathbf{j}}.$$

Evaluating the cross-product yields

$$(E_{Ni} + E_{Nr} - E_{Nt}) \hat{\mathbf{i}} - (E_{Pi} \cos \theta_i - E_{Pr} \cos \theta_i - E_{Pt} \cos \theta_t) \hat{\mathbf{j}} = 0.$$

Each vector component must independently be equal to 0; thus,

$$E_{Ni} + E_{Nr} = E_{Nt}, \tag{3.15}$$

$$(E_{Pi} - E_{Pr}) \cos \theta_i = E_{Pt} \cos \theta_t. \tag{3.16}$$

(2) From the Maxwell equation containing $\nabla \times \mathbf{H}$, we have that the tangential component of **H** is continuous if there are no surface currents. The tangent component of **H** can be written in terms of the electric field

$$\mathbf{H} \times \hat{\mathbf{n}} = \frac{\mathbf{B}}{\mu} \times \hat{\mathbf{n}} = \frac{\sqrt{\mu \varepsilon}}{\mu k} \mathbf{k} \times \mathbf{E} \times \hat{\mathbf{n}}. \tag{3.17}$$

The boundary condition is then written

$$\left[\frac{1}{k_i} \sqrt{\frac{\varepsilon_i}{\mu_i}} (\mathbf{k}_i \times \mathbf{E}_i + \mathbf{k}_r \times \mathbf{E}_r) - \frac{1}{k_t} \sqrt{\frac{\varepsilon_t}{\mu_t}} (\mathbf{k}_t \times \mathbf{E}_t) \right] \times \hat{\mathbf{n}} = 0,$$

$$(\mathbf{k} \times \mathbf{E}) \times \hat{\mathbf{n}} = \left[\left(E_y k_z - E_z k_y \right) \hat{\mathbf{i}} + (E_z k_x - E_x k_z) \hat{\mathbf{j}} + \left(E_x k_y - E_y k_x \right) \hat{\mathbf{k}} \right] \times \hat{\mathbf{k}},$$

$$= (E_z k_x - E_x k_z) \hat{\mathbf{i}} - E_y k_z \hat{\mathbf{j}} = 0. \tag{3.18}$$

Each vector component must independently be equal to 0. The x component gives

$$\sqrt{\frac{\varepsilon_i}{\mu_i}} [E_{Pi} - E_{Pr}] = \sqrt{\frac{\varepsilon_t}{\mu_t}} E_{Pt}. \tag{3.19}$$

The y component of Eq. (3.18) gives

$$\sqrt{\frac{\varepsilon_i}{\mu_i}}(E_{Ni} - E_{Nr})\cos\theta_i = \sqrt{\frac{\varepsilon_i}{\mu_i}}E_{Nt}\cos\theta_t. \tag{3.20}$$

There are three unknowns but only two equations for each polarization; thus, the amplitudes of the reflected and transmitted light can only be found in terms of the incident amplitude.

σ-case (Perpendicular polarization)

For this component of the polarization, **E** is perpendicular to the plane of incidence, i.e., the x, z-plane. This means that **E** is everywhere normal to **n** and parallel to the boundary surface between the two media.[7]

The boundary conditions provide, through Eqs. (3.16) and (3.20), relationships between the various normal electric fields. The amplitude of reflected and transmitted light will be found, using these equations, as a ratio to the incident amplitude. We use Snell's law Eq. (3.4) to modify Eq. (3.20)

$$E_{Ni} - E_{Nr} = \frac{\mu_i \sin\theta_i}{\mu_t \cos\theta_i} \cdot \frac{\cos\theta_t}{\sin\theta_t} \cdot E_{Nt} = \frac{\mu_i \tan\theta_i}{\mu_t \tan\theta_t} E_{Nt}. \tag{3.21}$$

Adding Eq. (3.20) to Eq. (3.15) yields

$$2E_{Ni} = \left[1 + \frac{\mu_i \tan\theta_i}{\mu_t \tan\theta_t}\right] E_{Ni},$$

$$\frac{E_{Nt}}{E_{Ni}} = \frac{2}{1 + \dfrac{\mu_i \tan\theta_i}{\mu_t \tan\theta_t}}. \tag{3.22}$$

As was shown in Chapter 2, for the majority of optical materials at optical frequencies, $\mu_i \approx \mu_t$ and the equation simplifies to

$$t_N = \frac{E_{Nt}}{E_{Ni}} = \frac{2\sin\theta_t \cos\theta_i}{\sin(\theta_i + \theta_t)}. \tag{3.23}$$

The amplitude ratio, t_N, is called the *amplitude transmission coefficient for perpendicular polarization*.

Now Eq. (3.22) can be substituted back into Eq. (3.15) to obtain the amplitude ratio for the reflected light:

$$E_{Ni} + E_{Nr} = \frac{2E_{Ni}}{1 + \dfrac{\mu_i \tan\theta_i}{\mu_t \tan\theta_t}},$$

$$\frac{E_{Nr}}{E_{Ni}} = \frac{1 - \left(\dfrac{\mu_i \tan\theta_i}{\mu_t \tan\theta_t}\right)}{1 + \left(\dfrac{\mu_i \tan\theta_i}{\mu_t \tan\theta_t}\right)}. \tag{3.24}$$

Again when $\mu_i \approx \mu_t$ we obtain the ratio of reflected amplitude to the incident amplitude that is called the *amplitude reflection coefficient for perpendicular polarization*:

$$r_N = \frac{E_{Nr}}{E_{Ni}} = \frac{-\sin(\theta_i - \theta_t)}{\sin(\theta_i + \theta_t)}. \tag{3.25}$$

π-case (Parallel polarization)

For this component of polarization, **E** is everywhere parallel to the plane of incidence; however, **B** and **H** are everywhere normal to **n** and parallel to the boundary between the two media.[8] The second boundary condition provides Eq. (3.16) which can be written

$$E_{Pi} - E_{Pr} = \frac{\cos\theta_t}{\cos\theta_i} E_{Pt}. \tag{3.26}$$

Applying Snell's law, Eq. (3.4), to Eq. (3.19) yields

$$E_{Pi} + E_{Pr} = \sqrt{\frac{\mu_i \varepsilon_t}{\mu_t \varepsilon_i}} E_{Pt} = \frac{\mu_i \sin\theta_i}{\mu_t \sin\theta_t} E_{Pt}. \tag{3.27}$$

Adding this equation to Eq. (3.26) yields the desired ratio of amplitudes

$$\frac{E_{Pt}}{E_{Pi}} = \frac{2\cos\theta_i \sin\theta_t}{\cos\theta_t \sin\theta_t + \frac{\mu_i}{\mu_t}\cos\theta_i \sin\theta_i}. \tag{3.28}$$

For the usual situation of $\mu_I \approx \mu_t$, the *amplitude transmission coefficient for parallel polarization* is

$$t_P = \frac{E_{Pt}}{E_{Pi}} = \frac{2\cos\theta_i \sin\theta_t}{\sin(\theta_i + \theta_t)\cos(\theta_i - \theta_t)}. \tag{3.29}$$

Substituting Eq. (3.28) into Eq. (3.27) produces the amplitude reflection ratio

$$\frac{E_{Pr}}{E_{Pi}} = \frac{\left(\frac{\mu_i}{\mu_t}\right)\sin 2\theta_i - \sin 2\theta_t}{\sin 2\theta_t + \left(\frac{\mu_i}{\mu_t}\right)\sin 2\theta_i}. \tag{3.30}$$

The assumption of $\mu_i \approx \mu_t$ (from now on, this assumption will be used) results in a *reflection coefficient for parallel polarization* of

$$r_P = \frac{E_{Pr}}{E_{Pi}} = \frac{\tan(\theta_i - \theta_t)}{\tan(\theta_i + \theta_t)}. \tag{3.31}$$

---

[7] This case could be labeled the *transverse electric field* (TE) case; we do not use this notation because the label implies that the wave may not be a transverse electromagnetic wave (a TEM wave); instead we use the subscript N. The TE notation will be reserved for inhomogeneous waves in a light-guiding structure which we will discuss in Chapter 8.

[8] This case could be labeled the *transverse magnetic field* (TM) case but we will use the subscript P for the same reason we did not use the TE notation; see the discussion of guided waves, where the TM notation is utilized for inhomogeneous waves.

3.4 **Reflected and Transmitted Energy**

The fraction of the incident amplitude reflected and transmitted at a surface are not experimentally available. The parameter that can be measured is the energy. At first, we might think that we could simply square the ratios we have derived to obtain the energies but this would lead to erroneous results. The area of the transmitted beam, A_t, is different than that of the incident beam, A_i, because of refraction, as shown in Figure 3.10. The energy in the beam is the intensity times the area of the beam. The circular incident beam of radius L in Figure 3.10 forms an ellipse on the surface of the interface with a minor axis of L and a major axis of D. The major axis and thus the area of the beam changes with incident angle;[9] the minor axis is a constant. The areas of the incident and transmitted beams are

$$A_i = \pi\,(D\cos\theta_i)\,L \ A_t = \pi\,(D\cos\theta_t)\,L.$$

The correct way to proceed is to use the average Poynting vector incident on a unit area, given by

$$\langle \mathbf{S}\rangle \bullet \hat{\mathbf{n}} = |\langle \mathbf{S}\rangle|\ \cos\theta,$$

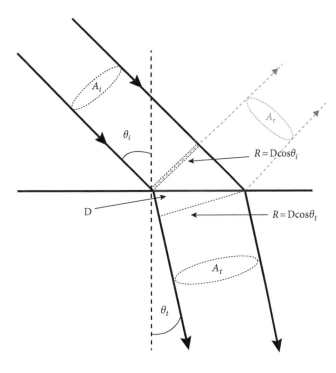

Figure 3.10 Change in area of light beam from incident to transmitted and the ratio is $A_i/A_t = \cos\theta_i/\cos\theta_t$

[9] The area of an ellipse is given by the product of the semi-major and semi-minor axis $A = \pi \bullet (Major) \bullet (minor)$

where the expression for the average Poynting vector is obtained from Eq. (2.31).[10]

The energy flow across the boundary (see Figure 3.10) can be obtained from Eq. (2.33):

$$v_i \langle U_i \rangle = \langle S_i \rangle \cos \theta_i = \frac{n_i}{2} \sqrt{\frac{\varepsilon_0}{\mu_0}} \mid E_i^2 \mid \cos \theta_i, \tag{3.32}$$

$$v_r \langle U_r \rangle = \langle S_r \rangle \cos \theta_i = \frac{n_i}{2} \sqrt{\frac{\varepsilon_0}{\mu_0}} \mid E_r^2 \mid \cos \theta_i, \tag{3.33}$$

$$v_t \langle U_t \rangle = \langle S_t \rangle \cos \theta_t = \frac{n_t}{2} \sqrt{\frac{\varepsilon_0}{\mu_0}} \mid E_t^2 \mid \cos \theta_t, \tag{3.34}$$

Each of these three equations applies separately to the normal and parallel components of polarization, resulting in six equations for the description of reflected and transmitted energy at a boundary.

We define[11] *reflectivity* as

$$R = \frac{\langle U_r \rangle}{\langle U_i \rangle} = \frac{\mid E_r^2 \mid}{\mid E_i^2 \mid}, \tag{3.35}$$

and *transmissivity* as

$$T = \frac{v_t \langle U_t \rangle}{v_i \langle U_i \rangle} \left(\frac{n_t}{n_i} \right) \frac{\cos \theta_t}{\cos \theta_i} \cdot \frac{\mid E_t^2 \mid}{\mid E_i^2 \mid}. \tag{3.36}$$

The expressions for the reflectivity and the transmissivity can be used to demonstrate that energy is conserved, $T + R = 1$, when light encounters a boundary.

Figure 3.11 displays the dependence of the reflectivity on the angle of incidence for both polarizations. Here the condition of $n_i < n_t$ and $n_i > n_t$ are shown. The special cases when $r^2_N = r^2_P = 1$, beyond the point labeled critical angle, and $r^2_P = 0$ (labeled Brewster's angle) are quite noticeable. After first discussing reflection and transmission when a light wave is incident nearly normal to the boundary, we will examine these special cases.

3.4.1 Reflection of unpolarized light

What happens if the incident wave electric field does not lie in the plane or perpendicular to the plane of incidence? It is not too difficult to generate the equations to handle such an event. If you have a light wave that is plane polarized at an angle γ with respect to the plane of incidence, you can decompose the incident wave so that the reflectivity can be calculated using the Fresnel

---

[10] We assume that $\mu \approx \mu_0$, resulting in $\mu c = \sqrt{\mu_0/\varepsilon_0} \simeq 377 \ ohms$, defined as the impedance of vacuum, Eq. (2.15).

[11] The quantities defined by Eqs. (3.35) and (3.36) are ratios of the Poynting vectors and therefore assume a wave of known frequency and phase. Experimentally, the ratios of the incident flux to the reflected and transmitted fluxes are determined using radiometric units called the *reflectance* and the *transmittance*. For homogenous materials reflectance and reflectivity are the same.

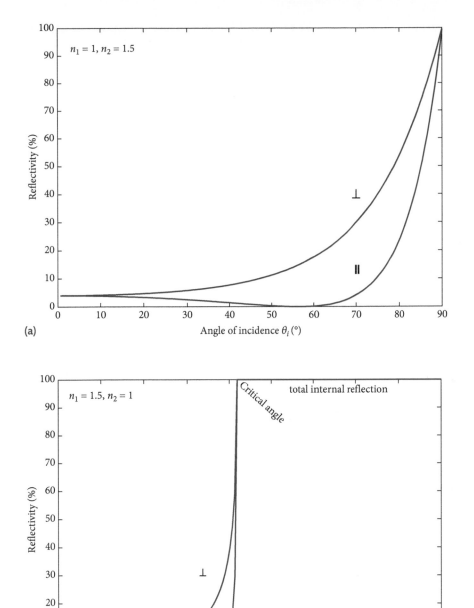

Figure 3.11 (a) Reflectivity as a function of incident angle for the case of $n_i = 1.0$ and $n_t = 1.5$. (b) Reflectivity as a function of incident angle for the case of $n_i = 1.5$ and $n_t = 1$.

formula from Eq. (3.35) with Eqs. (3.25) and (3.31). The incident wave's electric field can be decomposed as

$$E_{iP} = E_i \cos\gamma \quad E_{iN} = E_i \sin\gamma.$$

The reflectivity of the wave is the combination of the reflectivity for each polarization

$$R = \frac{I_r}{I_i} = \frac{I_{rP} + I_{rN}}{I_i},$$

where

$$I_{rP} = E_{rP}^2 \quad and \quad I_{rN} = E_{rN}^2.$$

The reflectivity for each polarization is given by

$$R_P = \frac{I_{rP}}{I_{iP}} = \frac{I_{rP}}{E_i^2 \cos^2\gamma} \quad R_N = \frac{I_{rN}}{E_i^2 \sin^2\gamma}.$$

The reflectivity for the incident wave is thus

$$R = R_P \cos^2\gamma + R_N \sin^2\gamma.$$

This result allows us to calculate the reflectivity of unpolarized light by treating γ as a random variable and calculating the time average

$$\langle \cos^2\gamma \rangle = \langle \sin^2\gamma \rangle = 1/2$$

$$R_{\text{unpolarized}} = \frac{R_P + R_N}{2}.$$

As an example, we will calculate the reflectivity of unpolarized light incident at 70° in air from a glass plate with an index of 1.5. We need to calculate in angle of transmission to be used in the reflection coefficient equations: Eqs. (3.25) and (3.31),

$$\sin\theta_t = \frac{n_i}{n_t} \sin 70^\circ = \frac{0.94}{1.5} \quad \Rightarrow \quad \theta_t = 38.8^\circ$$

$$\theta_i - \theta_t = 31.2^\circ \quad \sin(\theta_i - \theta_t) = 0.52 \quad \tan(\theta_i - \theta_t) = 0.61$$

$$\theta_i + \theta_t = 108.8^\circ \quad \sin(\theta_i + \theta_t) = 0.95 \quad \tan(\theta_i + \theta_t) = -2.94.$$

We can now use these values in our equation for the reflectivity

$$R_P = \frac{\tan^2(\theta_i - \theta_t)}{\tan^2(\theta_i + \theta_t)} = 0.04 \quad R_N = \frac{\sin^2(\theta_i - \theta_t)}{\tan^2(\theta_i + \theta_t)} = 0.30$$

$$R = 1/2\,(R_P + R_N) = 0.17$$

17% of unpolarized light is reflected.

We have generated partially polarized light by the reflection. In the chapter on polarizers (Chapter 5) we will find the Stokes's parameters for the reflected light.

3.4.2 Normal incidence

The meaning of a plane of incidence is lost for the incident angle of $\theta \approx 0°$, since the two vectors that are supposed to define the plane are parallel and thus do not define a plane.[12] Upon examining Figure 3.11, it can be seen that $R_N \simeq R_P$ when $\theta \lesssim 10°$ and we do not need to distinguish between the two types of polarization.

The reflection and transmission coefficients at $\theta = 0°$ can be found if the limits of the coefficients are taken as $\theta \to 0°$. The sine found in the denominator of Eq. (3.25) can be eliminated by using Snell's law in combination with the two trigonometric identities

$$\sin(\theta_i + \theta_t) = \sin\theta_i \cos\theta_t + \cos\theta_i \sin\theta_t,$$
$$\cos(\theta_i - \theta_t) = \cos\theta_i \cos\theta_t + \sin\theta_i \sin\theta_t.$$

$$\frac{E_{Pt}}{E_{Pi}} = \frac{2\cos\theta_i \sin\theta_t}{\sin\theta_t \left[\left(\frac{n_t}{n_i}\right)\cos\theta_t + \cos\theta_i\right][\cos\theta_i \cos\theta_t + \sin\theta_i \sin\theta_t]}.$$

Sin θ_t cancels out of the numerator and denominator so that the limit of the ratio as θ_i and $\theta_t \to 0°$. is found to be

$$\frac{E_{Pt}}{E_{Pi}} = \frac{2n_i}{n_t + n_i}. \tag{3.37}$$

In the same way, the limit of Eq. (3.31) as $\theta_i \to 0°$. is

$$\frac{E_{Nt}}{E_{Ni}} = \frac{2n_i}{n_t + n_i},$$

yielding the same transmission coefficient for the two polarizations.

The limit of Eq. (3.23) as $\theta_i \to 0°$. can also be obtained using the above trig identities:

$$\frac{E_{Nr}}{E_{Ni}} = \frac{-\sin\theta_t\left[\left(\frac{n_t}{n_i}\right)\cos\theta_t - \cos\theta_i\right]}{\sin\theta_t\left[\left(\frac{n_t}{n_i}\right)\cos\theta_t + \cos\theta_i\right]} = \frac{-(n_t - n_i)}{n_t + n_i}. \tag{3.38}$$

The limit of Eq. (3.31) is a little more difficult to obtain because terms that go to zero as θ approaches zero do not cancel out of the equation:

$$\frac{E_{Pr}}{E_{Pi}} = \frac{\dfrac{\sin(\theta_i - \theta_t)}{\cos(\theta_i - \theta_t)}}{\dfrac{\sin(\theta_i + \theta_t)}{\cos(\theta_i + \theta_t)}}.$$

Near normal incidence, $\theta_i \approx \theta_t \approx 0$ and we have $\cos\theta_i = \cos\theta_t \approx 1$ and $\sin\theta_t \approx \theta_t$, leading to

$$\sin\theta_i = \left(\frac{n_t}{n_i}\right)\sin\theta_t = \left(\frac{n_t}{n_i}\right)\theta_t.$$

[12] Mathematically, the cross product of the two vectors defines the normal to the surface. Since the two vectors are parallel, their cross product is zero and there is no normal.

We can write

$$\frac{\sin(\theta_i \pm \theta_t)}{\cos(\theta_i \pm \theta_t)} \approx \frac{\sin\theta_i\cos\theta_t \pm \cos\theta_i\sin\theta_t}{\cos\theta_i\cos\theta_t - \sin\theta_i\sin\theta_t},$$

$$\approx \frac{\left(\dfrac{n_t}{n_i}\right)\theta_t \pm \theta_t}{1 - \left(\dfrac{n_t}{n_i}\right)\theta_t^2}.$$

Keeping only terms linear in θ_t gives

$$\frac{E_{Pr}}{E_{Pi}} = \frac{n_t - n_i}{n_t + n_i}. \tag{3.39}$$

The reflection coefficient is directly proportional to the difference between the propagation velocities of the two media forming the interface. Experimental proof of this result is shown in Figure 3.12. A beaker containing water with a layer of xylene floating on the surface has a glass rod resting within. A glass rod, with an index of refraction of about 1.5, is seen passing through air ($n = 1.0$), xylene ($n = 1.5$), and water ($n = 1.3$). The rod is easily seen in air and water because of light reflected from its surface. The glass rod is nearly invisible in the xylene because the indexes of refraction of the rod and the liquid are nearly equal.

The reflection coefficient for the two polarizations have the same magnitude; the difference in sign between Eq. (3.38) and Eq. (3.39) is due to the geometry and sign convention used in Figure 3.9.[13]

Figure 3.12 A glass rod rests in a beaker containing water with $n = 1.33$. Xylene, $n = 1.49582$, is floating on its surface. The Pyrex glass rod with an index of 1.475 is invisible in the xylene because little light is reflected from its surface, $R = 0.004$.

---

[13] Upon reflection the normal polarization, along the y-axis, is not affected by the change in propagation direction but the parallel polarization, in the x, z-plane, has its x-component reversed in sign.

Most optical glasses have an index of refraction of about 1.5 and air has an index of 1.0; therefore, $n_t/n_i = 1.5$ and

$$R = \left(\frac{n_t - n_i}{n_t + n_i}\right)^2 = \left(\frac{1 - 1.5}{1 + 1.5}\right)^2 = (-0.2)^2 = 0.04.$$

This means that at each glass–air interface, 4% of the incident light is reflected. If we calculate the transmissivity for a glass–air interface, we find that $T = 0.96$; thus, $T + R = 1.0$ and energy is conserved.

Modern eyewear incorporates high index glass ($n \simeq 1.74$) for people who want lightweight eyewear but need strong corrections. Reflections from this glass are nearly that of double conventional glass

$$R = \left(\frac{n_t - n_i}{n_t + n_i}\right)^2 = \left(\frac{1 - 1.74}{1 + 1.74}\right)^2 = 0.07,$$

and antireflection coatings have become a popular enhancement. We will discuss these anti-reflection coatings in the next chapter.

The efficiency of solar collectors is affected by reflective losses, and a trade-off between thermal conduction loss to the surrounding air, reduced by multiple glass panes, and reflective loss due to multiple surfaces must be made when designing solar collectors.

Early camera lenses used optical designs that required few elements because of the loss of intensity due to reflections at air–glass interfaces.

Cell phone and computer screens battle the problem of unwanted reflection from the screen. The cell phone is a particular problem because it is used in both a horizontal and vertical orientation. Thin film anti-reflection coatings, which we will discuss later, have been used to reduce the reflections to 0.23%.

In any optical system the loss in transmitted energy due to reflective loss is called Fresnel loss. In fiber communications this loss is usually specified in terms of dB:

$$dB = 10 \log_{10}(1 - R).$$

For our assumed index for glass of $n = 1.5$, the one surface loss is -0.18 dB. For a typical fiber core of $n = 1.48$ the loss would be -0.166 dB.

In electromagnetic theory, the ability to transmit energy from a source, say an oscillator, to a load is affected by the impedance of the source, the transmission medium, and the load. To maximize the power transfer from the source or to minimize the power reflection from the load, the design step of impedance matching is undertaken. The same formalism can be used to describe the transmission of light. The equations for reflection and refraction can be reformulated in terms of the impedance of each medium, yielding new definitions for the reflectivity and transmissivity:

$$R = \left(\frac{Z_t - Z_i}{Z_t + Z_i}\right)^2, \tag{3.40}$$

$$T = \frac{4Z_i Z_t}{(Z_t + Z_i)^2}. \tag{3.41}$$

In microwave design, impedance is the preferred formulation and the concept of index is never discussed, though propagation times are considered when transmission lines are considered. The choice of using the impedance formulation or the index of refraction formulation for optics problems is one of personal preference.

We will limit our future discussion to the reflection from surfaces with the knowledge that the transmission can easily be discovered by applying conservation of energy.

3.5 **Polarization by Reflection**

Note in Figure 3.11 the reflectivity of a wave polarized in the plane of incidence has an angle where the amplitude of the reflected wave goes to 0. For the component of polarization parallel to the plane of incidence, the angle of incidence for which there is no reflected wave is named after *Sir David Brewster* (1781–1868), the inventor of the kaleidoscope.

From the equation for the reflection coefficient, Eq. (3.31), we see that the reflection coefficient is zero whenever $\tan(\theta_i + \theta_t) = \infty$. This occurs if the sum of the angles is 90°, i.e., when $\theta_i + \theta_t = \pi/2$. To calculate *Brewster's angle*, we use Snell's law:

$$\frac{n_t}{n_i} = \frac{\sin \theta_i}{\sin \theta_t} = \frac{\sin \theta_i}{\sin (\pi/2 - \theta_i)} = \frac{\sin \theta_i}{\cos \theta_i} = \tan \theta_B.$$

Brewster's angle is therefore given by

$$\theta_B = \tan^{-1} \left(\frac{n_t}{n_i} \right) \tag{3.42}$$

In Figure 3.11, when $n_t > n_i$, the ratio of indices is $n_t/n_i = 1.5$ and Brewster's angle is $\theta_B = 56.3°$. When the reverse is true and $n_i > n_t$, the ratio of the indices is $n_t/n_i = 0.67$ and Brewster's angle is the complement of the previous angle, $\theta_B = 33.7°$. As can be seen in Figure 3.13, the reflectivity remains near 0 over a large range of angles, making the effect quite easy to observe.

Some laser designs use windows or crystal surfaces set at Brewster's angle to reduce the Fresnel reflective losses in the laser cavity for one polarization. The result is that the laser produces polarized light. Some brands of sunglasses are designed to take advantage of the fact that light of one polarization is not reflected from dielectric surfaces as strongly as the perpendicular component (see Figure 3.11a). These sunglasses use a material (Polaroid) that transmits light polarized in one direction and absorb light polarized normal to that direction. We will discuss this topic later.

A simple physical explanation can be given for Brewster's angle [3]. In Figure 3.14, the angle between $\mathbf{k}_r$ and $\mathbf{k}_t$ is

$$\left(\frac{\pi}{2} - \theta_i \right) - \left(-\frac{\pi}{2} + \theta_t \right) = \pi - (\theta_i + \theta_t).$$

When the incident beam is oriented at Brewster's angle, $(\theta_I + \theta_t) = \pi/2$ and the angle between $\mathbf{k}_r$ and $\mathbf{k}_t$ is $\pi/2$. This results in $\mathbf{E}_t$ oriented parallel to $\mathbf{k}_r$. When light is incident on a medium, the electric field causes the charge distribution of the electrons, shown in Figure 2.1, to vibrate in the direction of the field of the transmitted wave, $\mathbf{E}_t$. The vibrating electrons radiate an

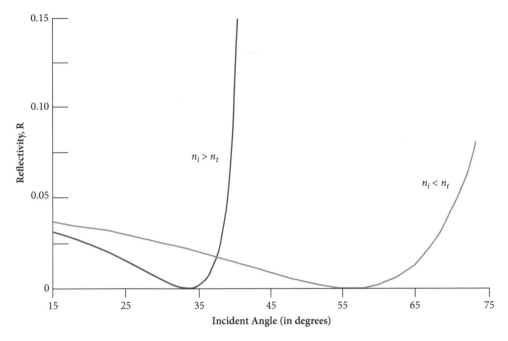

Figure 3.13 The reflectivity of a wave polarized in the plane of incidence as a function of the incident angle around the Brewster's angle. The two indices are 1.0 and 1.5.

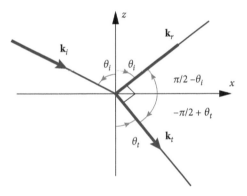

Figure 3.14 Geometry used to explain the occurrence of Brewster's angle.

electromagnetic wave that propagates back into the first medium; this is the origin of the reflected wave.[14]

There is no radiation produced in the direction of vibration [4] of the electrons, along the vertical axis in Figure 3.15. Thus, when the reflected and transmitted waves are propagating at right angles to each other, the reflected wave does not receive any energy from oscillations in the plane of incidence. Based on the geometry of the waves at Brewster's angle, shown in

---

[14] This radiation is not due to atomic transitions but rather a modulation of a charge distribution as discussed following Figure 2.1.

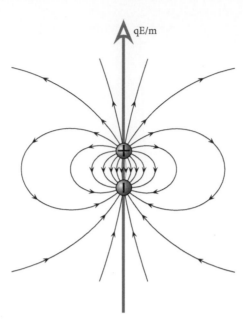

Figure 3.15 Reradiating electric field generated by the incident field acceleration of a molecular charge distribution (the plus and minus dots). It is assumed that relativistic velocities are not present so that the two radiation lobs are symmetric.
By Geek3 - Own work, CC BY-SA 3.0, https://commons.wikimedia.org/w/index.php?curid=11621756.

Figure 3.14, no reflection will be observed. Magnetic materials complicate the problem [5] but the mathematics needed to describe the effects is a straightforward extension of this derivation.

3.6 Total Reflection

From Figure 3.11b, it is apparent that when light is incident on a boundary from a more dense (larger value of index, n) medium toward a less dense medium, then the reflectivity for the perpendicular polarization is a monotonically increasing function of the incident angle. The reflectivity for the parallel polarization decreases to 0 at Brewster's angle and then exhibits the same behavior as the reflectivity for the perpendicular polarization. At an angle we call the critical angle, the reflectivity becomes one and total reflection occurs for all larger angles. For angles beyond this critical angle, the light is said to undergo *total reflection* and there is no transmission across the interface . If you are underwater (where $n = 1.33$) and look toward the surface (into air where $n = 1.0$), you will see mainly reflection from items under the water; see Figure 3.16.

A schematic representation of refraction and total reflection is shown in Figure 3.17. When $n_i > n_t$, Snell's law states that $\theta_t > \theta_i$ because, for θ between $0°$ and $\pi/2$, $\sin\theta$ is a monotonically increasing function. The angle θ_t reaches $\pi/2$ and $\sin\theta_t$ reaches 1 when the incident angle is equal to θ_c ($< \pi/2$) obtained from Snell's law:

$$\theta_c = \sin^{-1}\left(\frac{n_t}{n_i}\right). \tag{3.43}$$

Figure 3.16 Weedy sea dragon (*Phyllopteryx taeniolatus*) images reflected from the water surface of an aquarium. This image was recorded in the Georgia Aquarium.

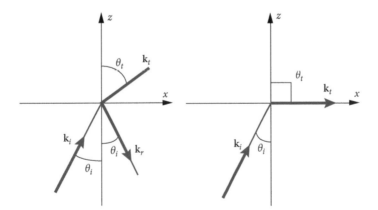

Figure 3.17 Example of a wave incident at an angle well below the critical angle is shown on the left; this is normal refraction. On the right is shown a wave incident at the critical angle when total reflection occurs.

For example, in Figure 3.16 two different critical angles occur. At a glass–air interface

$$\theta_c = \arcsin(1/1.5) \approx 41.8°$$

and at a glass–water interface

$$\theta_c = \arcsin(1.33/1.5) \approx 62.5°.$$

Figure 3.18 Total reflection occurring at a water–xylene interface. Light from a HeCd laser is incident on the interface from the xylene. The water contains the dye Rhodamine 6G in solution. If the blue laser light propagates in the water it is absorbed by the Rhodamine that then reradiates a red fluorescence; see Figure 3.23. At angles of incidence greater than the critical angle, no radiation propagates into the water and the red light is not visible.

Figure 3.18 shows an experimental demonstration of total reflection of blue light from a HeCd laser. In the experiment, a layer of xylene floating on water forms a water–xylene interface. When light is incident at an angle less than the critical angle

$$\theta_c \approx \sin^{-1}\left(\frac{1.33}{1.505}\right) = 62.1°,$$

normal reflection and refraction take place; see Figure 3.23. When light is incident at an angle greater than the critical angle, total internal reflection takes place, as is shown in Figure 3.18. This fact allows us to construct a perfectly reflecting mirror.

Total reflection occurs when $\theta_c < \theta_i \leq \pi/2$. Beyond the critical angle, θ_c, the transmission angle θ_t becomes imaginary. To understand the physical significance of the imaginary angle, recall that

$$\cos\theta_t = \sqrt{1 - \sin^2\theta_t}.$$

Snell's law can be used to rewrite this identity

$$\cos\theta_t = \sqrt{1 - \left(\frac{n_i}{n_t}\right)^2 \sin^2\theta_t} = \sqrt{1 - \left(\frac{\sin\theta_i}{\sin\theta_c}\right)^2}.$$

Over the interval $0 \leq \theta \leq \pi/2$, we have $0 \leq \sin\theta \leq 1$. When $\theta_i > \theta_c$ we have $\sin\theta_i / \sin\theta_c > 1$, leading to a negative number under the radical, making $\cos\theta_t$ an imaginary function:

$$\cos\theta_t = i\sqrt{\left(\frac{\sin\theta_i}{\sin\theta_c}\right)^2 - 1} \qquad \theta_c < \theta_i \leq \frac{\pi}{2}.$$

For notational convenience define $-i\alpha = \cos\theta_t$. The transmitted wave, represented in Figure 3.17 by its propagation vector, $\mathbf{k}_t$, can be written in this notation as

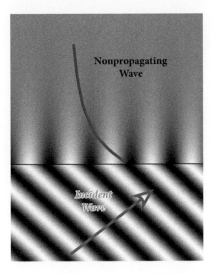

Figure 3.19 Inhomogeneous plane wave incident from bottom onto dielectric junction; the wave vector is indicated by red arrow. Evanescent field with its decay in amplitude indicated by exponential. The reflected wave is not shown. By Aluminium (adapted by Gavin R Putland). [CC BY-SA 3.0 (https://creativecommons.org/licenses/by-sa/3.0)], via Wikimedia Commons.

$$E_t \propto e^{-i\mathbf{k}_t \cdot \mathbf{r}} = e^{-ik_t(x\,\sin\theta_t + z\cos\theta_t)},$$

$$E_t = \underbrace{e^{-k_t a z}}_{attenuation}\underbrace{e^{-ik_t x\sqrt{1+\alpha^2}}}_{x-propagation}.$$

The transmitted wave propagates parallel to the surface (i.e., along the x-axis) at a velocity given by

$$V_{TR} = \frac{c}{n_i \sin\theta_i}$$

that is a function of the incident wave velocity. The wave propagating along the surface is polarized in the direction of propagation, the x-direction. There is no propagation of the transmitted wave in the z-direction. The amplitude of this non-propagating component decays exponentially in the z-direction with the decay constant:

$$\gamma^2 = (kn_t\alpha)^2 \tag{3.44}$$

This is called an inhomogeneous plane wave or a surface wave (Figure 3.19). The phase of the two components differs in phase by $\pi/2$ and energy is interchanged periodically between the x- and z-components as the wave propagates along the x-direction.

The reflectivity for the two polarizations is

$$R_P = \frac{\tan^2(\theta_i-\theta_t)}{\tan^2(\theta_i+\theta_t)} = \frac{\tan^2\left(\theta_i-\frac{\pi}{2}\right)}{\tan^2\left(\theta_i+\frac{\pi}{2}\right)} = \frac{ctn^2\theta_i}{ctn^2\theta_i} = 1,$$

$$R_N = \frac{\sin^2\ (\theta_i - \theta_t)}{\sin^2\ (\theta_i + \theta_t)} = \frac{\sin^2\ \left(\theta_i - \dfrac{\pi}{2}\right)}{\sin^2\ \left(\theta_i + \dfrac{\pi}{2}\right)} = \frac{\cos^2 \theta_i}{\cos^2 \theta_i} = 1.$$

Since the reflectivity is one, there can be no net energy flow across the boundary. Calculation of the z-component of the Poynting vector will verify this.

To calculate the Poynting vector, we use the geometry of Figure 3.17 and assume that the wave is incident with its polarization, i.e., $\mathbf{E}$, parallel to the y-axis,

$$E_{yt} \propto e^{-k_t \alpha z} e^{-i k_t x \sqrt{1+\alpha^2}}$$

$$\mathbf{H} \propto \mathbf{k} \times \mathbf{E} = -E_y k_z\, \hat{\mathbf{i}} + E_y k_x \hat{\mathbf{k}},$$

$$H_{zt} \propto \sqrt{1 + \alpha^2}\ \sqrt{\frac{\varepsilon_t}{\mu_t}} E_{yt},$$

$$H_{xt} \propto i\alpha E\sqrt{\frac{\varepsilon_t}{\mu_t}}.$$

The Poynting vector is

$$\mathbf{S} = \mathbf{E} \times \mathbf{H} = E_y H_z \hat{\mathbf{i}} - E_y H_x \hat{\mathbf{k}},$$

$$S_x \propto E_{yt}^2 \sqrt{1 + \alpha^2}\sqrt{\frac{\varepsilon_t}{\mu_t}}.$$

$$S_z \propto -E_{yt}^2 i\alpha \sqrt{\frac{\varepsilon_t}{\mu_t}}.$$

Energy flows along the x-direction, cycling back and forth across the border into the decaying component in the z-direction until it reaches the edge of the illumination spot. This leads to what is called the *Goos–Hänchen shift*; see Figure 3.20. If the decaying component does not interact with anything, all of the incident energy ends up in the reflected wave. This inhomogeneous wave attached to the surface with amplitude decaying exponentially normal to the surface is called an *evanescent wave*.

The penetration depth $z = 1/\gamma$ of the totally reflected wave is the value of z where the amplitude of the non-propagating component of the wave drops to $1/e$ of its original value, called the decay constant, γ. We can detect this exponentially decaying wave by bringing another high index material within a few penetration depths of the first material. This is called *frustrated total reflection* and is analogous to quantum mechanical tunneling [6]. Figure 3.21 shows two applications of frustrated total reflection, a variable attenuator that can be used as a transducer to convert ultrasound to a visible signal [7] and a prism coupler for launching a wave into an optical waveguide. The prism coupler is also used to construct a total internal reflection fluorescence microscope, or launching surface plasmons on the surface of a metal film [8, 9].

We have replaced the incandescent light bulb for lighting with light emitting diodes (LEDs). Much of the light emitted at the diode junction of the first LEDs was trapped in the diode material by total reflections producing a light source that was not competitive as a light source with incandescent bulbs and fluorescent lamps. Using a variety of optical techniques developed

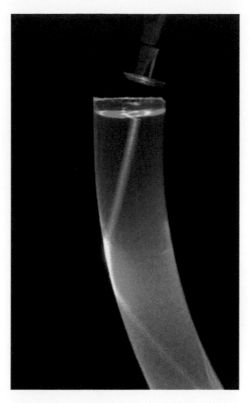

Figure 3.20 Goos–Hänchen effect: There is a small lateral shift by a light beam when undergoing total reflection. This is due to the exchange of energy between the x- and z-components we talked about in the text.

(Copyright 2017 Emilio Gomez-Gonzalez, Universidad de Sevilla, Spain.)

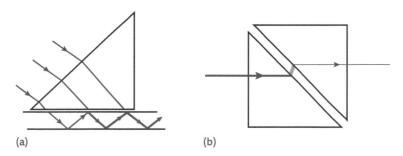

(a)   (b)

Figure 3.21 Two examples of frustrated total reflection. (a) The evanescent wave from a light beam undergoing total internal reflection in a prism is coupled into a waveguide mode of an optical waveguide. (b) Two identical prisms are used to make a variable attenuator by varying the amount of evanescent wave coupled into the second prism. The intensity of the beam is indicated by the breath of the line.

using photonic crystal theory, the total reflection has been defeated [10], producing LED with an energy cost $\frac{1}{6}$ of traditional incandescent and 70% of the cost of a compact fluorescent lamp (CFL).

Total reflection has had an enormous impact on the use of optics in modern technology. The fiber optics used in communications depends on total reflection of a wave trapped in the core of an optical fiber, allowing long distant communications of signals with a large bandwidth such as streaming video. We also use total reflection in microscopy in the total internal reflecting microscope (TIRM)[15] and for imaging of internal organs and tissue with endoscopes.

While the amplitude of the wave is not modified when the light undergoes total reflection, the phase of the light is modified. This is important for guided waves, as we will see in Chapter 8. We can examine this phase change by looking at the reflection coefficients. We will rewrite Eqs. (3.38) and (3.39) to get

$$r_N = -\frac{\sin\theta_i \cos\theta_t - \cos\theta_i \sin\theta_t}{\sin\theta_i \cos\theta_t + \cos\theta_i \sin\theta_t} = \frac{\sqrt{1+\alpha^2}\cos\theta_i + i\alpha \sin\theta_i}{\sqrt{1+\alpha^2}\cos\theta_i - i\alpha \sin\theta_i} \tag{3.45a}$$

$$r_P = \frac{\sin\theta_i \cos\theta_i - \cos\theta_t \sin\theta_t}{\sin\theta_i \cos\theta_i + \cos\theta_t \sin\theta_t} = \frac{\sin\theta_i \cos\theta_i + i\alpha\sqrt{1+\alpha^2}}{\sin\theta_i \cos\theta_i - i\alpha\sqrt{1+\alpha^2}} \tag{3.45b}$$

Both r_N and r_P are complex numbers of the form $(a+ib)/(a-ib)$. These complex variables can be written as

$$r = \frac{a+ib}{a-ib} = \frac{\cos\varphi + i\sin\varphi}{\cos\varphi - i\sin\varphi} = \frac{e^{i\varphi}}{e^{-i\varphi}} = e^{2i\varphi} = e^{i\delta}$$

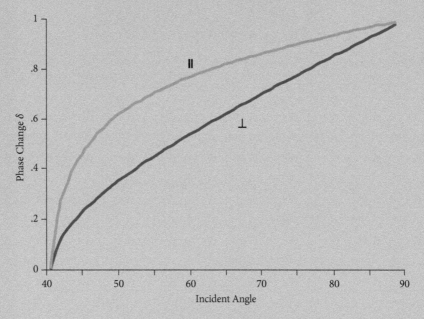

Figure 3.22 The phase change upon total reflection for each of the two polarizations. For this calculation we assumed that the index of the dense medium was 1.5 and the index of the less dense medium was 1.0.

---

[15] One of the suppliers of TIRM devices can be found at http://www.tirf-labs.com/lightguidetirf.html.

$$\frac{\sin\varphi}{\cos\varphi} = \tan\varphi = \frac{b}{a}.$$

We then define $\delta/2 = \varphi$ and $\tan(\delta/2) = b/a$. This leads to phase shifts of

$$\tan(\delta_P/2) = \frac{\alpha\sqrt{1+\alpha^2}}{\sin\theta_i\cos\theta_i} \tag{3.46}$$

$$\tan(\delta_N/2) = \frac{\alpha\sin\theta_i}{\sqrt{1+\alpha^2}\cos\theta_i} \tag{.3.47}$$

The phase shifts as a function of incident angle are shown in Figure 3.22 for the case of $n_i = 1.5$ and $n_t = 1.0$.

3.7 Reflection from a Conductor

Most mirrors that we use are made by coating a transparent piece of glass with a metal film like silver or aluminum. Our description of the propagating medium that we have been using contains the assumption that it is non-conducting:

$$\sigma = 0 \Rightarrow \quad \mathbf{J} = 0.$$

We now relax that assumption and allow $\sigma \neq 0$; i.e., the material is dissipative. This will allow us to include metal mirrors in our discussion of reflection. Maxwell's equations include a current density, $\mathbf{J}$, term:

$$\nabla \cdot \mathbf{D} = 0 \qquad \nabla \cdot \mathbf{B} = 0$$

$$\nabla \times \mathbf{H} = \mathbf{J} + \frac{\partial \mathbf{D}}{\partial t} \qquad \nabla \times \mathbf{E} = -\frac{\partial \mathbf{B}}{\partial t}.$$

We continue to neglect dynamic or resonant effects so that we may use the simple constitutive relations

$$\mathbf{J} = \sigma\mathbf{E} \qquad \mathbf{D} = \varepsilon\mathbf{E} \qquad \mathbf{B} = \mu\mathbf{H},$$

where ε, μ, and σ are scalars, independent of time and space. Maxwell's equations in a medium with dissipation can be rewritten using these constitutive relations as

$$\nabla \cdot \mathbf{E} = 0 \qquad \nabla \cdot \mathbf{H} = 0$$

$$\nabla \times \mathbf{H} = \sigma\mathbf{E} + \varepsilon\frac{\partial \mathbf{E}}{\partial t} \qquad \nabla \times \mathbf{E} = -\mu\frac{\partial \mathbf{H}}{\partial t}. \tag{3.48}$$

We now apply the same procedure used to derive the wave equation for free space:

$$\nabla \times (\nabla \times \mathbf{E}) = \nabla \times \left(-\mu\frac{\partial \mathbf{H}}{\partial t}\right) = -\mu\frac{\partial}{\partial t}(\nabla \times \mathbf{H}),$$

$$-\mu\frac{\partial}{\partial t}(\nabla \times \mathbf{H}) = -\mu\frac{\partial}{\partial t}\left(\sigma\mathbf{E} + \varepsilon\frac{\partial \mathbf{H}}{\partial t}\right),$$

$$\nabla \times (\nabla \times \mathbf{E}) = \nabla(\nabla \cdot \mathbf{E}) - \nabla^2\mathbf{E}.$$

Yielding the wave equation in a conducting medium

$$\nabla^2 \mathbf{E} = \mu\sigma\frac{\partial \mathbf{E}}{\partial t} + \mu\varepsilon\frac{\partial^2 \mathbf{E}}{\partial t^2}. \tag{3.49}$$

We can derive a similar equation for the magnetic field

$$\nabla^2 \mathbf{B} = \mu\sigma\frac{\partial \mathbf{B}}{\partial t} + \mu\varepsilon\frac{\partial^2 \mathbf{B}}{\partial t^2}. \tag{3.50}$$

Equations (3.49) and (3.50) are called the *telegraph equations* and are wave equations developed by *Oliver Heaviside* (1850–1925) to explain the propagation of pulses on telegraph lines.

Propagation in a Conducting Media

Remember in Chapter 1 we assume that a point on the wave underwent simple harmonic motion, Eq. (1.3). With the addition of a loss to the wave equation, the simple harmonic motion becomes damped harmonic motion

$$m\frac{d^2x}{dt^2} = -sx - b\frac{dx}{dt},$$

where b is called the damping constant and $\gamma = \dfrac{b}{m}$ is the resistance per unit time. In a conducting medium, the wave equation, Eq. (3.49), now contains a damping term, $\partial \mathbf{E}/\partial t$, and the damping constant for Eq. (3.49) is $\mu\sigma$. The dissipative wave equation will be an electromagnetic wave that will experience attenuation proportional to $\mu\sigma$ as it propagates. Using Eqs. (2.20) and (2.18), we may rewrite Eq. (3.49), for plane wave solutions, as

$$\begin{aligned} \nabla \times \mathbf{E} = -\mu\frac{\partial \mathbf{H}}{\partial t} &\Rightarrow \nabla \times \mathbf{E} = -i\omega\mu\mathbf{H} \\ \nabla \times \mathbf{E} = \sigma\mathbf{E} + \varepsilon\frac{\partial \mathbf{E}}{\partial t} &\Rightarrow \nabla \times \mathbf{E} = i\omega\left[\varepsilon - i\frac{\sigma}{\omega}\right]\mathbf{E} \end{aligned} \tag{3.51}$$

We rewrite Eq. (3.49) in terms of these expressions for the curl of **E** and **H**:

$$\nabla^2 \mathbf{E} + \omega^2\mu\left[\varepsilon - i\frac{\sigma}{\omega}\right]\mathbf{E} = 0 \tag{3.52}$$

This has the form of the Helmholtz equation, (1.14), if we replace k^2 with the complex function (denoted by the ~)

$$\tilde{k}^2 = \omega^2\mu\left[\varepsilon - i\frac{\sigma}{\omega}\right] \tag{3.53}$$

We use the identity

$$k = \frac{n\omega}{c} = \omega\sqrt{\mu\varepsilon},$$

to reinforce the fact that the equations for conducting media are identical to those derived for non-conducting media if the dielectric constant ε is replaced by a complex dielectric constant

$$\tilde{\varepsilon} = \varepsilon' - \varepsilon'' = \varepsilon' - i\left(\frac{\sigma}{\omega}\right). \tag{3.54}$$

This equation suggests that σ may be frequency dependent (in fact, in the cgs system the units of σ are s^{-1}; for copper in cgs units, $\sigma = 5.14 \times 10^{17}$/s). In condensed matter physics one finds that the mobility of the electrons contains the frequency dependence that shows up in σ.[16]

Since we have replaced k by the complex quantity,

$$\tilde{k} = \omega \sqrt{\mu \left[\varepsilon - i \frac{\sigma}{\omega} \right]}.$$

[16] The mobility of a carrier in a metal or a semiconductor describes how quickly a charge is pulled through the material when in an electric field and has units of cm^2/(V·s).

We must replace the index of refraction by a complex index when we have a conductivity. In the literature this is accomplished in two ways:

$$\begin{aligned} \tilde{n} &= n - i\kappa \\ \tilde{n} &= n_1 - in_2 \end{aligned} \tag{3.55}$$

where κ is the *extinction coefficient* or mass attenuation coefficient. We will use the notation displayed in Eq. (3.55). The complex index is related to the complex dielectric constant by the equations

$$\begin{aligned} \mathcal{R}e\,(\tilde{\varepsilon}) &= \varepsilon' = n^2 - \kappa^2 \\ \mathcal{I}m\,(\tilde{\varepsilon}) &= \varepsilon'' = 2n\kappa \end{aligned} .$$

To find out how the plane wave propagates in this conductive medium, we simply replace the propagation constant, k, by

$$\tilde{k} = \tilde{n}\frac{\omega}{c} = \left(\frac{\omega}{c}\right)(n - i\kappa).$$

If we assume the wave vector $\mathbf{k}$ is parallel to the z-axis, then $\mathbf{k} \cdot \mathbf{r} = kz$, and the plane wave is

$$\mathbf{E} = \mathbf{E}_0 e^{-\omega\kappa z/c} \left[e^{-i\omega(t-z/c)} \right] \tag{3.56}$$

$$\mathrm{Re}\,\{\mathbf{E}\} = \mathbf{E}_0 e^{-(\omega/c)\kappa z} \cos(\omega t - kz) \tag{3.57}$$

The wave described by Eq. (3.57) is a plane wave, with an amplitude attenuated by the exponent

$$e^{-(\omega/c)\kappa z}. \tag{3.58}$$

Figure 3.23 displays the exponential decay of a light wave propagating in an absorbing medium. The light wave is blue radiation from a HeCd laser and the absorbing medium is water containing the dye Rhodamine 6G in solution.

In Figure 3.23, a layer of xylene floats upon water with Rhodamine 6G in solution. The blue light from a HeCd laser (440 nm) passes through the xylene without any perceptible change in intensity. When it enters the water with Rhodamine in solution, the blue light is rapidly

Figure 3.23 Blue laser light is shown propagating in xylene (above) and water (below). The water contains the dye Rhodamine 6G in solution. The red Rhodamine dye absorbs the blue light and the blue beam rapidly decays to 0. Some of the energy absorbed by the dye is re-emitted in the yellow to red region of the spectrum. This re-emitted light caused the diffuse appearance of the light as it propagates in the water because the dye molecules are moving in the solution.

absorbed by the Rhodamine. Some of the energy absorbed by the Rhodamine is re-emitted at longer red wavelengths. The re-emitted light travels in all directions, as it has no memory of the direction traveled by the blue light. For this reason, the reddish beam of light in the water appears diffuse. The red light intensity is proportional to the amount of blue light absorbed so as the blue light is attenuated the brightness of the red fluorescence decreases. Because of the recording process of the film, the red light masks the blue radiation. The bending tail at the end of the red beam is due to the diffusion of the Rhodamine molecules in the water before the fluorescence occurs.

To evaluate the absorption coefficient κ, in terms of electromagnetic properties of the medium, we will derive a relationship between κ and σ. We rewrite Eq. (3.55) as

$$\tilde{n}^2 = n^2 - \kappa^2 - 2i\kappa = \frac{c^2}{\omega^2}\tilde{k}^2.$$

Equation (3.53) can be used to express $\tilde{n}^2$ in terms of the constants of the material

$$\tilde{n}^2 = c^2\mu\left(\varepsilon - i\frac{\sigma}{\omega}\right). \tag{3.59}$$

Equating real and imaginary terms, we obtain

$$n^2 - k^2 = c^2 \mu \varepsilon \qquad\qquad 2n\kappa = c^2 \frac{\mu\sigma}{\omega}.$$

Note that when $\sigma = 0$, $\kappa = 0$, and we obtain the free space result, Eq. (2.15),

$$n^2 = \frac{\mu\varepsilon}{\mu_0 \varepsilon_0}.$$

An estimate of the magnitude of the real and imaginary parts can be obtained by using values for copper where in the MKS units $\sigma = 5.8 \times 10^7$ mhos/m and $n = 0.62$ at $\lambda = 589.3$ nm:[17]

$$\frac{\mu\sigma}{\omega} = \frac{\left[\left(4\pi \times 10^{-7}\right)\left(5.8 \times 10^7\right)\left(5.893 \times 10^{-7}\right)\right]}{(2\pi)\left(3 \times 10^8\right)} = 2.3 \times 10^{-14} \frac{s^2}{m^2},$$

$$\mu\varepsilon = \mu_0 \varepsilon_0 n^2 = \left(4\pi \times 10^{-7}\right)\left(8.8542 \times 10^{-12}\right)(0.62)^2 = 4.3 \times 10^{-18} \frac{s^2}{m^2}.$$

By comparing the relative magnitude of these two terms, we are justified in assuming that $\sigma/\omega \gg \varepsilon$ and the loss term dominates

$$n^2\kappa^2 \approx \frac{c^2\mu\sigma}{2\omega}$$
$$\sqrt{n\kappa} \approx c\sqrt{\frac{\mu\sigma}{2\omega}} \qquad\qquad\qquad (3.60)$$

We use Eq. (3.60) to find the depth at which an electromagnetic wave is attenuated, to $1/e$ of its original energy, when propagating into a conductor. At that depth, denoted by d, the exponent in Eq. (3.57) will equal 1; thus,

$$\left(\frac{\omega}{c}\right)n\kappa d = \frac{2\pi}{\lambda_0} n\kappa d = 1$$

$$d = \frac{\lambda_0}{2\pi n\kappa} \approx \frac{\lambda_0}{2\pi c}\sqrt{\frac{2\omega}{\mu\sigma}}.$$

The depth d is called the *skin depth* (see Figure 3.24):

$$d = \sqrt{\frac{2}{\mu\sigma\omega}}. \qquad\qquad\qquad (3.61)$$

The transmitted optical wave therefore does not penetrate very far into any material with an appreciable conductivity.

The Fresnel equations derived for reflection and transmission from a boundary apply to materials with a complex index of refraction as well as a material with a real index. To use the derived equations for materials with complex indices, simply replace the real index with the equivalent complex one. We will demonstrate the procedure by finding the reflectivity of a metal that has a complex index of refraction. We will limit our discussion to the near-normal reflection from an air–metal interface where

$$n_t = \tilde{n}$$

---

[17] The index of refraction is less than one that implies that the phase velocity is greater than the speed of light. This typically occurs near a resonance in an absorbing medium.

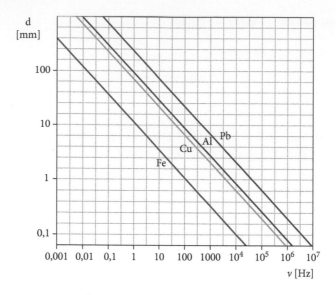

Figure 3.24 Skin depth for a few metals.

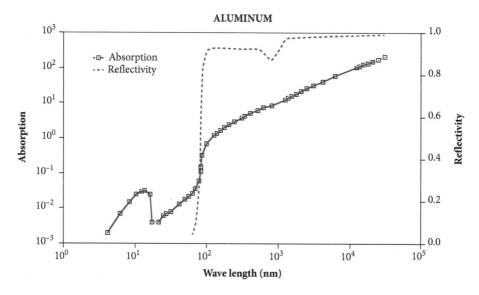

Figure 3.25 Reflectivity and absorption as a function of wavelength in aluminum.

(a complex index of refraction) and $n_i = 1$. Replacing the real index in Eq. (3.39) by a complex index

$$R = \frac{(\tilde{n} - 1)(\tilde{n}^* - 1)}{(\tilde{n} + 1)(\tilde{n}^* + 1)} = \frac{(n - 1 + in\kappa)(n - 1 - in\kappa)}{(n + 1 + in\kappa)(n + 1 - in\kappa)} = \frac{(n-1)^2 + (n\kappa)^2}{(n+1)^2 + (n\kappa)^2},$$

$$R = 1 - \frac{4n}{(n+1)^2 + (n\kappa)^2} \tag{3.62}$$

The reflectivity, R, and the absorption, $n\kappa$, in a material can be observed in Figure 3.25, where $n\kappa$ and R are plotted as a function of wavelength for the metal aluminum. Note in Figure 3.25

that the reflectivity rises with increasing absorption as our simple theory predicts. If the index of refraction were purely imaginary, then

$$\tilde{n} = in\kappa,$$

$$R = \frac{(in\kappa - 1)(-in\kappa - 1)}{(in\kappa + 1)(-in\kappa + 1)} = \frac{-(in\kappa - 1)(in\kappa + 1)}{-(in\kappa + 1)(in\kappa - 1)} = 1,$$

and the material would be a perfect reflector.

Let's look at a real metal mirror made out of aluminum. The values for n and κ as a function of wavelength are shown in Figure 3.26 over the wavelength range from the near IR to the UV. Light is incident on an aluminum mirror from air at an angle near normal. The reflection coefficient is

$$r_1 = \frac{n_0 - (n_1 - in_1\kappa_1)}{n_0 + (n_1 - in_1\kappa_1)}.$$

We will pick a wavelength of light of 546 nm yield $n_1 \approx 0.8$ and $\kappa_1 = 6$. This give us $n_1\kappa_1 = 4.91$:

$$r_1 = \frac{1 - (.82 - i4.91)}{1 + (.82 - i4.91)} = -0.87 + i0.36$$

$$r_1 = 0.94e^{i(\pi - 39)} = -0.94e^{-.39i}$$

$$R_1 = 0.88.$$

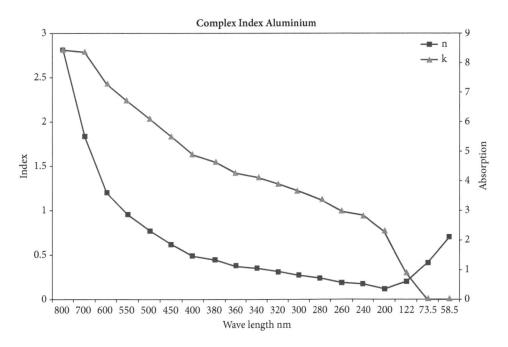

Figure 3.26 Complex index of refraction for aluminum.

Aluminum oxidizes in air so it might be interesting to see what would happen if we put a glass overcoat over the aluminum to prevent oxidization. Now

$$r_2 = \frac{105 - 0.82 + i4.91}{105 + 0.82 - i4.91} = -0.76 + i0.5$$

$$r_2 = -0.91e^{-.58i}$$

$$R_2 = 0.83.$$

Some of the light also reflects from the glass–air surface so that the measured reflectivity is

$$\frac{I_f}{I_0} = (1 - R_2) R_2 (1 - R_1)$$

$$= (1 - 0.04)(0.83)(1 - 0.04)$$

$$= 0.76.$$

We have a much lower reflectivity when we put the protective coating on the aluminum. It is possible by the use of dielectric coatings to beat this loss in reflectivity and still protect the metal surface.

3.8 Problem Set 3

(1) The path of light in air incident on and transmitted through a glass plate. The angle of the incident ray to the normal is 45° and equals that of the reflected ray. The transmitted ray is refracted at an angle of 28° to the normal and exits the glass at an angle of 45° to the normal, an angle equal to that of the incident ray. What is the index of refraction of the glass?

(2) An unpolarized beam of light is one whose Stokes' vectors are $s_1 = s_2 = s_3 = 0$. If such a beam in air is incident at an angle of 30° on glass with an index of 1.50. What is the percentage of light energy refracted in the normal and parallel polarized components? What is the degree of polarization?

(3) Calculate the critical angle and Brewster's angle for the image in Figure 3.14. The index of water ($n = 1.33$) and we assume the aquarium glass is dense flint glass ($n = 1.75$). The water interfaces with air and the glass walls of the aquarium.

(4) A tank of water is covered with a 1-cm-thick layer of oil ($n_0 = 1.48$); above the oil is air. If a beam of light originates in the water, what angle must the light beam make at the water–oil interface if no light is to escape into the air?

(5) Find the skin-depth for seawater with a resistivity of $\rho = 0.20/m$ for $\omega = 30$ KHz and 30 MHz. What frequency should we use to communicate with a submarine that will not be deeper than 100 meters?

(6) The index of refraction of germanium at $\lambda = 500$ nm is complex: $3.47 - i(1.4)$. What is the reflection coefficient of a polished germanium surface at normal incidence?

(7)* Derive an expression for the transmittance of light normally through a stack of N glass plates, each separated by a small air space. Assume no absorption and that all the plates have an index of refraction, n. If we did this calculation for R, we would have initiated the calculation needed to design a multilayer mirror.

(8) Determine the value of the amplitude reflection coefficients for unpolarized light incident at 30° on an air–glass interface where $n_i = 1.0$ and $n_t = 1.5$. Write the Stokes vector for the reflected light.

(9) Mirrors used in x-ray optics are often based upon total reflection. The index of refraction for x-rays of wavelength 0.15 nm in a vacuum is 1.0. For silver, we will treat the index as real and equal to 0.99998. What is the angle for total reflection?

(10) What is the refractive index of a glass plate that polarizes light reflected at 57.5°? This is a good technique for measuring the index of refraction of an unknown material.

(11) A beam of light passes normally through one side and is totally reflected off the hypotenuse of a 45°–90°–45° glass prism of index 1.6. What is the decay constant for the evanescent wave at the point of total reflection, assuming the prism is in air? What happens if the prism is in water (assume the index of water is real and equal to 1.3)?

(12) A silver sample has complex refractive index and damping constant equal to 0.135 and 3.985, respectively, when a light of 632.8 nm ($\nu = 4.74 \times 10^{14}$ Hz) is focused on the sample.

   (a) Calculate the reflectance (R) for this sample.

   (b) With the value of reflectance obtained in (a), calculate the expected conductivity for silver.

(13) Prove that $T + R = 1$ using Eqs. (3.35) and (3.36).

REFERENCES

1. J. M. Enoch, *Optometry & Vision Science* **83**(10), 775–81 (2006).

2. R. P. Feynman, R. B. Leighton, and M. Sands, in *The Feynman Lectures on Physics*, Vol. II, edited by M. A. Gottlieb and R. Pfeiffer (California Institute of Technology, 1963).

3. W. T. Doyle, *Am. J. Phys.* **53**, 463–8 (1985).

4. R. K. Wangsness, *Electromagnetic Fields*, 2nd edn (Wiley, 1986).

5. C. L. Giles and W. J. Wild, *International J. Infra. and Millimeter Waves* **6**(3), 187–97 (1985).

6. S. Zhu, A. W. Yu, D. Hawley, and R. Roy, *Am J. Phys.* **54**, 601–7 (1986).

7. P. J. Phillips, O. T. v. Ramm, J. C. Swartz, and B. D. Guenther, *J. Acoust. Soc. Am.* **93**, 1182–91 (1993).

8. P. K. Tien and R. Ulrich, *J. Opt. Soc. Am.* **60**, 1325–37 (1970).

9. J. R. Sambles, G. W. Bradbery, and F. Yang, *Contemporary Physics* **32**(3), 173–83 (1991).

10. Y.-C. Lee and S.-H. Tu, in *Recent Advances in Nanofabrication Techniques and Applications*, edited by B. Cui (IntechOpen, 2011).

4 Interference and Coherence

The superposition principle applied to wave theory states that for all linear systems the sum of solutions of the wave equation is also a solution to the wave equation. Linearity forbids scattering of one photon by another and prevents the observation of the interaction between overlapping waves unless a square law detector[1] is used to record bright and dark bands of light called *fringes* produced by the overlap. These fringes are commonly observed in soap bubbles or on oil films on a wet roadway. The bright regions occur when a number of waves add together to produce an intensity maximum of the resultant wave; this is called *constructive interference.Destructive interference* occurs when a number of waves add together to produce an intensity minimum of the resultant wave.

Collectively the distribution of fringes is called an *interference pattern.* It is interesting to note that if you do not detect the waves in the overlap region, then there is no record carried by any of the waves to indicate outside of the overlap region that the waves ever commingled. Our discussion must be then confined to the region of overlap and include the detection process before we are able to record an interference pattern. Quantum electrodynamics does allow photons to interact through the pair production process at fairly high photon energies but this interaction has only been observed once [1].

Everyday experience suggests that the observation of interference must be subject to very restrictive conditions for there are no indications of fringes in most illumination. When interference is observed, it is usually associated with small dimensions; for example, two slits separated by only a few millimeters, or thin films much less than a millimeter thick, are required to produce interference fringes in natural light. The need for small dimensions is not, however, associated with the wavelength of light. When laser illumination is used, interference is quite easy to produce. In fact, interference is a major noise source in images produced with laser illumination. The ability to observe interference is associated with a property of a wave called coherence. The mathematical description of interference is the correlation operation. We will develop the connection between coherence theory and correlation and discuss the experimental techniques that duplicate the correlation operation. We will discover that a temporal correlation of a wave with itself can be performed using a Michelson interferometer and that a spatial correlation can be performed using Young's two-slit configuration.

The first observation of interference was made by *Robert Boyle* (1627–91) in 1663. *Robert Hooke* (1635–1703) was a co-discoverer of the interference pattern of concentric rings but it is Newton who has his name associated with the interference pattern because he performed a number of experiments on the effect. *Thomas Young* (1773–1829) in 1802 conducted experiments [2] that could only be explained if light was a wave phenomenon. His experiments

---

[1] A detector whose output is proportional to the square of the input amplitude, for example, the human eye.

Modern Optics Simplified. B. D. Guenther. © B. D. Guenther 2020.
Published in 2020 by Oxford University Press. DOI: 10.1093/oso/9780198842859.001.0001

disagreed with the then-accepted particle theory of light developed by Newton, and were rejected by most of the scientific community, sometimes rudely. Young's experiments were not decisive in establishing the wave theory because it was possible for there to be other origins for the fringes. Ten years after Young's experiments, Fresnel performed experiments that confirmed Young's results and eliminated all other possible sources of the observed patterns. These experiments led to the rejection of the particle theory of Newton and its replacement by the wave theory of light.

4.1 Addition of Waves

We will utilize two related methods for adding waves based on the use of the complex notation introduced in Chapter 1.

4.1.1 Complex approach

N scalar waves of the form given by

$$y_1 = Y_1 \cos(\omega t + \phi_1), y_2 = Y_2 \cos(\omega t + \phi_2), \cdots y_N = Y_N \cos(\omega t + \phi_N) \tag{4.1}$$

can be rewritten in complex notation as

$$\tilde{y}_1 = Y_1 e^{i(\omega t + \phi_1)}, \ldots, \tilde{y}_i = Y_i e^{i(\omega t + \phi_i)}, \ldots, \tilde{y}_N = Y_N e^{i(\omega t + \phi_N)}. \tag{4.2}$$

To simplify the discussion, assume that the phase of each wave (in radians) is ϕ_0 different from waves adjacent in index ($i-1, i, i+1$) and that the amplitude of each wave is identical and equal to the value Y_0. The sum of N waves of this type is written as

$$\tilde{y} = \sum_{j=1}^{N} \tilde{y}_j = Y_0 e^{i\omega t} \left[e^{i\phi_0} + e^{2i\phi_0} + \cdots + e^{Ni\phi_0} \right] = Y_0 e^{i\omega t} \sum_{j=1}^{N} e^{j(i\phi_0)} \tag{4.3}$$

The summation is a geometric series[2] with a value given by the formula

$$\sum_{n=0}^{N-1} r^n = \frac{1 - r^N}{1 - r}.$$

Equation (4.1) becomes

$$\tilde{y} = Y_0 e^{i(\omega t + \phi_0)} \frac{1 - e^{Ni\phi_0}}{1 - e^{i\phi_0}}. \tag{4.4}$$

We stated when we introduced the complex notation that upon completion of the evaluation of the relationship we should retain only the real part of the solution.

To obtain the real part of $\tilde{y}$, Eq. (4.4), we use the identities for sine and cosine in exponential form[3]

---

[2] A geometric series is a series with a constant ratio between successive terms.

[3] Euler's formula $e^{i\theta} = \cos\theta + i\sin\theta$ provides a connection between the trigonometric functions and the exponential function.

$$\cos\theta = 1/2\left(e^{+i\theta} + e^{-i\theta}\right) \qquad \sin\theta = 1/2i\left(e^{+i\theta} - e^{-i\theta}\right)$$

$$\tilde{y} = Y_0 e^{i(\omega t + \phi_0)}\frac{-e^{iN\phi_0/2}\left[e^{-iN\phi_0/2} - e^{iN\phi_0/2}\right]}{-e^{i\phi_0/2}\left[e^{-i\phi_0/2} - e^{i\phi_0/2}\right]} = Y_0 e^{i(\omega t + \phi_0)}e^{i(N-1)\phi_0/2}\frac{\sin\left(N\phi_0/2\right)}{\sin\left(\phi_0/2\right)}.$$

Now the real part of $\tilde{y}$ can be separated from the imaginary part and placed in the form

$$\mathrm{Re}\{\tilde{y}\} = \frac{Y_0 \sin\left(N\phi_0/2\right)}{\sin\left(\phi_0/2\right)}\cos\left[\omega t + (N+1)\phi_0/2\right].$$

Note how the use of complex notation made it easy to obtain the phase of the cosine wave. We can generalize this result to state that the sum of a number of harmonic waves of the same frequency, ω, leads to a resultant wave which is also a sinusoidal wave with the frequency ω and with an amplitude and phase given by

$$Y = Y_0 \frac{\sin\left(N\phi_0/2\right)}{\sin\left(\phi_0/2\right)} \qquad \delta = (N+1)\phi_0/2. \tag{4.5}$$

4.1.2 **Graphic approach**

Waves described in complex notation in Eq. (4.3) can be represented, at a time t_1, by vectors. The length of each vector is equal to the maximum amplitude of the wave, and the angle that each vector makes with the abscissa is equal to the phase $\omega t_1 + \phi_n$. A similar graphical representation was used in our discussion of polarization; see Figure 2.7. If these vectors are drawn in a coordinate system rotating about the axis normal to the plane of the figure at a frequency ω, the vectors will appear stationary. Figure 4.1 shows the vector representation of

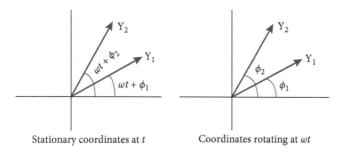

Stationary coordinates at t Coordinates rotating at ωt

Figure 4.1 Vector representation of two waves with the same frequency and different phase. The coordinate system on the left is stationary while that on the right is rotating at an angular frequency equal to the waves' frequency so that the vectors appear stationary in this frame.

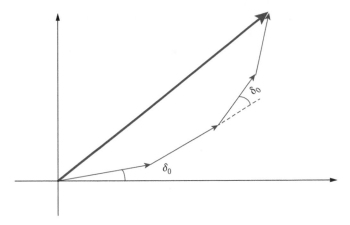

Figure 4.2 Vector representation of four waves with equal amplitudes and phases in an arithmetic progression starting at δ_0. The coordinate system is rotating at a frequency ω.

two waves in a fixed coordinate system, on the left, and in a coordinate system rotating at an angular velocity, ω, on the right.

Vectors are added graphically by placing the vectors tail to head, as shown in Figure 4.2. The resultant wave is then drawn from the tail of the first wave to the head of the last wave. The resultant vector shown in Figure 4.2 is stationary in the rotating frame and rotating at a frequency ω in the fixed coordinate system; therefore, the resultant wave is a harmonic wave of frequency ω. This graphic implementation of the complex approach provides visual insight of the interference process.

4.2 **Polarization and Interference**

In introducing the mathematics of wave addition, we have been treating the wave as a scalar. Before continuing we need to evaluate the effect that the vector amplitude (polarization) of the electromagnetic wave has on the treatment of interference. We will represent the addition of two electromagnetic plane waves in its most general form. We will then discover, if we limit our attention to plane electromagnetic waves propagating in free space, that orthogonal polarizations will not interfere due to the fact that light waves are transverse.

We will represent two electromagnetic waves by

$$\tilde{\mathbf{E}}_1 = \tilde{\mathbf{A}}e^{i\omega t} \text{ and } \tilde{\mathbf{E}}_2 = \tilde{\mathbf{B}}e^{i\omega t}.$$

We will assume that the waves have the same frequency and thus will need only to concern ourselves with the *complex amplitudes* of the waves. The components of the complex amplitudes of the two waves are

$$\tilde{A}_x = a_1 e^{-ig_1}, \tilde{A}_y = a_2 e^{-ig_2}, \tilde{A}_z = a_3 e^{-ig_3}, \tag{4.6}$$

$$\tilde{B}_x = b_1 e^{-ih_1}, \tilde{B}_y = b_2 e^{-ih_2}, \tilde{B}_z = b_3 e^{-ih_3}, \tag{4.7}$$

where g and h are of the form $(\mathbf{k} \cdot \mathbf{r} + \phi)$. The vector $\mathbf{r}$ represents the position of the point in space where the wave amplitudes are added and $\mathbf{k}$ is the wave vector indicating the direction of motion. Since we can only measure the intensity of the light wave, the time average of the square of the sum of the two waves must be calculated before the theoretical result can be compared with experiment,

$$\langle E^2 \rangle = \langle (E_1 + E_2) \cdot (E_1 + E_2) \rangle = \langle E_1^2 \rangle + \langle E_2^2 \rangle + 2 \langle E_1 \cdot E_2 \rangle. \tag{4.8}$$

Equation (2.31) defines the intensity of the wave as proportional to the square of the electric field of the light wave. We are only interested in relative intensities so we will replace $\langle E^2 \rangle$ by I and neglect the constants of proportionality.

The resultant wave's intensity is given by

$$I = I_1 + I_2 + 2 \langle E_1 \cdot E_2 \rangle.$$

I_1 and I_2 are the intensities of each wave, independent of the other. All information about interference is contained in the third term of this equation. If the third term is zero at all positions, the waves do not interfere and are said to be *incoherent*. We will evaluate the third, interference, term utilizing a useful relationship that makes it possible to calculate the time average of the product of two waves, A and B, when complex notation is used.

$$A = \mathcal{R}e\{\tilde{A}\} = \mathcal{R}e\{A_0 e^{i(\omega t + \phi_1)}\}$$
$$B = \mathcal{R}e\{\tilde{B}\} = \mathcal{R}e\{B_0 e^{i(\omega t + \phi_2)}\}$$

use

$$\mathfrak{R}e\{\tilde{z}\} = x = r\cos\phi = \frac{\tilde{z} + \tilde{z}^*}{2}$$

$$\mathfrak{I}m\{\tilde{z}\} = y = r\sin\phi = \frac{\tilde{z} - \tilde{z}^*}{2i}$$

to write the average over one period as

$$\langle AB \rangle = \frac{1}{T} \int_0^T \left(\frac{\tilde{A} + \tilde{A}^*}{2} \right) \left(\frac{\tilde{B} + \tilde{B}^*}{2} \right) dt,$$

$$\left(\tilde{A} + \tilde{A}^* \right) \left(\tilde{B} + \tilde{B}^* \right) = \tilde{A}\tilde{B} + \tilde{A}^* \tilde{B}^* + \tilde{A}\tilde{B}^* + \tilde{A}^* \tilde{B},$$

where

$$\tilde{A}\tilde{B} = A_0 B_0 e^{i(2\omega t + \phi_1 + \phi_2)}$$

and

$$\tilde{A}^* \tilde{B}^* = A_0 B_0 e^{-i(2\omega t + \phi_1 + \phi_2)}.$$

The time averages of these two terms are zero and we are left with

$$\langle AB \rangle = \frac{1}{T} \int_0^T \frac{\tilde{A}\tilde{B}^* + \tilde{A}^* \tilde{B}}{4} dt.$$

We may rewrite this as

$$\langle AB \rangle = \frac{1}{2} \mathfrak{Re} \left\{ \tilde{A} \tilde{B}^* \right\}. \tag{4.9}$$

$$2 \langle E_1 \cdot E_2 \rangle = \mathfrak{Re} \left\{ \tilde{E}_1 \cdot \tilde{E}_2^* \right\} = \frac{1}{2} \left(\tilde{E}_1 \cdot \tilde{E}_2^* + \tilde{E}_1^* \cdot \tilde{E}_2 \right),$$

$$= a_1 b_1 \cos \left(g_1 - h_1 \right) + a_2 b_2 \cos \left(g_2 - h_2 \right) + a_3 b_3 \cos \left(g_3 - h_3 \right).$$

Assume that the two waves to be added together are plane waves, of the same frequency, orthogonally polarized, and propagating parallel to one another along the z-axis. One wave is assumed to have its $\mathbf{E}$ vector located in the x,z-plane so that $a_2 = 0$ and the other is assumed to have its $\mathbf{E}$ vector in the y,z-plane so that $b_1 = 0$. With these assumptions, the interference term is

$$2 \langle \mathbf{E}_1 \cdot \mathbf{E}_2 \rangle = a_3 b_3 \cos \left(g_3 - h_3 \right).$$

Up to now the electromagnetic properties of the waves have not been utilized. Now we will use the physical fact that light waves are transverse in free space. The transverse nature of light requires that $a_3 = b_3 = 0$ for the waves represented by $\mathbf{E}_1$ and $\mathbf{E}_2$ and results in the interference term being equal to 0 for orthogonal polarizations.

The interference between two sources, a and b, can also be calculated by calculating the cross-correlation of $a(t)$ and $b(t)$. The mathematical expression for the cross-correlation is

$$h(\tau) = a \oplus b = \int_{-\infty}^{\infty} a^*(t) \, b(t+\tau) \, dt$$

$$a(t) = A \cos(\omega_0 t + \theta) \qquad b(t) = B \cos(\omega_0 t + \phi).$$

The cross-correlation function is

$$h(\tau) = \frac{AB}{2} \cos(\omega_0 \tau + \theta - \phi).$$

This mathematical operation describes the interference process where the fringes are due to the phase difference $(\theta - \phi) = (g_3 - h_3)$, as we just derived.

Maxwell's equations provide the key result for interference that light waves propagating in free space and polarized at right angles to each other will not interfere. The key assumptions applied to Maxwell's equations leading to this result are that the medium is isotropic and free of charge; thus, the result applies to many simple dielectrics.[4]

Since only parallel polarized waves interfere with one another, we can simplify the notation without loss of generality by assuming that all light waves are linearly polarized in the y-direction and the waves are propagating in the x,z-plane

---

[4] There are materials in which an electromagnetic wave can have a longitudinal component and for those special cases (viz., nonlinear optics), the above result must be modified.

$$a_1 = a_3 = b_1 = b_3 = 0$$

$$I_1 = \frac{a_2^2}{2}, \ I_2 = \frac{b_2^2}{2}$$

$$2 \langle \mathbf{E}_1 \cdot \mathbf{E}_2 \rangle = a_2 b_2 \cos(g_2 - h_2) = 2\sqrt{I_1 I_2} \cos \delta,$$

allowing the relation for the intensity of the resultant wave to be written as

$$I = I_1 + I_2 + 2\sqrt{I_1 I_2} \cos \delta, \tag{4.10}$$

where

$$\delta = [(\mathbf{k}_1 \cdot \mathbf{r}_1 - \mathbf{k}_2 \cdot \mathbf{r}_2) + (\phi_1 - \phi_2)] = \Delta + \Delta\phi. \tag{4.11}$$

Two electromagnetic waves have been added using the complex approach and have been used to demonstrate that only parallel components of the electric field contribute to the interference term. The interference term has been shown to be a function of the amplitudes of the two waves and a harmonic function of the phase difference, δ, between the two waves.

In Eq. (4.11), we have separated the phase difference, δ, into two components: $\Delta\phi = \phi_1 - \phi_2$, having nothing to do with propagation (often this is a result of reflection) and a component, $\Delta = (\mathbf{k}_1 \cdot \mathbf{r}_1 - \mathbf{k}_2 \cdot \mathbf{r}_2)$, equal to the phase difference due to optical path propagation differences. The propagation paths for the two waves are measured in units of wavelength since $|\mathbf{k}| = 2\pi/\lambda$.

4.2.1 Optical path length

We know that the wavelength of light depends upon the propagation velocity in the medium; see Figure 3.4. In order to allow the light to propagate along paths in media with different indices of refraction and still evaluate their phase differences, all path lengths are converted to an equivalent path length in a vacuum. The equivalent path length, the *optical path length*, between points A and B, in a medium with an index of refraction, n, is defined in Chapter 3 by Eq. (3.5) and repeated here:

$$c\tau = \frac{cr}{v} = nr.$$

The phase difference, Δ, can now be expressed in terms of the optical path length. Both waves were assumed to have the same frequency; therefore, the propagation constant of wave i is

$$|\mathbf{k}_i| = \frac{n_i 2\pi}{\lambda_0},$$

where λ_0 is the wavelength of the waves in a vacuum and n_i is the index of refraction associated with the path of the ith wave. With the definitions just introduced, it is apparent that Δ is the difference in optical path lengths of the two waves:

$$\Delta = \mathbf{k}_1 \cdot \mathbf{r}_1 - \mathbf{k}_2 \cdot \mathbf{r}_2 = k_0 \cdot (n_1 \mathbf{r}_1 - n_2 \mathbf{r}_2),$$

where the subscript 0 on $\mathbf{k}$ denotes the propagation vector in a vacuum.

When the two waves have the same amplitude, Eq. (4.10) can be rewritten as

$$I = 4I_1 \cos^2 \frac{\delta}{2}. \tag{4.12}$$

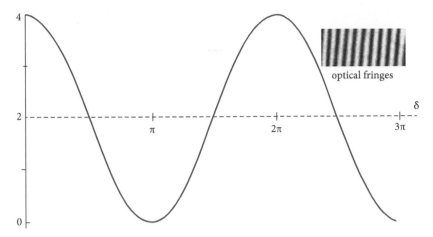

Figure 4.3 Interference of two waves of equal intensity as described by Eq. (4.12). The insert is a photo of optically generated fringes matching the theoretical result shown in the plotted curve.

Figure 4.3 shows a plot of the intensity distribution that is described by Eq. (4.12).

When constructive interference occurs, the maximum intensity is $4I_1$ because the phase difference between the two waves is $\delta = m \cdot 2\pi$. The minimum intensity is zero when destructive interference occurs because the phase difference between the two waves is $\delta = (2m + 1)\pi$. The integer m is equal to the optical path length measured in wavelengths. If the energy in each beam is I_1, conservation of energy requires the energy in the combined beams to be $2I_1$. Is conservation of energy violated when the observed intensity of the interference reaches a value of $4I_1$? The answer is no; the energy has had its spatial variation redistributed by interference but over the entire cross section of the resultant wave the energy per unit area is still $2I_1$ (the spatial average of $\cos^2 \delta$ is equal to ½).

Equation (4.8) requires that the observed light distribution be a temporal average. The importance of the time average in the equations for interference cannot be overemphasized. Temporal variations of δ must be small during the period, T, over which we take the temporal average, if the interference is to be observed. From an operational viewpoint, the detector used to observe interference must respond in a time T that is faster than the time for any intensity variations of the interference process to occur. We define a *coherence time*, τ_c, to characterize the temporal variations of the phase; if $\tau_c > T$, then the waves are said to be *coherent* and we can track any intensity variation. If we can temporally track any intensity variation of the interference term we can observe interference.[5] We will develop some other experimental definitions of coherence later.

[5] Tracking of the intensity variation is what occurs when your radio uses its detector to extract the audio signal that is carried by the radiofrequency wave transmitted by the radio station.

The expected fringe structure can be predicted mathematically by looking at a simple experimental configuration for the generation of a fringe pattern. We assume that two plane waves overlap as shown in Figure 4.4. The fringes that occur along the plane of intersection, **OB**, are shown in Figure 4.5. The intensity distribution described by Eq. (4.10) can be connected to the experimental arrangement in Figure 4.4 by evaluating the phase change in terms of a dimension in the experimental configuration. At point **O** the two waves are assumed to be out of phase and we have the dark band in Figure 4.5. They are out of phase again at point **B** where the wave traveling at an angle has traveled an additional full wavelength. The spacing between dark bands[6] if the plane waves have a wavelength of 632.8 nm (the wavelength of a HeNe laser) and intersect at an angle of 5° is given by

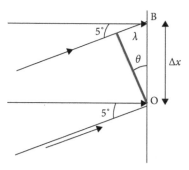

Figure 4.4 Two plane waves intersecting on the plane **OB**. *Note:* The angle as drawn is four times larger (20°) than the 5° labeled in the drawing.

Figure 4.5 Interference fringes occurring on the plane **OB.**

[6] Experimentally the minimum dark fringe is easier to locate accurately than is the peak of the bright fringe and the good experimentalist measures dimensions with respect to dark fringes.

$$\sin \theta = \frac{\overline{AP}}{\overline{OP}} = \frac{\lambda}{\Delta x}$$

$$\Delta x = \frac{\lambda}{\sin \theta} \approx \frac{\lambda}{\theta}$$

$$\Delta x = 7 \, \mu m.$$

Our eye cannot resolve fringe spacing this small but could easily resolve a fringe spacing of 1 mm.[7] The angle between the two beams that would produce fringes similar to Figure 4.5 with 1-mm spacing would be

$$\sin \theta = \frac{632.8 \times 10^{-9}}{10^{-3}}.$$

$$\theta \approx 0.04°$$

To produce the required two plane waves, we could simply reflect a plane wave from two glass plates—one tilted at an angle θ with respect to the other. The plates create an air gap that can be thought of as a dielectric film. To produce the small angle required to see the fringes, we could place a single tissue between the edges along the far right of Figure 4.6. Each fringe represents a gap of constant thickness.

The parameter m in Figure 4.6 is equal to the number of total wavelengths contained in the optical path. It should be noted that the value of m is not needed to obtain the angle between the two waves. We only need the change in m, Δm, the number of fringes we cross while measuring the distance, Δx, in the **OB** plane containing the fringes. The second observation we need to make is that by generating fringes, we have produced a ruler with wavelength accuracy that allows us to accurately measure very small physical dimensions. In the remaining part of this chapter, we will develop physical models that will allow the use of the wavelength accuracy provided by interference to measure useful properties of physical objects.

4.2.1.1 Newton's rings

Newton's rings are the first useful application of interference that we will discuss. The interference fringes are produced not by two flat plates as in our initial example but rather by a flat plate and a spherical surface. Newton's rings are called Fizeau fringes or fringes of constant

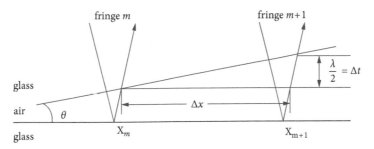

Figure 4.6 Two glass plates with a wedge of air between them will match the geometrical construct of Figure 4.4. The Δt $\lambda/2$ is due to the fact that the light must make a round trip so the actual thickness must be multiplied by 2.

---

[7] An unaided eye can see 0.1 mm at reading distances, 400 mm. By using a factor of 10 more, we ease the observation task.

thickness[8] and can be obtained by placing a convex lens in contact with a flat glass plate, forming an air wedge. Circular interference fringes are formed about the point of contact—point O in Figure 4.7a. The radius of a dark or light band of the fringes can be calculated in terms of the radius of curvature of the lens.

In Figure 4.7a the radius of curvature of the lens is R and the radius of a circular contour forming one of Newton's rings is ρ. The Pythagorean theorem can relate these two parameters

$$R^2 = \rho^2 + (R - d)^2,$$

where d is the thickness of the air gap at the fringe,

$$\rho^2 = d(2R - d),$$

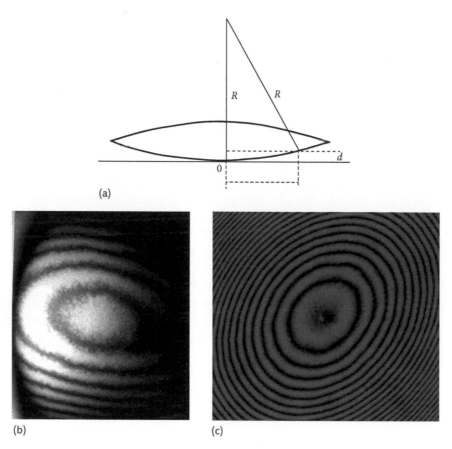

(a)

(b)   (c)

Figure 4.7 (a) Experimental setup for the observation of Newton's rings. (b) Newton's rings seen in transmission and (c) Newton's rings seen in reflection
(by Robert D. Anderson (CC BY-SA 3.0 (https://creativecommons.org/licenses/by-sa/3.0)), from Wikimedia Commons). Both fringe patterns were produced by the air film between a lens and an optical flat. Note that in transmission the center fringe is bright but in reflection it is dark.

---

[8] A dark fringe produces a contour line mapping out points where the film thickness is the same.

since $d \ll R$ we can write

$$d = \frac{\rho^2}{2R}.$$ (4.13)

We assume normal incidence so that $\cos(\theta_t) \approx 1$ and get from Eq. (4.13) that the dark bands will be rings with a radius of

$$\rho^2 = \frac{m\lambda_0 R}{n_2} \quad m = 0, 1, 2, \ldots,$$ (4.14)

and the bright bands will be rings with a radius of

$$\rho^2 = \frac{\left(m + \frac{1}{2}\right)\lambda_0 R}{n_2} \quad m = 0, 1, 2, \ldots.$$ (4.15)

Usually the gap between the test plate and the lens is air so that $n_2 = 1$. Note that here, because we see the contact point of the lens and the test plate, we actually know the position of the fringe with $m = 0$, i.e., the place where there is no air gap.

Modern lens grinders use Newton's rings to monitor their work [3, 4]. Newton's rings can provide not only a qualitative idea of the quality of the lens but also a quantitative measure by providing the lens designer a measurable parameter, ρ, that will relate to the radius of curvature of the lens under construction. Through the use of either Eq. (4.14) or (4.15), the optician can measure the radius of a Newton's ring to determine whether the lens being ground has reached the proper curvature. If the Newton's rings are not symmetric then errors in the figure[9] of the lens can be identified and corrected. This is a special case of thin film interference which we will discuss next.

4.2.1.2 Dielectric layer

Nature produces interference in a configuration that we encounter in our everyday lives arising from waves reflecting from two parallel surfaces created by a thin, dielectric film. This is a generalization of Newton's rings to cover nearly parallel surfaces. The two interfering waves come from the same transverse position on a wavefront, resulting in this type of interference being labeled *interference by division of amplitude*.

The geometrical model that will be used to discuss interference by reflection from a thin, dielectric layer is shown in Figure 4.8. In Figure 4.8, light enters a thin, dielectric film of uniform thickness, d, at point A. A portion of the amplitude of the wave is reflected toward D, remaining in the medium with index, n_1, while another portion proceeds from A to B and then to C in the medium with index n_2. A lens (for example, the eye) is assumed to gather the waves at C and D and bring them together on a viewing screen (your retina), as shown in Figure 4.9. The paths taken from C and D to the viewing screen are the same for both waves;[10] therefore, this segment of the two paths will not contribute to the phase difference, δ, between the two waves and may be ignored. Using the geometry in Figure 4.8, the optical path lengths for the waves

---

[9] The *figure* of an optical surface refers to how well the actual shape of the surface conforms to the desired shape and is measured in fractions of a wavelength.

[10] This is based on Fermat's principal.

from A to D and from A to C via B will be obtained. These optical path lengths will then be used to find δ for use in Eq. (4.10).

The optical path, as light travels from A to B in the thin film, is

$$\frac{k_2 d}{\cos \theta_t} = \frac{\omega n_2 d}{c \cos \theta_t}.$$

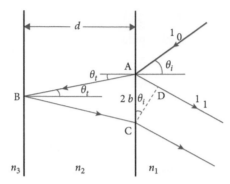

Figure 4.8 Dielectric layer of index n_2 and thickness d separating a medium of index n_1 and a medium of index n_3.

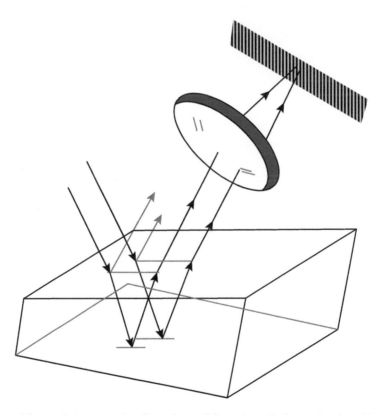

Figure 4.9 Use of a lens to produce fringes from a dielectric layer. The lens is often that of the eye.

Light is reflected off the back surface of the dielectric film at point B and travels an equivalent optical path from B to C. Thus, the optical path from A to B to C in the medium with index n_2 is

$$\overline{ABC} = \frac{2\omega n_2 d}{c \cos \theta_t}.$$ (4.16)

The optical path length from A to D in the medium with index n_1 is

$$\overline{AD} = k_1 2b \sin \theta_i = 2k_1 d \tan \theta_t \sin \theta_i$$
$$= \frac{2\omega n_1 d}{c} \frac{\sin \theta_i \sin \theta_t}{\cos \theta_t}.$$ (4.17)

Applying Snell's law, we can rewrite Eq. (4.17) as

$$\overline{AD} = 2\omega d n_2 \frac{\sin^2 \theta_t}{c \cos \theta_t}.$$

Since the source for the two waves is identical (point A, Figure 4.8), the starting phase of the two waves is the same, $\Delta\phi$, in Eq. (4.11). The phase difference between the waves at points C and D are determined by the optical path difference, $\Delta = \delta$,

$$\delta = \overline{ABC} - \overline{AD} = \frac{4\pi n_2 d}{\lambda_0} \cos \theta_t.$$ (4.18)

It is important to remember that we will assume the polarization is normal to the plane of incidence for all interference problems so polarization is not an issue. There could be an additional phase change of π upon reflection from a dense medium (see Eq. (3.25)). If $n_2 > n_1$, the reflection at A will experience this additional phase change and if $n_3 > n_2$, the reflection at B will experience a phase change. We will assume $n_3 < n_2 > n_1$ for our discussion so that the only phase change due to reflection occurs at A; we denote this as $\Delta\phi = \pi$. A bright band will occur when $\delta = 2\pi m$:

$$\delta = 2\pi m = \frac{2n_2 d}{\lambda_0} 2\pi \cos \theta_t + \pi.$$ (4.19)

Three of the parameters in this equation could independently cause δ to vary and produce a set of interference fringes. These parameters are d, λ_0, and θ_t. Fringes produced by varying each of the parameters are given special names but the names are only of historical interest.

4.2.1.2.1 Fizeau fringes

If we illuminate the dielectric layer with a plane wave, with a single frequency, then $\cos \theta_t$ is constant over the layer and each bright or dark band will show the region of constant optical thickness. These are the *Fizeau fringes* we saw with Newton's rings configuration.

Another geometry that could produce Fizeau fringes is shown schematically in Figure 4.10a where a monochromatic plane wave illuminates a dielectric film with a non-uniform thickness, the tilted glass plate in Figure 4.6. The fringes, as observed by the eye, are produced by reflections of the wave in the film. Another example of fringes in a wedge-shaped dielectric film is seen in Figure 4.10b where fringes are produced in a vertical soap film illuminated by sodium light. The soap film's thickness increases as your eye moves down the film's surface because gravity drags the water in the film downward. The fringe spacing decreases as you move down the film since the film thickness grows.

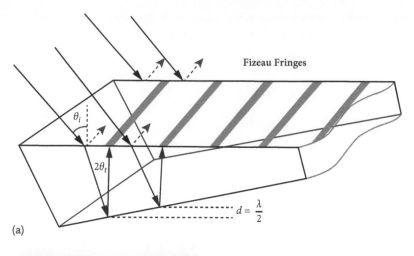

(b)

Figure 4.10 (a) The production of Fizeau fringes in a dielectric layer with a nonuniform thickness. The fringes appear to be located at the dielectric layer but originate because of the interference of light collected by a lens such as our eye. (b) Fizeau fringes produced by illuminating a vertical soap film with sodium light. Gravity creates a nearly linear thickness change in the film.

4.2.1.2.2 *Color fringes*

The thickness of the dielectric film could be a constant, set by surface tension, and the illumination could be from a source producing a plane wave containing a range of wavelengths. An example of this configuration is an oil film, floating on the water surface of a wet road, illuminated by sunlight, Figure 4.11a, or the vertical soap film in Figure 4.11b, illuminated by an sodium vapor lamp.

4.2.1.3 Dielectric thin film coating

The destructive interference resulting from light reflected by the front and back surfaces of a dielectric layer can reduce the reflectivity of the substrate supporting the layer [5]. This is accomplished by selecting a layer thickness that results in a phase shift of 180° between the

(a)

(b)

Figure 4.11 (a) A roadside oil film of uniform thickness created by diesel fuel floating on water. John [CC BY-SA 2.5 (*http://creativecommons.org/licenses/by-sa/2.5*)], v.W.C., *Diesel spill on a road*, Dieselrainbow.jpg, Editor. 2007: Wikimedia Commons. (b) A wedge-shaped soap film similar to that shown in Figure 4.10b but here the soap film is illuminated by white light from an incandescent bulb. The color pattern repeats when the thickness changes by one wavelength; thus, each cycle in an identical color corresponds to one order, *m*, in the interference pattern.

waves reflected by the two surfaces of the dielectric film. Assume that light is incident normal to the surface of the dielectric so that $\theta_t \approx 0$ and $\cos(\theta_t) \approx 1$; also assume that $n_1 < n_2 < n_3$ so that there is an equal phase change for reflections from each of the two surfaces of the dielectric film. To observe a dark band, we require Eq. (4.18) to fulfill the condition[11]

$$\delta = (2m+1)\pi = \frac{2n_2 d \cdot 2\pi \cos \theta_t}{\lambda_0}.$$

Thus, the film thickness required to produce destructive interference for light incident normal to the surface is

$$d = \frac{(2m+1)}{4} \cdot \frac{\lambda_0}{n_2} = (2m+1)\frac{\lambda}{4}.$$

This equation is identical to Eq. (4.19) but now the bands are dark bands. The film layer that meets our design criteria is usually called a ¼ wave layer and it serves as an antireflection coating; see Figure 4.14. The index, n_2, of the film appears in the equation to correct the wavelength for the velocity of propagation in the medium.

We will utilize the vector approach to design multilayer interference filters. The vector approach will provide us a means of visualizing the operation of the multilayer stack. A single layer will support multiple reflections from the top and bottom of the layer.

We will now only pay attention to the two reflections from the front surface and one from the back surface of the dielectric, including the phase shift for the wave that is reflected from the back surface.[12] From Figure 4.12, the reflection coefficient is

$$\rho = \frac{1}{A}\left(Ar_0 + At_0 t_0' r_1 e^{-i\delta_1}\right).$$

From energy conservation and Eq. (3.38) we can write

$$t_0 = 1 - r_0 \qquad t_0' = 1 - r_0' = 1 + r_0,$$

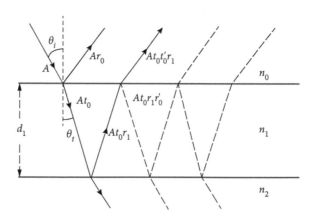

Figure 4.12 Reflection from a dielectric layer. We ignore the multiple reflections shown by the dotted lines.

[11] The phase units are radians. If you want to see it in degrees you must multiply by $180/\pi$.

[12] Later we will use Figure 4.12 to design an optical resonant structure called a Fabry–Pérot etalon or interferometer.

thus

$$t_0 t_0' = 1 - r_0^2,$$

and we may rewrite the reflection coefficient as the series

$$\rho = r_0 + r_1 e^{-i\delta_1} - r_0^2 r_1 e^{-i\delta_1}. \ldots$$

We will limit our analysis to a single reflection from each interface; thus, only the first two terms of this expansion are needed:

$$\rho = r_0 + r_1 e^{-i\delta_1}$$

From Eq. (4.18) the phase shift experienced by the wave reflected from the back surface is

$$\delta_1 = \frac{4\pi d_1 n_1 \cos \theta_i}{\lambda_0} \tag{4.20}$$

Table 4.1 lists some of the coating materials we can use to design an antireflective coating. The index of refraction of many of the films depends upon how the films were prepared; the table gives average values.

Table 4.1 Thin Film Materials

MATERIAL	Index of refraction	Wavelength range
Cryolite (Na_3AlF_6)	1.35	0.15–14
Magnesium fluoride (MgF_2)	1.38	0.12–8
Silicon dioxide (SiO_2)	1.46	0.17–8
Thorium fluoride (ThF_4)	1.52	0.15–13
Aluminum oxide (Al_2O_3)	1.62	0.15–6
Silicon monoxide (SiO)	1.9	0.5–8
Zirconium dioxide (ZrO_2)	2.00	0.3–7
Cerium dioxide (CeO_2)	2.2	0.4–16
Titanium dioxide (TiO_2)	2.3	0.4–12
Zinc sulfide (ZnS)	2.3	0.4–12
Zinc selenide (ZnSe)	2.44	0.5–20
Cadmium Telluride (CdTe)	2.69	1.0–30
Silicon (Si)	3.5	1.1–10
Germanium (Ge)	4.05	1.5–20
Lead telluride (PbTe)	5.1	3.9–20+

First, we will graphically design a single-layer antireflection coating for a glass substrate of index 1.70[13] using MgF_2 as the dielectric layer. The glass is to be used in air so $n_0 = 1.0$, $n_1 = 1.38$, and $n_2 = 1.7$. The amount of light reflected at each interface is

$$r_0 = \frac{n_0 - n_1}{n_0 + n_1} = \frac{1 - 1.38}{1 + 1.38} = -0.16 \qquad r_1 = \frac{1.38 - 1.7}{1.38 + 1.7} = -0.1.$$

Here we see that different amounts of light are reflected from the two surfaces. If we do not have equal amounts of light reflected by the two surfaces, then total cancellation of the reflected component cannot occur. Fresnel's formula can be used to determine the condition for which equal fractions of light will be reflected from the two surfaces:

$$r_0 = \frac{n_1 - n_0}{n_1 + n_0} = \frac{n_2 - n_1}{n_2 + n_1}.$$

We can assume $n_1 = 1$ since most optical surfaces are used in air. With this assumption, the equation reduces to

$$n_1 = \sqrt{n_2}$$

For a heavy flint glass

$$n_2 = 1.7 \; and \; \sqrt{n_2} = 1.304.$$

Magnesium fluoride has an index of $n = 1.38$, which is reasonably close to the desired index, n_2, and should act as a good antireflection coating for glasses with $n \approx 1.7$. For many years, magnesium fluoride was used as an antireflection coating for glasses of all indices because it was easy to apply, low cost, and relatively rugged. For crown glasses with an index of about 1.52, magnesium fluoride reduces the reflectivity from 4% to about 1%.

We have shown that destructive interference will occur when the film thickness is some integer multiple of $n_1 d_1 = \lambda/4$. We will design for the wavelength $\lambda = 500$ nm, resulting in $d_1 = 90.6$ nm. The phase change upon passing through the film will be calculated, assuming normal incidence, for $\lambda = 500$ nm and $\lambda = 650$ nm to allow an estimate of the bandpass response of the antireflection coating:

$$\delta_1(500) = \frac{4\pi d_1 n_1}{\lambda_0} = \left[\frac{4\pi \left(9.1 \times 10^{-8}\right)(1.38)}{5 \times 10^{-7}} \right] \cdot \frac{180}{\pi} = 180°$$

$$\delta_1(500) = \frac{4\pi d_1 n_1}{\lambda_0} = \left[\frac{4\pi \left(9.1 \times 10^{-8}\right)(1.38)}{6.5 \times 10^{-7}} \right] \cdot \frac{180}{\pi} = 140°$$

The vector diagrams for these two cases, following the procedure for vector addition outlined in Figure 4.2, are shown in Figure 4.13.

When the film's thickness is set to produce a phase shift of 180°, as it was for the wavelength of $\lambda = 500$ nm, then the vectors associated with the reflection coefficients, r_0 and r_1, are antiparallel. For this case, it is possible to make the resultant reflection coefficient, ρ, equal

---

[13] This glass is used by people needing an eyeglass or contact lens with a strong correction and provides correction with reduced distortion and less curvature.

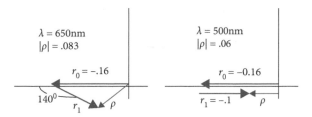

Figure 4.13 Design of an antireflection coating for a glass of index 1.7 using a film of index 1.38. The design wavelength that was used to select the film thickness was 500 nm.

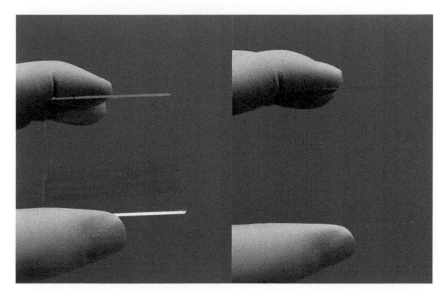

Figure 4.14 Use of multilayer antireflective coating. The coating was designed to give >0.5% reflectivity across the visible wavelength range at normal incidence shown on the right. Rotating the glass plate to 45° tunes the ¼ wave effective coating thickness into the red.

(By Zaereth (Own work) [CC0], via Wikimedia Commons.)

to zero if the reflection coefficients from the two interfaces can be made equal (right side of Figure 4.13). When the phase shift produced by propagation through the dielectric film is not 180°, then there is no possibility of obtaining a resultant reflection coefficient that is zero (left side of Figure 4.13). The index of refraction of glasses varies with wavelength. For example, BK7 glass increases from $n = 1.514$ at a wavelength of 656 nm to $n = 1.527$ for a wavelength of 435 nm. This results in a higher reflectance in the blue than in the red. The coating is designed around the green so a mismatch will occur in both the red and the blue and the coating will appear purple, a mix of red and blue reflected light.

The chances of getting a zero reflectivity with a single dielectric coating are slight because the desired indices of the films available do not match well with the glasses and plastics used for optics. For this reason, several layers are used to reduce the reflectivity, similar to the use of multiple quarter-wave sections in impedance-matching networks in electronics.

4.2.1.4 Multilayer dielectric coating

It is quite simple to extend Eq. (4.20) to allow the use of multilayers to reduce the reflectivity (Figure 4.14). An m layer stack is

$$\rho = r_0 + \sum_{j=1}^{m} r_j e^{\sum_{k=1}^{j} \delta_k} \tag{4.21}$$

We can perform the summation graphically according to the procedure shown in Figure 4.2. We will limit the design of an antireflection coating to two layers to simplify the mathematics and to allow the use of a graphically generated solution. In Figure 4.15 a graphic addition of the reflection coefficients for a two-layer stack is shown.

The glass will be used in air, $n_0 = 1$, and the top most layer, 1, will be SiO_2, $n_1 = 1.46$, the second layer, 2, will be TiO_2, $n_2 = 2.3$, and the films are to be deposited on a glass substrate, 3, called BK-7 ($n_3 = 1.517$), as shown in Figure 4.16. Using these index values in Eq. (3.37) yields

$$r_0 = -0.187 \qquad r_1 = -0.213 \qquad r_2 = 0.195.$$

We don't know the proper thickness for the two layers so we will draw two circles of radii r_1 and r_2, as shown in Figure 4.17. We center the circle of radius $|r_1|$ at the head of r_0 and the circle

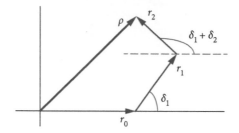

Figure 4.15 Graphic formulation of Eq. (4.21) with $m = 3$.

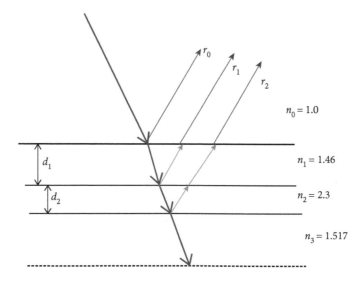

Figure 4.16 A antireflection coating using two dielectric layers.

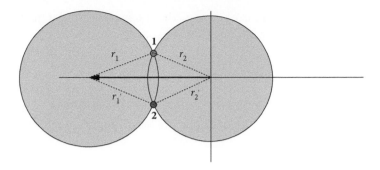

Figure 4.17 Design of a two-layer antireflection coating on a glass of index 1.517 using films with indices 1.46 and 2.3.

of radius $|r_2|$ at the tail of r_0, to ensure complete reflection cancelation.[14] The intersections of the two circles, labeled 1 and 2, represent two solutions to the design problem. Point 1 has

$$\delta (SiO_2) = 118° \qquad \delta (TiO_2) = 26.5°$$

and Point 2 has

$$\delta' (SiO_2) = 61.5° \qquad \delta' (TiO_2) = 153.5°$$

The solution associated with Point 2 is easier to manufacture because it is thicker, but the solution associated with Point 1 is less sensitive to wavelength changes.

Three-layer designs are also used for antireflection coating but one must balance the cost of the coatings against the application value.

Olexander Smakula (1900–83) (in English the first name is Alexander) patented the antireflection coating in 1935 while at Zeiss. Full utilization of the concept of interference filters had to wait until the end of World War II when the technology developed by Zeiss was globally available.

There are multilayer applications other than antireflection coatings. You can also use dielectric layers to produce a high reflectivity at a design wavelength or over a desired wavelength range. Cold mirrors designed to reflect the visible wavelengths and transmit the infrared wavelengths were one of the first products of this technology and today are found in every dentist's lamp to keep the heat from the lamps from falling on the patients. In the 1970s the multilayer coating technology had developed to a point where the mass production of laser mirrors with very low absorption became an off-the-shelf commodity. Coating technology is now being used to produce durable mirrors for copy machines and conductive coatings to provide frost-free aircraft windshields.

4.3 **Interferometry and Temporal Coherence**

The application of thin films to control the reflectivity of a surface is but one of a number of applications of interference. We will now explore the use of interference for the measurement

---

[14] Do not worry about conservation of energy; when you cancel the reflective energy, the missing energy shows up in the transmitted energy.

of small displacements and surface irregularities. The design of devices for this type of purpose contains these basic components:

1. A single light source and a method for dividing it into multiple beams;
2. Two separate optical paths;[15]
3. A means for recombining the two beams to produce interference; and
4. A means for analyzing the resulting interference beams to obtain information about the optical properties of the two paths.

4.3.1 Michelson interferometer

Albert Abraham Michelson (1852–1931), the first American scientist to win the Nobel Prize in physics, developed the design for an interferometer using the basic components listed above that has found a number of useful applications. The interferometer will work with a large source, giving a set of bright interference fringes. A beam splitter (a semitransparent mirror) is used to divide the light into two beams, as shown in Figure 4.18a.

The two beams generated by the beam splitter are directed along orthogonal paths, usually called the arms of the interferometer, where they strike two mirrors, M_1 and M_2, and then return to the beam splitter where they interfere. Looking at the beam splitter from the detector, we see an image of mirror M_2 near mirror M_1. The image, M_2', and the mirror, M_1, form a dielectric layer of thickness d allowing the use of our theory for thin films in the understanding of this interferometer. The interferometer is assumed to be in air so that the optically generated dielectric layer has an index of refraction, $n_2 = 1$. Also light reflected from M_1 and M_2 experience the same phase change upon reflection; thus, no additional phase shift needs to be added to δ in calculating the phase difference between the two waves. The total phase difference for a bright band is

$$\delta = 2\pi m = \frac{2d2\pi \cos \theta_t}{\lambda_0} \tag{4.22}$$

The source shown in Figure 4.18b is nearly a point source, generating spherical waves that produce fringes of constant inclination, also called *Haidinger's fringes*. Bright bands will occur at an angle given by

$$\cos \theta_t = \frac{(2m-1)\lambda_0}{4n_2d}.$$

We can rewrite this in terms of the angle of incidence by using Snell's law

$$\sin \theta_i = \frac{1}{n_1}\sqrt{n_2^2 - \frac{n_2(2m-1)\lambda_0}{3d}}.$$

[15] There is a class of interferometers called *common-path interferometers* where the reference beam and sample beam travel the same path. They are very sensitive to optical path changes but insensitive to vibrations. They may not measure length changes but rather novel changes such as the shearing interferometer to test the collimation of light beams and the Sagnac rotation sensor used in inertial guidance.

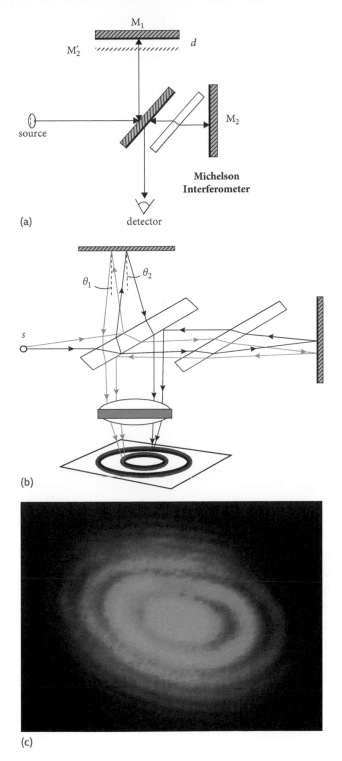

Figure 4.18 (a) Michelson Interferometer. The diagonal plate with one solid surface is a beam splitter, and the solid surface represents a partially reflective coating. It creates two beams traveling toward mirrors M_1 and M_2. The second diagonal plate is called the compensating plate. The compensating plate is made of the same material and thickness as the beam splitter. Its purpose is to equalize the optical path length of the two arms of the interferometer and was a requirement before coherent sources were available. (b) Haidinger's fringes produced in a Michelson interferometer. Both mirror surfaces are assumed to be normal to the axis of the interferometer. The light source is a point source producing spherical waves. The light-gray line has a smaller angle of incidence on the mirrors than has the dark line, $\theta_1 < \theta_2$. (c) Interferogram from a Michelson interferometer.

The fringes are circularly symmetric, as shown in Figure 4.18c, if d is a constant across the aperture. As we stated earlier, we really don't need to know the value of m but we can figure it out. The maximum value of m, the order of the fringe, occurs at the center of the set of fringe rings where $\theta_t = 0$:

$$m_{max} = \frac{2d}{\lambda_0},\tag{4.23}$$

and the order of the fringes decreases as we move out from the center fringe in the viewing plane. The order of the fringe is shown by Eq. (4.23) to be equal to the difference in lengths of the two interferometer arms, d, expressed as the number of wavelengths of light contained in d.

If we increase d, a bright band will move out from the center of the aperture and a new bright band of higher order (larger value for m) will take its place at the center. If we place a detector at the position of the center fringe and monitor the intensity as we move one of the mirrors, and thereby change d, we measure the intensity

$$I = I_1 + I_2 + 2\sqrt{I_1 I_2}\cos\left(\frac{2\omega d}{c}\right),\tag{4.24}$$

where we have substituted

$$\frac{\omega}{c} = \frac{2\pi}{\lambda_0}.$$

If the beam splitter is a 50:50 splitter, then the intensity of light in the two arms will be the same and we can write

$$I = I_0\left[1 + \cos\left(\frac{2\omega d}{c}\right)\right].\tag{4.25}$$

The fringe pattern is a sinusoidal relationship of the difference in path length of the two arms modulating the intensity maxima.

The Michelson interferometer has a number of shortcomings that have been overcome by modifying its design. The Twyman–Green interferometer (Figure 4.19) is a Michelson interferometer that utilizes a plane wave, produced by a collimating lens. The simple fringe pattern makes the Twyman–Green interferometer useful for the evaluation of optical components. Without the collimating lens, a large number of fringes (Haidinger's fringes) would be present, making interpretation of the data difficult. By the addition of a collimating lens, θ_t is converted from a variable to a constant. This means that, since the angle of incidence is a constant, the fringes are of equal thickness (Fizeau fringes) and Eq. (4.19) applies.

The Mach–Zehnder interferometer (Figure 4.20) also uses collimated light but further simplifies the understanding of the observed fringes by passing light through the test area only once. By separating the two optical paths, very large objects such as wind tunnels can be tested. Either arm of the interferometer can contain the test area.

Measurements using interferometers are expressed in fringes per unit length. The spacing of one fringe equals the distance between adjacent dark bands.

4.3.2 Interference spectroscopy

We can also treat the interference pattern produced by the Michelson interferometer as a comparison between waves at two different times. Using the physical arrangement shown in

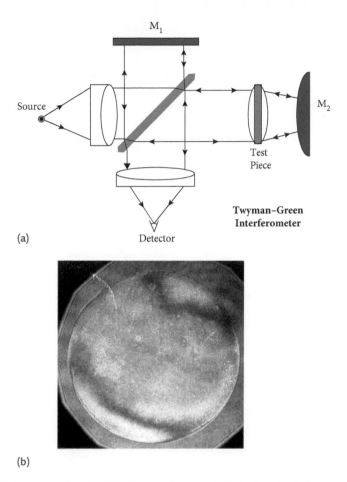

(a)

(b)

Figure 4.19 (a) An improvement on the Michelson interferometer for the testing of optical components. If the test piece is perfect, the waves interfering on the beam splitter are plane waves. M_2 can be either a plane mirror or spherical mirror. The spherical mirror has several experimental advantages. (b) The Twyman–Green interferometer can be used to test a mirror surface by replacing the "test piece" and M_2, shown in (a), by the test mirror. Here is shown a mirror in just such a configuration. The mirror was designed for use at 10 μm, but the wavelength used in the interferometer was 0.44 μm. Thus, the fringes in the interferogram correspond to 1/25 of a fringe at the mirror's operating wavelength.

Figure 4.18a, the light in the M_1 arm of the interferometer travels an extra distance, $2d$, to reach the detector. This means that two waves that originated at different times are added together; the difference in time between the origination of the two waves is

$$\tau = \frac{2d}{c}, \tag{4.26}$$

which is called the *retardation time*.[16] Using the definition Eq. (4.26), we can rewrite Eq. (4.25) as

$$I = I_0 \left[1 + \cos \omega \tau \right]. \tag{4.27}$$

---

[16] The light wave in arm M_1 has been delayed or retarded with respect to the wave in arm M_2 so that M_1 light is older than M_2 light.

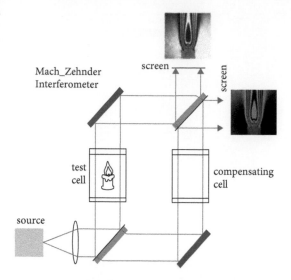

Figure 4.20 A Mach–Zehnder interferometer configured to use plane waves. Either of the arms can be used as the test area. Shown is the interferogram of the air density distribution created by a burning candle. Note that the interferogram in one port is the negative of the interferogram in the second port due to the conservation of energy.
By Stigmatella aurantiaca—own work, CC BY-SA 3.0, https://commons.wikimedia.org/w/index.php?curid= 25125975.

The signal is made up of a constant term plus an oscillatory term. The oscillatory term will provide information about the coherence properties of the light.

4.3.3 Temporal Coherence

The Michelson Interferometer, Figure 4.18a, can be used to characterize the coherence of a wave. To discover the origin of this characterization, we will calculate the form of the interference produced by a Michelson interferometer when the input wave contains two frequencies.

Assume two co-propagating waves of equal amplitude but different frequencies and propagating constants

$$\omega_1 = \omega + \Delta\omega \qquad k_1 = k + \Delta k$$
$$\omega_2 = \omega - \Delta\omega \qquad k_2 = k - \Delta k.$$

The resulting wave is obtained by simple algebraic addition of the two waves:

$$f(x, t) = A \cos(\omega_1 t - k_1 x) + A \cos(\omega_2 t - k_2 x),$$

$$f(x, t) = A \cos[(\omega t - kx) + (\Delta\omega t - \Delta k x)] + A \cos[(\omega t - kx) - (\Delta\omega t - \Delta k x)].$$

Using the trig identity

$$\cos(\alpha \pm \beta) = \cos\alpha \cos\beta - \sin\alpha \sin\beta,$$

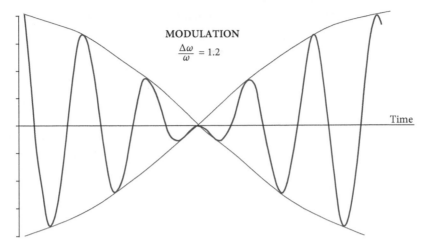

Figure 4.21 A traveling wave group with a ratio of the carrier frequency to the modulation frequency of 1.2. We could also identify this wave as an amplitude-modulated wave or as an example of beating between two frequencies. These descriptions are equivalent.

the resultant wave becomes

$$f(x,t) = 2A \cos\left(\Delta\omega t - \Delta k x\right) \cos\left(\omega t - k x\right).$$

This wave is shown in Figure 4.21.

The arms of the Michelson interferometer act as delay lines, allowing light waves that were generated at different times to interfere. Because the two waves are at slightly different frequencies, their relative phase difference is a function of time. The time-dependent phase results in the two waves alternately adding constructively and destructively, creating an amplitude modulation that is a periodic series of maxima due to constructive interference. We will call these maxima *groups*.

The amplitude modulation is on a carrier wave of frequency

$$\omega = \frac{1}{2}\left(\omega_1 + \omega_2\right)$$

and propagation constant k traveling at a velocity

$$v = \frac{\omega}{k}. \tag{4.28}$$

This velocity we recognize as the phase velocity Eq. (1.14).

The amplitude modulation of the carrier wave is a wave of frequency $\Delta\omega$, called the *beat frequency*, and propagation constant Δk propagating at a velocity of

$$u = \frac{\Delta\omega}{\Delta k}. \tag{4.29}$$

This velocity is called the *group velocity*. If the medium of propagation has a velocity that is independent of frequency, we say it is a nondispersive medium, and the group and phase velocity are identical.

We will assume that the detector cannot respond to the beat frequencies produced on the detector surface by the two waves. If we look at Eq. (4.8), we see that the assumption is the same as saying that the time average $\langle \mathbf{E}_1 \cdot \mathbf{E}_2 \rangle = 0$ and interference between the two frequencies is unobservable. Because of this assumption, the output intensity of the interferometer is obtained by adding the intensities associated with the interference pattern created by each frequency. We will find that measurement of the interferometer's output intensity, as we move one mirror, will generate information about the spectral content of the source.

Each frequency will produce an interference pattern at the detector whose intensity is given by Eq. (4.27). We add the two interference patterns

$$
\begin{aligned}
I_d &= I_{d1} + I_{d2}, \\
&= I_1 + I_2 + I_1 \cos \omega_1 \tau + I_2 \cos \omega_2 \tau, \\
&= I_0 \left(1 + \gamma\right).
\end{aligned}
\tag{4.30}
$$

The resultant intensity contains a constant term,

$$
I_0 = I_1 + I_2,
$$

and an oscillating term

$$
I_0 \gamma \left(\tau\right).
$$

For the current example of two frequencies, the normalized oscillating term is

$$
\gamma \left(\tau\right) = \frac{I_1}{I_0} \cos \omega_1 \tau + \frac{I_2}{I_0} \cos \omega_2 \tau.
$$

If the two arms of the Michelson interferometer are the same length, we have $\tau = 0$ and γ has its maximum value,

$$
\gamma \left(\tau\right) = \frac{I_1 + I_2}{I_0} = 1.
$$

The minimum value that γ can have is zero. Whenever $\gamma = 0$, there is no interference and the resultant observed intensity is

$$
I_d = I_1 + I_2 = I_0.
$$

An example of the output intensity of the interferometer as the one mirror is moved is given in Figure 4.22 for three representative values of γ. The intensities plotted in Figure 4.22 are measured on axis as one mirror is moved to vary d. Note the variation we observe is not in the spacing of the fringes but in their contrast.[17]

By measuring γ, we can measure the separation between the two wavelengths present in the source illumination. To see how this measurement is accomplished, first note that the two wavelengths of the source can prevent the observation of interference. The loss of interference fringes is due to a filling in of the minimum intensity fringe associated with one frequency by the maximum intensity fringe of another frequency. The loss of visible fringes occurs when

[17] Contrast is defined in Eq. (4.34).

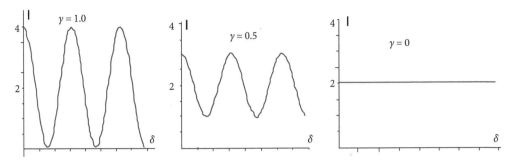

Figure 4.22 The intensity output of a Michelson interferometer as a function of mirror spacing for three values of the degree of coherence function.

$$\gamma(\tau) = \frac{I_1}{I_0}\cos\omega_1\tau + \frac{I_2}{I_0}\cos\omega_2\tau = 0.$$

If

$$I_1 = I_2 = \frac{1}{2}I_0,$$

the function $\gamma(\tau)$ for the two frequencies can be rewritten as

$$\gamma(\tau) = \cos\left(\frac{\omega_1 + \omega_2}{2}\right)\tau \cos\left(\frac{\omega_1 - \omega_2}{2}\right)\tau,$$

which equals zero whenever

$$\frac{\omega_2 - \omega_1}{2}\tau = \pi(\nu_2 - \nu_1)\tau = \frac{\pi}{2},$$

$$\delta\nu = \frac{1}{2\tau}.$$

Physically, γ is zero when the maximum intensity of the fringe associated with one frequency occurs at the same mirror separation as the minimum intensity of the fringe associated with the second frequency.

If we had a continuous distribution of frequencies rather than just two discrete frequencies, then γ would be a continuous variable and we would characterize the coherence in terms of a parameter, τ_c, called *coherence time,* which characterizes the function γ; for example, we might use the half-width at half-maximum if γ were a Gaussian function. As we saw earlier, the coherence time is inversely proportional to the spectral width of the source. The broader the spectral source, the shorter the coherence time.

We now have a technique for measuring the separation, in wavelength, of two monochromatic sources. We begin by adjusting the two arms of a Michelson interferometer so that they have equal optical path lengths. With equal path lengths in the two arms we observe a bright fringe, called the zero fringe, that is independent of wavelength.[18] We introduce light from the two monochromatic sources into the Michelson interferometer. A detector monitors the

---

[18] The zero fringe is hard to find, so we usually cheat by using a narrow band light source in the preliminary alignment. That detail does not affect the theory.

central fringe at the output of the interferometer as one mirror is moved, increasing d; see Figure 4.22a. The intensity measured at the central fringe will rapidly vary, in intensity as shown in Figure 4.22a. When the difference in optical path length of the two arms becomes large, the output intensity will no longer show any fringes but will remain constant as d changes ($\gamma = 0$), Figure 4.22c. The value of d where the variation in intensity ceases can be used to calculate the wavelength separation between the two sources,

$$\Delta v = \frac{c}{4d}.$$

We see that for $\gamma = 0$, the output intensity of the interferometer is a constant, independent of the difference in length between the two interferometer arms, $d = \tau_c/2$. For this condition, the wave is said to be incoherent. We find that the degree of coherence, $\gamma(\tau)$, has a total range of values,

$$0 \leq \mid \gamma(\tau) \mid \leq 1.$$

Natural occurring sources contain a distribution of frequencies; therefore, our simple theory must be expanded to include a continuous distribution of frequencies. The resultant intensity at the output of a Michelson interferometer for a source with a continuous spectral distribution of intensity, $I(\omega)$, is obtained by analogy with the two-frequency result. $I(\omega)$ equals the fraction of light contained in an interval between ω and $\omega + d$. The form of the resultant intensity is given by

$$I_d(\tau) = \int_0^\infty I(\omega)(1 + \cos\omega\tau)\,d\omega. \tag{4.31}$$

This integral contains the sum of constant and oscillatory terms. The constant term is

$$I_0 = \int_0^\infty I(\omega)\,d\omega$$

and the oscillatory term is

$$\int_0^\infty I(\omega)\cos\omega\tau\,d\omega.$$

By defining a normalized intensity distribution of frequencies called the *spectral distribution function* or the *power spectrum* of the light source,

$$P(\omega) = \frac{I(\omega)}{\int_0^\infty I(\omega)\,d\omega} = \frac{I(\omega)}{I_0}, \tag{4.32}$$

the oscillating term can be written in a form independent of the incident intensity

$$\gamma(\tau) = \int_0^\infty P(\omega)\cos\omega\tau\,d\omega. \tag{4.33}$$

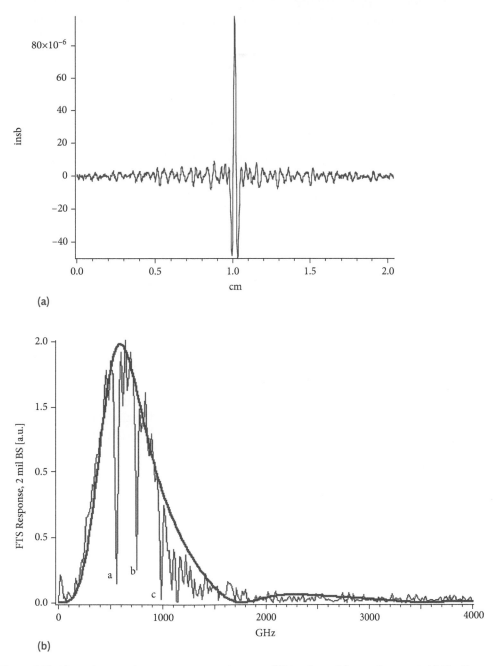

Figure 4.23 (a) Interferogram from a Michelson interferometer. (b) Transform of the interferogram in (a). The dips labeled a–c are absorption dips due to rotational spectral lines of water vapor in the atmosphere.

The oscillatory function, Eq. (4.33), is called the *degree of coherence*. The integral, Eq. (4.33), is the cosine transform of $P(\omega)$[19]:

$$I_d(\tau) = \int_0^\infty I(\omega)\, d\omega = \int_0^\infty I_0 P(\omega)\, d\omega = I_0.$$

Because γ describes the interference part of the intensity distribution of the Michelson interferometer, and involves the frequency distribution of the source it is called *temporal coherence*. Our discussion suggests that we should be able to measure the coherence function of any light source with the Michelson interferometer. From Eq. (4.33), we infer that by taking the Fourier transform of the measured coherence function, the spectral distribution of a light source can be recovered. Michelson used this procedure to determine the shape of a number of spectral emission lines such as the hydrogen line at $\lambda = 656.3$ nm. The approach developed by Michelson was ignored for many years because the Fabry–Perot interferometer was much easier to use and computers did not exist to allow the rapid calculation of the Fourier transform. Now, with the availability of small computers and the fast Fourier transform algorithm, the technique is very popular in the far-infrared and terahertz regions of the spectrum.

A typical output from a Michelson interferometer operating around 300 GHz ($\lambda = 1$ mm) is shown in Figure 4.23a. The big spike is the fringe at $\tau = 0$ called the white light (or zero) fringe. As we mentioned earlier, to align the interferometer you must start by obtaining the zero fringe. This type of spectrometer is often called a *Fourier transform spectrometer* since it depends on performing a Fourier transform to generate the spectrum collected by scanning one of the arms of a Michelson interferometer.

By taking the Fourier transform of the interferogram shown in Figure 4.23a you generate the spectrum of the gigahertz source that is under evaluation. The absorption peaks, a, b, c, shown in the spectrum of Figure 23b, are rotational absorption lines of water vapor and here were used to help calibrate the frequency spectrum.

4.3.4 Optical coherence imaging

A recent imaging modality called optical coherence tomography (OCT) [6] utilizes the temporal coherence of a super luminous diode[20] to map a third dimension (depth) onto a spatially resolved microscopic image for medical applications, especially in ophthalmology. Figure 4.24 illustrates a simplified version of an optical coherence imaging system to generate an image of a human retina. The vertical displacement of one mirror allows the measurement of the coherence function of the diode modulated by the depth profile of the retina. If we know the coherence function of the diode, then the measurement of the fringe visibility can be used to determine the third dimension. The x- and y-coordinates needed to generate the 2d surface image are created by scanning a mirror in a second interferometer arm; see Figure 4.25 displays a typical OCT image, here of a human retina.

---

[19] Here the degree of coherence is shown as the real part of the Fourier transform of $P(\omega)$; see Eq. (1.22). We would expect in general to find $\gamma(\tau)$ to be a complex function but in our discussion we will ignore the phase and use only real functions.

[20] A forward biased pn-junction similar to a laser diode but designed with insufficient optical feedback to lase usually by tilting the low reflective end mirrors.

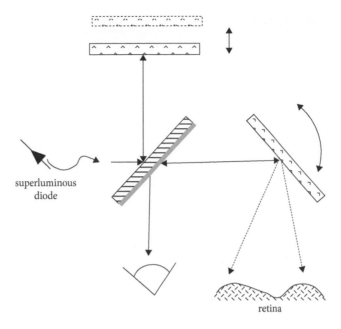

Figure 4.24 Optical coherence tomography configuration that uses a Michelson interferometer with a superluminous diode for a light source. One mirror is scanned along the interferometer axis to measure the depth of the object (here a human retina). In the other arm of the interferometer the beam is scanned over the retina to generate a 2d surface image.

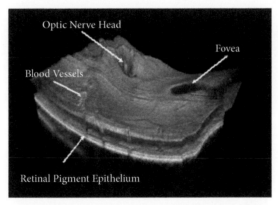

Figure 4.25 An optical coherence tomographic image of the retina of the human eye [7]. Three-dimensional image provided by Joseph A. Izatt, Duke University.

The coherence function of a typical superluminous diode of know linewidth (as large as 100 nm to measure d with resolution better than 10 μm) is displayed in Figure 4.26 as a scope trace of the intensity from a Michelson interferometer with a flat mirror replacing the retina. The modulation, $\cos \omega_0 \tau$, contained within the envelope shown in Figure 4.26 is made up of the interference fringes produced by the light in the two arms of the interferometer. The equation that describes the waveform is

$$I_d(\tau) = I_1 + I_2 + 2\sqrt{I_1 I_2}\vartheta(\tau)\cos\omega_0\tau.$$

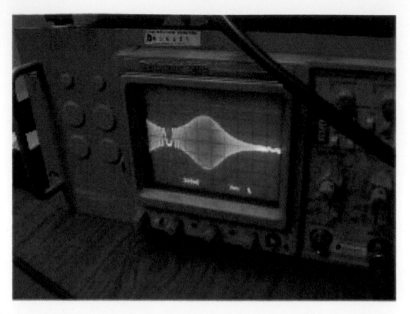

Figure 4.26 Coherence function of a superluminous diode from a Michelson interferometer configured as an OCT spectrometer. To generate this function, the OCT illuminates a plane mirror to remove any variation in the z-dimension. The wiggle on the left of the trace is due to the reversal in motion of the moving mirror. The same glitch is on the right side of the trace but because it occurs when there is little interference, it is hard to see.

We define a parameter, called the fringe visibility, by the equation

$$v = \frac{I_{MAX} - I_{\min}}{I_{MAX} + I_{\min}}. \tag{4.34}$$

The maximum intensity of the output of the Michelson interferometer is

$$I_{MAX}(\tau) = I_1 + I_2 + 2\sqrt{I_1 I_2}\vartheta(\tau)$$

and the minimum intensity is

$$I_{\min}(\tau) = I_1 + I_2 - 2\sqrt{I_1 I_2}\vartheta(\tau).$$

The fringe visibility is thus

$$v = \frac{2\sqrt{I_1 I_2}}{I_1 + I_2}\vartheta(\tau) \tag{4.35}$$

When $I_1 = I_2$ then $V = \vartheta(\tau)$ and the degree of coherence is simply the visibility of the fringes.

4.4 Interference by Multiple Reflection

In our previous discussion of a thin dielectric film, we neglected multiple reflections in the film. We now wish to consider what effect multiple reflections have on interference. We will

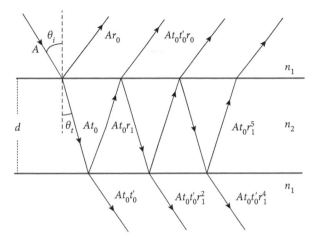

Figure 4.27 Reflection of a plane wave in a plane parallel dielectric layer.

find that the dielectric layer forms a resonant cavity. If the losses at each reflection are not too great, a set of standing waves is created in the dielectric layer similar to a set of standing waves that forms on a guitar string. The mathematics become somewhat involved but once completed they demonstrate that the dielectric layer acts like a wavelength filter, transmitting some frequencies and rejecting others.

In a dielectric layer (see Figure 4.27), the light waves reflecting back and forth between the boundaries of the dielectric layer will result in solutions that cause us to view the dielectric layer as a resonant structure. Artistic license has been used to spread out the multiple reflections to make them easy to see in our derivation. Actually, the plane wave illumination is adjusted so that all of the reflections are approximately collinear.

We will ignore the fields in the dielectric layer and direct our attention toward the light transmitted through the layer. Assume that a wave of amplitude A is incident upon the layer. By replacing the index of refraction used to calculate reflection and refraction with an effective index for normal polarization of a material as

$$N_j = \tilde{n}_j \cos\theta_j$$

and the effective index for parallel polarization as

$$N_j = \frac{\tilde{n}_j}{\cos\theta_j},$$

we can write a generalized set of Fresnel equations for reflection and transmission at the two interfaces based on the effective index we have just defined:

$$t_0 = \frac{2N_1}{N_1 + N_2} \qquad t_0' = \frac{2N_1}{N_1 + N_2} \qquad r_1 = \pm\frac{N_2 - N_1}{N_1 + N_2}$$

We will determine the interference of the transmitted waves by calculating the amplitude and phase of each transmitted wave. The waves will then be added using the complex approach. As shown in Figure 4.9, a lens will be used to collect all of the transmitted waves and bring them together at a viewing screen. All of the waves are assumed to travel the same optical path length upon exit from the dielectric so that the wave-combining process can be neglected.

We will represent the first transmitted wave by

$$\tilde{E} = A t_0 t_0' e^{i(\omega t - \mathbf{k} \cdot \mathbf{r})}.$$

The reflectivity of the layer is defined as r_1^2 but we will suppress the subscript since it is always 1 in our example. The transmitted amplitudes for the various reflected waves form a geometric progression (sequence) in terms of the reflectivity:

$$\tilde{E}, \tilde{E} r^2, \tilde{E} r^4, \tilde{E} r^6 \dots .$$

Each transmitted wave will have a constant phase difference, relative to its neighbor, given by Eq. (4.22):

$$\delta = \frac{4\pi d n_2 \cos \theta_t}{\lambda_0} \tag{4.36}$$

The general, nth, transmitted wave is of the form

$$\tilde{E} r^{2n} e^{i\delta} = r^{2n} A t_0 t_0' e^{i(\omega t - \mathbf{k} \cdot \mathbf{r})} e^{in\delta}, \tag{4.37}$$

where $n = 0,1,2,\dots$. In adding the transmitted waves, we can factor E outside the summation because all of the transmitted waves are at the same frequency and all are propagating in a parallel direction:

$$\tilde{A} = \tilde{E} \left(1 + r^2 e^{i\delta} + r^4 e^{i2\delta} + r^6 e^{i3\delta} + \dots \right)$$
$$= \tilde{E} \left(1 + r^2 \cos \delta + r^4 \cos 2\delta + \dots \right) + i\tilde{E} \left(r^2 \sin \delta + r^4 \sin 2\delta + r^5 \sin 3\delta + \dots \right).$$

To evaluate this equation, we use the geometric series

$$\frac{1}{1+x} = 1 - x + x^2 - x^3 + x^4 - \dots, \quad -1 < x < 1,$$

to write

$$\frac{1}{1 - r^2 e^{i\delta}} = 1 + r^2 e^{i\delta} + r^4 e^{i2\delta} + r^6 e^{i3\delta} \dots r^2 < 1$$
$$= \frac{1}{1 - r^2 e^{i\delta}} \frac{1 - r^2 e^{-i\delta}}{1 - r^2 e^{-i\delta}} = \frac{1 - r^2 e^{-i\delta}}{1 - r^2 \left(e^{i\delta} + e^{-i\delta}\right) + r^4}. \tag{4.38}$$

Using the identity

$$e^{\pm i\delta} = \cos \delta \pm i \sin \delta,$$

$$\frac{1}{1 - r^2 e^{i\delta}} = \frac{1 - r^2 \cos \delta + i r^2 \sin \delta}{1 - 2r^2 \cos \delta + r^4}. \tag{4.39}$$

Equating real and imaginary parts of Eq. (4.39), we obtain for the real part

$$\frac{1 - r^2 \cos \delta}{1 - 2r^2 \cos \delta + r^4} = 1 + r^2 \cos \delta + r^4 \cos 2\delta + \dots \tag{4.40}$$

and for the imaginary part

$$\frac{r^2 \sin \delta}{1 - 2r^2 \cos \delta + r^4} = r^2 \sin \delta + r^4 \sin 2\delta + r^6 \sin 3\delta + \dots . \tag{4.41}$$

The intensity of the transmitted light is

$$I = (A_{\mathcal{R}} + iA_{\mathcal{I}}) \cdot (A_{\mathcal{R}} - iA_{\mathcal{I}}) = A_{\mathcal{R}}^2 + A_{\mathcal{I}}^2,$$

which is the sum of the squares of the real and the imaginary terms we have just calculated,

$$I\alpha \frac{\tilde{E}^2 \left[\left(1 - r^2 \cos \delta\right)^2 + r^4 \sin^2 \delta \right]}{\left(1 - 2r^2 \cos \delta + r^4\right)^2} = \frac{\tilde{E}^2}{1 - 2r^2 \cos \delta + r^4} \tag{4.42}$$

To obtain the form of the transmitted wave normally used, the denominator of Eq. (4.42) is rewritten

$$1 - 2r^2 \cos \delta + r^4 = \left(1 - 2r^2 \cos \delta + r^4\right) - 2r^2 + 2r^2$$

$$= 2r^2 \left(1 - \cos \delta\right) + \left(1 - 2r^2 + r^4\right).$$

This new expression allows Eq. (4.42) to be written in the form

$$I\alpha \frac{\tilde{E}^2}{\left(1 - r^2\right)^2 + 4r^2 \sin^2 \delta/2} \tag{4.43}$$

When $\delta = 0, 2\pi, 4\pi, \dots$, we get a maximum in transmission

$$I_{MAX} \propto \left(\frac{\tilde{E}}{1 - r^2} \right)^2 \tag{4.44}$$

When $\delta = \pi, 3\pi, 5\pi, \dots$, we get a minimum in transmission

$$I_{min} \alpha \left(\frac{\tilde{E}}{1 + r^2} \right)^2 . \tag{4.45}$$

We recall the definition of the fringe visibility

$$\vartheta = \frac{I_{MAX} - I_{min}}{I_{MAX} + I_{min}}. \tag{4.46}$$

The larger the value of fringe visibility, the easier it is to observe the fringe pattern. The maximum value of ϑ is 1 and occurs when $I_{min} = 0$, the minimum value is 0, because $I_{MAX} = I_{min}$. Using Eqs. (4.44) and (4.45) in Eq. (4.46), we obtain the fringe visibility in terms of the reflectivity of the film surface

$$\vartheta = \frac{2r^2}{1 + r^4}.$$

As the reflectivity, r^2, approaches 1, the fringe visibility approaches its maximum value of 1.

We can write the intensity transmitted through the dielectric layer in terms of I_{max}

$$I = \frac{I_{max}}{1 + \frac{4r^2}{\left(1 - r^2\right)} \sin^2 \frac{\delta}{2}}.$$

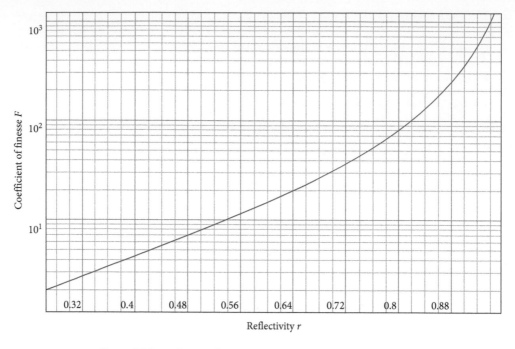

Figure 4.28 Coefficient of finesse, F, as a function of reflection coefficient.

The factor

$$\frac{4r^2}{(1-r^2)^2} = F \tag{4.47}$$

is called the *coefficient of finesse*,[21] F (Figure 4.28), and is proportional to the fringe visibility, ϑ. Its use allows the relative transmission of the dielectric layer to be expressed in a format called the Airy function:

$$\frac{I}{I_{max}} = \frac{1}{1 + F\sin^2\frac{\delta}{2}}. \tag{4.48}$$

If we plot the Airy function for different values of the reflectivity, r^2, we obtain Figure 4.29, which displays a periodic maximum in the transmission of the dielectric layer as d (or equivalently δ) is varied. The peak in transmission occurs when d is equal to a multiple of $\lambda/2$, where λ is the illuminating wavelength. Thus, the maximum transmission of the film occurs when the film thickness will support standing waves in the layer; i.e., the transmission maxima occur for eigenvalues of the layer. At this resonant condition the transmission is 1.

---

[21] Unfortunately, the term finesse is also given to figure of merit $\mathcal{F}$ that we will introduce as Eq. (4.57). With this warning it should be possible to avoid confusion.

Assume we illuminate the Fabry–Perot interferometer with a 1-mW laser; our results say that the output is also 1 mW. If we have a mirror reflectivity of 0.94, how can we get 100% transmission? This can occur only if 24 mW is stored in the resonant structure of the Fabry–Perot interferometer. The mirrors normally used in a Fabry–Perot interferometer to reach 100% transmission are constructed with dielectric layers that have no loss. If we used mirrors with a loss, A, for example through the use of metal mirrors, then we replace the numerator of Eq. (4.43) with $1 - r^2 - A$:

$$\frac{I_t}{I_i} = \left(1 - \frac{A}{1 - r^2}\right)^2 \frac{1}{1 + F\sin^2\left(\delta/2\right)}.$$

Assume $r^2 = 0.997$ and $A = 0.002$ the maximum transmission is 11%, but if A is just a little larger, say $A = 0.0029$, then the maximum transmission is only 0.11%. The large numbers of reflections cause large losses.

The peak in transmission in Figure 4.29 narrows as the reflectivity, r^2, increases. The value of δ over which I goes from $I_{\max}$ to $I_{\max}/2$ is a measure of the fringe sharpness and can be obtained by noting $I = I_{\max}$ when $\delta = 0$. When $I = I_{\max}/2$,

$$\frac{4r^2\sin^2\dfrac{\delta}{2}}{\left(1 - r^2\right)^2} = 1.$$

The fringe sharpness is given by

$$\delta_{\frac{1}{2}} = 2\sin^{-1}\left(\frac{1 - r^2}{2r}\right) = 2\sin^{-1}\left(\frac{1}{\sqrt{F}}\right). \tag{4.49}$$

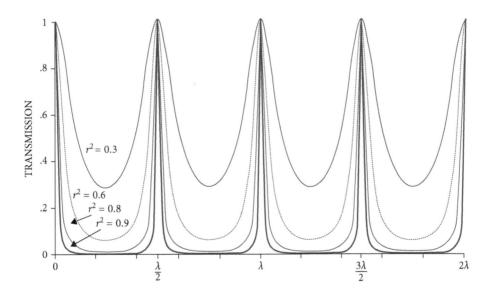

Figure 4.29 This is a plot of the fraction of transmitted light as a function of the optical path, *nd*, for reflectivities of 0.3, 0.6, 0.8, and 0.9. The distance between adjacent peaks is the free spectral range. The fringes become narrower as the reflectivity increases but the spacing does not change. A more detailed discussion of the widths of the transmission peaks is found near Figure 4.37.

The sharpness of the fringes increases and $\delta_{1/2}$ decreases with increasing reflectivity. If the reflectivity is $r^2 = 0.9$, then

$$\delta_{\frac{1}{2}} = 0.211 \text{ radians.}$$

The multiple reflections are the cause of the sharp fringes as we can see by comparing the fringe sharpness of the dielectric layer with multiple reflections to a Michelson interferometer that was modeled as a dielectric layer with one reflection. Rewriting Eq. (4.29) as

$$I = I_{max}\cos^2\frac{\delta}{2},$$

the sharpness of the fringes for a Michelson interferometer can be obtained by noting that $I = I_{max}$ whenever $\delta = 0$ and $I = I_{max}/2$ whenever

$$\delta_{\frac{1}{2}} = 2\cos^{-1}\left(\frac{1}{\sqrt{2}}\right) = \frac{\pi}{2} = 1.57 \text{ radians.}$$

This value for the fringe sharpness is much larger when multiple reflections occur than the value when only one reflection occurs; thus, the fringes of a Michelson interferometer are much broader than those in a Fabry–Perot interferometer.

It is obvious that a device constructed as a dielectric layer could be used to accurately measure wavelength. Such a device was first constructed by *Marie Paul Auguste Charles Fabry* (1867–1945) and Jean Baptiste Gaspard Gustave Alfred Perot (1863–1925).

4.5 Fabry–Perot Interferometer

The Fabry–Perot interferometer is constructed using two highly reflective surfaces usually separated by air. In Figure 4.30 is displayed a typical experimental arrangement. Two plane glass plates separated by a distance, d, have reflective dielectric mirrors on their facing surfaces.[22] The waves exiting the plates, after multiple reflections, are collected by the lens and imaged onto an observation screen. Only one propagation vector, incident at an angle θ, is followed through the system in Figure 4.30. Other incident propagation vectors will result in a bright fringe if $\delta = 2\pi m$. The incident angle of the propagation vectors forming the bright fringes must satisfy the equation

$$m\lambda = 2nd \cos\theta.$$

The observed fringes, such as those in Figure 4.30, are circularly symmetric if the illumination is symmetric about the symmetry axis of the optical system.

To use the Fabry–Perot interferometer as a wavelength-measuring instrument, a detector behind an aperture in the screen measures the light passing through the screen as the optical path length, n_2d, is varied either by varying the pressure of the air between the mirrors of the interferometer or by using piezoelectric translators to move one mirror relative to the other. If

---

[22] We will limit our attention here to the interferometers constructed using plane, parallel plates but interferometers are also constructed using spherical mirrors.

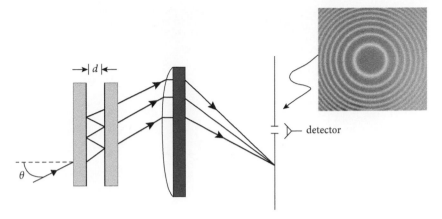

Figure 4.30 The experimental arrangement of a Fabry-Perot interferometer. Shown on the right are the fringes observed on the screen with laser illumination. A detector measures the intensity through the pinhole of diameter *a* as *d* is varied.

a single wavelength illuminates, the mirrors then the signal recorded by the detector will look like one of the curves in Figure 4.29. If multiple wavelengths are present then an interferogram like Figure 4.31 might be observed.

The accuracy with which an interferometer can measure the wavelength of its illumination is called *chromatic resolving power*, R, and is defined as $\lambda/\Delta\lambda$, where λ is the mean wavelength of the illumination and $\Delta\lambda$ is the wavelength difference that can be resolved. We need a criterion for resolution and the one we will use assumes that two wavelengths, λ_1 and λ_2, of equal intensity are present. The criterion states that the two wavelengths are just resolved if the half-maximum intensity of a fringe produced by λ_1 falls on the half-maximum intensity of a fringe produced by λ_2, as shown in Figure 4.32.

When this occurs, the transmitted intensity is a constant as d is varied from the resonant condition of λ_1 to the resonant condition of λ_2. The phase shift going from the intensity maximum for λ_1 to the intensity maximum for λ_2 is then $\Delta\delta = 2\delta_{1/2}$. The fringes are assumed to be narrow so that this is a small value and we may make the approximation

$$\sin\delta_{\frac{1}{2}} = \sin\frac{\Delta\delta}{2} \approx \frac{\Delta\delta}{2}$$

and

$$\sin\left[\frac{\frac{\delta_1}{2}}{2}\right] = \frac{\Delta\delta}{4}.$$

Using Eq. (4.46)

$$\frac{\Delta\delta}{4} \approx \frac{1}{\sqrt{F}} = \frac{1-r^2}{2r}. \tag{4.50}$$

Differentiating Eq. (4.34) to get a relationship between $\Delta\delta$ and $\Delta\theta$

$$\Delta\delta = -4\pi n_2 d \sin\theta_t \frac{\Delta\theta_t}{\lambda_0}. \tag{4.51}$$

Figure 4.31 The output fringes from a Fabry–Perot interferometer. Several sets of fringes due to multiple colors in the light source are present in this photograph.
(Photograph by Frank E. Barmore.)

A bright fringe will occur whenever

$$2n_2 d \cos \theta_t = m\lambda. \tag{4.52}$$

If we differentiate this equation we will obtain a relationship between $\Delta\theta$ and $\Delta\lambda$:

$$-2n_2 d \sin \theta_t \Delta\theta_t = m\Delta\lambda,$$

$$-\sin \theta_t \Delta\theta_t = \frac{m\Delta\lambda}{2n_2 d}. \tag{4.53}$$

Using Eq. (4.53) in Eq. (4.51) and then equating with Eq. (4.50), we get

$$\Delta\delta = 2\pi m \frac{\Delta\lambda}{\lambda} = \frac{2\left(1 - r^2\right)}{r}.$$

The resolving power is thus

$$\mathcal{R} = \lambda \Big/ \Delta\lambda = \frac{m\pi r}{1 - r^2} = \frac{m\pi}{2}\sqrt{F} \tag{4.54}$$

We see from Eq. (4.52) that the order number, m, has a maximum value whenever $\cos \theta_t = 1$; i.e., m is a maximum at the center of the Fabry–Perot fringe pattern and this maximum is given by

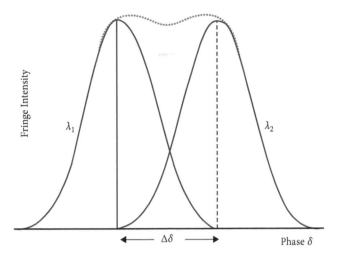

Figure 4.32 Resolving power of the Fabry–Perot interferometer. From Eq. (4.47) the width of the fringe is $4/\sqrt{F}$. The dotted curve is the sum of the two fringes and displays the output intensity of the interferometer with two wavelengths that are just resolved.

$$m_{\max} = \frac{2n_2 d}{\lambda_0}. \tag{4.55}$$

Physically, m corresponds to the number of wavelengths of light that can fit between the two mirrors.

As an example of the chromatic resolving power of a Fabry–Perot interferometer, assume that $r^2 = R = 0.9$, the reflecting surfaces are separated by $d = 1$ cm, the dielectric between the two reflecting surfaces is air, $n_2 = 1.0$, and the illuminating wavelength is $\lambda_0 = 500$ nm. The resolving power is

$$\frac{\lambda}{\Delta\lambda} = 1.2 \times 10^6$$

and the smallest wavelength difference that can be measured by the instrument is

$$\Delta\lambda = 4.2 \times 10^{-4} nm.$$

The peaks in the transmitted intensity of two adjacent red fringes shown in Figure 4.31 occur when the spacing, d, is a multiple of $\lambda/2$, where λ is the illuminating red wavelength. The wavelength difference, $(\Delta\lambda)_{SR}$, corresponding to a change in d of $\lambda/2$, or a change in m of 1, is called the *free spectral range* of the interferometer. This parameter is the maximum wavelength difference that can be unambiguously measured by the interferometer. When we change the order, m, by 1 we change the phase $\Delta\delta = 2\pi$. Using Eq. (4.53) we have just derived the relationship

$$\Delta\delta = 2\pi m \frac{\Delta\lambda}{\lambda}.$$

This allows us to write for the free spectral range

$$(\Delta \lambda)_{SR} = \frac{\lambda}{m}.$$

Substituting the maximum value for m from Eq. (4.52) into the equation will give us the minimum free spectral range

$$(\Delta \lambda)_{SR} = \frac{\lambda^2}{2n_2 d} \qquad (4.56)$$

or in terms of frequency

$$(\Delta \nu)_{SR} = \frac{c}{2n_2 d}.$$

If we substitute Eq. (4.55) into Eq. (4.54), we see that increasing the separation, d, increases the resolving power of the Fabry–Perot interferometer, but accompanying that increase is a decrease in the free spectral range, as shown by Eq. (4.56). If the separation of two wavelengths exceeds the free spectral range, we will obtain an incorrect value for the wavelength difference, as would be the case in Figure 4.31 if we were not able to use color to identify the fringes.

Our eye can identify fringes of increasing order by the fringe color in the spectrum shown in Figure 4.31. It acts as a post-filter to the spectral data. If we had to simply use the current response of a detector to the intensity variation, we would be dependent on the angular spacing to identify the spectral value. The overlap occurs because the order, m, in Fabry–Perot interferometers is very large and there is no tag labeling the order of the fringe. If we cannot use our eye to discriminate the spectral color we cannot discriminate between, say, the fringe associated with $(m + 1)\lambda_1$ and the fringe associated with $(m - 1)\lambda_2$. This means that the separation between fringes may yield a multiple of the true wavelength separation.

In Figure 4.33 we show a Fabry–Perot output of the sodium doublet. Our eye cannot discriminate between the two wavelengths. Where should we measure the distance between the observed fringe maxima? Is the proper distance the narrow spacing or the large spacing between fringes? To ensure that we know where we are, another wavelength selective device is often used as a pre-filter of the light incident on the Fabry–Perot interferometer in replace of the eye serving as a post-filter.

The ratio of the free spectral range to the minimum resolvable wavelength is called the *finesse*, $\mathcal{F}$. Using Eqs. (4.54) and (4.56) we can write that

$$\mathcal{F} = \frac{(\Delta \lambda)}{\Delta \lambda} = \frac{\pi \sqrt{F}}{2}. \qquad (4.57)$$

The finesse is the key measure of performance of the interferometer and as it should be, it is independent of the spacing, d.

In terms of frequency, the finesse is

$$\mathcal{F} = \frac{(\Delta \nu)_{SR}}{\Delta \nu}.$$

Since the finesse is proportional to the reciprocal of the minimum resolvable bandwidth of the instrument, the finesse can be thought of as proportional to the decay time of the Fabry–Perot interferometer, i.e., the time taken for the optical fields associated with the

Figure 4.33 Fabry–Perot output of the sodium doublet spectrum at 588.9950 and 589.5924 nm. The pairs of lines disappear in the center of the image because the exposure saturates.
(From HyperPhysics by Rod Nave, Ga. State University. http://hyperphysics.phy-astr.gsu.edu/hbase/hframe.html)

standing waves in the Fabry–Perot interferometer to fall from their steady-state value to 0 after light is removed from the interferometer. This is sometimes called the photon lifetime.[23] The relaxation time is given by

$$\tau_D = \frac{1}{2\pi\Delta v} = \frac{d/c}{\delta_{loss}},$$

where we assume all losses are due to reflective losses,

$$\delta_{loss} = (1 - R).$$

This allows us to associate the finesse of the Fabry–Perot cavity with the Q of a classical oscillator:

$$Q = \frac{v}{\Delta v} = \frac{\text{Energy stored}}{\text{Energy loss/s}}$$

$$Q = \frac{c}{\lambda} \cdot 2\pi \left(\frac{d}{c\delta_{loss}}\right) = \frac{2\pi d}{\lambda\delta_{loss}} = \frac{2\pi d}{\lambda(1 - R)}.$$

From Eq. (4.55) we can write the reflective finesse as

$$\mathcal{F} = \pi r \big/ (1 - r^2) = \frac{\lambda r}{2d}Q.$$

The resolution in terms of the cavity Q is thus

$$\mathcal{R} = \frac{n_2 \lambda r}{4} Q.$$

The resolution for several values of finesse are listed here:

$$\mathcal{R} = 0.9 \quad \mathcal{F} = \frac{\pi\sqrt{0.9}}{0.1} \approx 30$$

$$\mathcal{R} = 0.95 \quad \mathcal{F} = \frac{\pi\sqrt{0.9}}{0.05} \approx 61$$

$$\mathcal{R} = 0.99 \quad \mathcal{F} = \frac{\pi}{0.01} \approx 314.$$

[23] Photon lifetime is the time constant associated with the decay (or growth) of energy in the Fabry–Perot cavity and not the conversion of a photon to another particle. If each mirror has a reflectivity given by R_i, and L is the spacing between the mirrors, then

$$\tau_{photon} = \frac{\tau_{rndtrp}}{(1 - R_1 R_2)} = \frac{2nL}{c(1 - R_1 R_2)}.$$

Rather than using two highly reflective mirrors separated by some gas, a solid rectangular block, called an etalon, is often used to serve as a narrow band filter for applications in lasers and telecom.

As an example of the design of an etalon, assume we have a 2-cm-thick ZnS plate to be used as a wavelength locking device in a dye laser operating at a wavelength of 600 nm. The index of refraction of the ZnS in $n_p = 2.3$. The high index of the ZnS determines the reflectivity

$$r^2 = \left(\frac{n_p - n_i}{n_p + n_i}\right)^2 = \left(\frac{2.3 - 1}{2.3 + 1}\right)^2 = 0.155$$

The coefficient of finesse, F, due to the reflectivity is

$$F = \frac{4r^2}{(1 - r^2)^2} = \frac{4(0.155)}{(1 - 0.155)^2} = 0.87$$

The resolving power of the etalon is given by

$$\mathcal{R} = \frac{m\pi}{2}\sqrt{F} \leq \frac{2n_p d}{\lambda}\sqrt{F} = \frac{2(2.3)\sqrt{0.87}}{6x10^{-7}}d = 7.15 \times 10^6 d$$

The thickness of the etalon is $d = 2$ cm, giving a resolving power of

$$\mathcal{R} \leq 1.43 \times 10^5.$$

The finesse of this etalon is

$$\mathcal{F} = \frac{\pi}{2}\sqrt{F} = \frac{\pi}{2}\sqrt{0.87} = 1.47$$

It would be very difficult to machine an etalon to an accuracy needed to hit the resonance at 600 nm. For the 2-cm etalon the correct thickness is given by

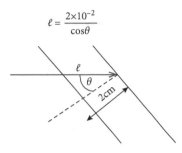

Figure 4.34 Rotation of etalon.

$$n_p d = \frac{m\lambda}{2}$$

$$m = \frac{2(2.3)\left(2 \times 10^{-2}\right)}{6 \times 10^{-7}} = 153,333.3.$$

We need a technique to correct errors in matching the thickness to the wavelength. We can change the effective thickness of the etalon to match the desired wavelength by rotation of the etalon, giving us a tunable optical filter, as we can see in Figure 4.34. We use Figure 4.34 to calculate the rotation needed to produce the next largest integer value for m:

$$\cos\theta = \frac{2nd}{m\lambda} = \frac{2(2.3)\left(2 \times 10^{-2}\right)}{(153334)\left(6 \times 10^{-7}\right)} = 0.17°.$$

We need to turn the etalon through 10 minutes of rotation.

We could improve the finesse, i.e., sharpen the resonance of the etalon without reducing the free spectral range by increasing the reflectivity. Applying a thin film dielectric stack can do this. If we created a reflectivity of 0.9, then we would obtain the following contrast, resolving power, and finesse:

$$F = \frac{4(0.9)}{(1 - 0.9)^2} = 360$$

$$\mathcal{R} = \frac{m\pi}{2}\sqrt{F} \le \frac{2n_p d}{\lambda}\sqrt{F} = 2.91 \times 10^6$$

$$\mathcal{F} = \frac{\pi}{2}\sqrt{F} = 29.8.$$

It would appear that by increasing the reflectivity, and thus the coefficient of finesse, F, we could continue to increase the finesse without limit; see Eq. (4.57). This is not the case; the figure on the mirrors, i.e., the flatness of the mirrors, will place an ultimate limit on the resolving power of the interferometer. To see why, return to the expression for the phase difference of each transmitted wave, Eq. (4.35), and assume the index in the interferometer is $n_2 = 1$, the operating wavelength is the vacuum wavelength, $\lambda = \lambda_0$, and the illumination is a plane wave so that $\cos\theta_t = 1$, then

$$\delta = \frac{4\pi d}{\lambda}.$$

The variation of δ with mirror spacing, d, is

$$\Delta\delta = \frac{4\pi\Delta d}{\lambda}.$$

We assume that, because the mirrors are not perfectly flat, d varies across the Fabry–Perot aperture by a fraction of a wavelength given by $\Delta d = \lambda/\mathcal{M}$. The phase variation due to the *mirror figure* (the variation in surface flatness) is given by

$$\Delta\delta = \frac{4\pi}{\mathcal{M}}.$$

Using the relation

$$\Delta\delta = 2\pi m \frac{\Delta\lambda}{\lambda}.$$

we can write the resolving power in terms of the mirror figure

$$\mathfrak{R} = \frac{\lambda}{\Delta\lambda} = \frac{m}{2}\mathcal{M}.$$

In calculating the resolution due to mirror reflectivity, we found that the variation in phase due to the mirror's reflectivity is

$$\Delta\delta < \frac{2\left(1 - r^2\right)}{r}.$$

If we have

$$\frac{1}{\mathcal{M}} < \frac{1 - r^2}{2\pi r} = \frac{1/\pi}{\sqrt{F}} = \frac{1}{2\mathcal{F}},$$

then the mirror figure, not its reflectivity, will determine the wavelength resolution.

We define a finesse measure called the figure finesse to characterize the performance of a Fabry–Perot limited by mirror flatness:

$$\mathcal{F}_F = \frac{\mathcal{M}}{2}.$$

It is difficult to create a mirror with an $\mathcal{M}$ value larger than 100 that would correspond to a surface that did not depart from a plane by more than 5 nm.[24] It is particularly difficult to reach a figure near 100 using the soft ZnS used in the above example.

The size, a, of the pinhole used in Figure 4.30 to limit the view of the detector can also determine the wavelength resolution. There is a pinhole finesse, defined as

$$\mathcal{F}_P = \frac{4\lambda f^2}{a^2 d},$$

where f is the focal length of the lens. (The origin of this relationship is diffraction theory that we will cover later.)

---

[24] A flatness of $\lambda/100$ can usually only be obtained using a liquid surface like mercury.

The net finesse due to the pinhole size, mirror figure, and reflectivity is called the *instrument finesse*, F_I,

$$\frac{1}{\mathcal{F}_I^2} = \frac{1}{\mathcal{F}^2} + \frac{1}{\mathcal{F}_F^2} + \frac{1}{\mathcal{F}_P^2}.$$

4.6 Young's Interference

There are a number of ways to produce two waves that exhibit interference. One way is to mix samples of a single wavefront. This approach was first used by *Francesco Maria Grimaldi* (1618–63), who used an extended source to illuminate a pair of pinholes. Thomas Young placed a pinhole between the source and the pair of pinholes to try to remove the effects of the source on the observed light distribution. Fresnel and Lloyd each developed an experimental arrangement (see Figure 4.35) that removed all question as to the origin of the observed dark and light bands.

The interference pattern from a Lloyd's mirror experiment is shown in Figure 4.36. Lloyd's mirror is used in uv photolithography and nanopatterning to produce diffraction gratings for such applications as surface encoders and medical implants without the need for expensive masks. Lloyd's mirror theory can also be used to explain fringes similar to those shown in Figure 4.36 to explain multipath interference in radio communications, ghosting in analog television reception, and signal fading.

We will limit our discussion to the experiment of Young's because of its simplicity. The law of reflection and Young's configuration can be combined to explain the experiments of Fresnel and Lloyd. The parameters shown in Figure 4.37 allow the math developed below for Young's two slit experiment to explain Lloyd's experiment. All discussion of diffraction in Young's experiment will be delayed until we develop diffraction theory.

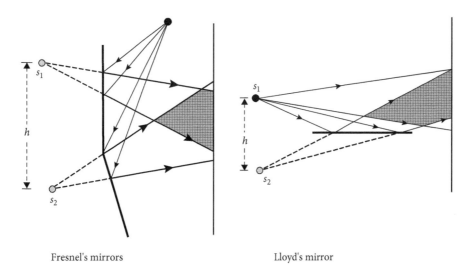

Fresnel's mirrors   Lloyd's mirror

Figure 4.35 Fresnel's mirrors and Lloyd's mirror. The shaded area is the region where fringes are observed.

Figure 4.36 Interference fringes from a Lloyd's mirror experiment shown in Figure 4.37. A HeNe laser was focused onto a pinhole and the light transmitted through the pinhole illuminated a flat mirror. The fringe at the plane established by the mirror surface is a dark fringe due to the phase shift of π upon reflection.

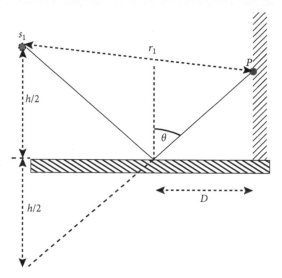

Figure 4.37 Configuration of Lloyd's experiment.

4.6.1 Young's two-slit experiment

The experimental geometry of Young's experiment is shown in Figure 4.38. The origin of the x,z-coordinate system is centered between two slits, labeled s_1 and s_2, separated by a distance h and extending out of the sheet of paper.[25] These slits are illuminated by a light source, s, with

---

[25] Slits make the interference pattern much easier to observe than the interference pattern produced by two pinholes by increasing the length of the interference fringes.

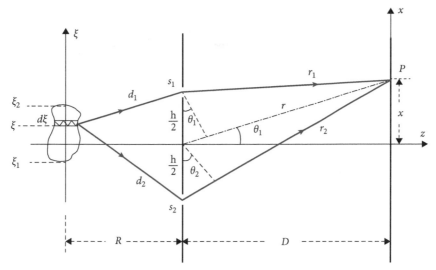

Figure 4.38 Young's two-slit experiment with an extended source.

a single frequency and having a shape whose extent is limited by a source pinhole of size $d\xi$. The light, reaching s_1 and s_2, differs in phase by $\phi_2 - \phi_1$.

(We can insure that this phase difference is zero, by centering the pinhole with respect to the two slits so that both values of d are the same.) The two slits collect samples of the wavefront that is emitted from the source pinhole. We observe the interference produced by the two wave samples after they have propagated through the distance D. A lens can be used to image the source pinhole on the observation screen and thereby forcing the overlap of the waves from s_1 and s_2.

The light from s_1 travels a path r_1 and the light from s_2 travels the path r_2 to the observation point P in the observation plane, a distance D from the slit plane. The intensity at P is determined by the total phase difference

$$\delta = \phi_2 - \phi_1 + k(r_2 - r_1).$$

The distances r_1 and r_2 are not easy to measure; however, the angles are experimentally accessible.

The angles shown in Figure 4.38 are defined by the equations

$$\sin\theta_1 = \frac{r - r_1}{h/2}, \sin\theta_2 = \frac{r_2 - r}{h/2}. \tag{4.58}$$

The angles are here defined using distances that are not easily measured. To define the angles in terms of easily measured parameters, we assume that the distance from the slits to the observation screen, D, is much larger than the height of the observation point above the z-axis, i.e., $x < D$. With this assumption, $\sin\theta_1 \approx \tan\theta_1$ and we may write the distances r_1 and r_2 in terms of the coordinates of the observation point and the slit spacing:

$$\sin \theta_1 = \frac{r - r_1}{\dfrac{h}{2}} \approx \tan \theta_1 = \frac{x}{D}$$

and

$$\sin \theta_2 = \frac{r_2 - r}{\dfrac{h}{2}} \approx \tan \theta_2 = \frac{x + \dfrac{h}{2}}{D}.$$

We further assume that the slit spacing is small, $h < x < D$, so that we may neglect terms that involve h^2,

$$r_2 - r \approx \frac{xh}{2D}.$$

Now adding $(r_2 - r) + (r - r_1)$, we obtain the difference in the path lengths from s_1 to P and s_2 to P in terms of easily measured experimental parameters,

$$r_2 - r_1 = \frac{xh}{D}. \tag{4.59}$$

The observed intensity distribution is given by Eq. (4.10). The interference term, allowing for different intensities passing through the two slits, is

$$2\sqrt{I_1 I_2} \cos \delta.$$

With our geometry the phase angle is given by

$$\delta = \phi_2 - \phi_1 + k(r_2 - r_1) = \Delta\phi + \frac{2\pi x h}{\lambda D}. \tag{4.60}$$

The resulting interference pattern consists of bright and dark bands perpendicular to the plane of Figure 4.38. An experimental example of Young's fringes, generated using white light, is shown in Figure 4.39. Note that the central fringe is white, independent of wavelength, since when $x = 0$, $\cos\delta = 1$.

As stated in our introductory arguments about interference, whenever the argument of the cosine changes by 2π the intensity on the screen passes through one cycle of intensity, say maximum to minimum and back to maximum

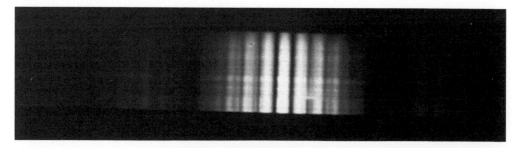

Figure 4.39 Fringes from a two-slit experiment, made with white light. Note that no color is visible in the center (zero order) fringe, often called the white light fringe.

$$\left[\frac{2\pi x_1 h}{\lambda D} + \Delta\phi\right] - \left[\frac{2\pi x_2 h}{\lambda D} + \Delta\phi\right] = 2\pi.$$

A single cycle in intensity that we have defined as a *fringe* corresponds to spacing between two bright or two dark bands. The spatial period of a fringe, i.e., its repeat distance, is given by

$$x_1 - x_2 = \frac{\lambda D}{h}. \tag{4.61}$$

From Eq. (4.61) we see that the spacing of the fringes on the viewing screen is inversely proportional to the slit spacing. Note also that the fringe spacing is a linear function of the wavelength. These results are supported by what can be observed in Figure 4.39.

The fringes are generated by adding waves originating from two samples of a wavefront obtained by using the two slits. For this reason, the term *interference by division of wavefront* is often used to describe this type of interference. We earlier pointed out that the interference process performs a comparison of the waves and provides a measure of the similarity (a correlation) of the two waves; here the comparison would be between two different parts of the same wavefront while in the Michelson interferometer the comparison is between a wave at different times.

4.7 **Spatial Coherence**

Temporal coherence is associated with the wave's properties along the direction of propagation. Spatial coherence we are now considering is associated with the wave's properties transverse to the direction of propagation. If the wave is a perfect plane wave, then there is a uniform phase in a plane perpendicular to the direction of propagation. The phase may exhibit fluctuations but all points in the plane will have identical fluctuations. The perfect plane wave is said to be *spatially coherent.*

Fourier theory gives us a mathematical structure to understand what is going on. The temporal coherence is associated with the frequency distribution of the source while the spatial coherence is associated with a distribution of propagation vectors, **k**, associated with the wave, i.e., with a departure of the wave from the ideal plane wave. We will see, in our later discussion of diffraction, that complete spatial coherence is as difficult to obtain as complete temporal coherence.

The theoretical development of the concept of spatial coherence is identical to the development for temporal coherence. By a simple redefinition of the retardation time, τ, and a generalization of the degree of coherence function, we can apply the same equations to the two classes of coherence.

In our discussion of spatial coherence, Young's two-slit experiment, as displayed in Figure 4.38, will play the same role as Michelson's interferometer did in the discussion of temporal coherence. The time to travel from the source to the observation point P by way of slit 1 is given by

$$t + \frac{d_1}{c} + \frac{r_1}{c}.$$

The time for light to travel from the source to point P by way of slit 2 is

$$t' = t + \frac{d_2}{c} + \frac{r_2}{c}.$$

The phase difference between the waves from the source to the slits is

$$\Delta\varphi = \frac{d_2 - d_1}{c},$$

and the difference in propagation time from the two slits to point P is

$$\tau = \frac{r_2 - r_1}{c}.$$

The difference in propagation time in Young's two-slit experiment, τ, is equivalent to the retardation time used in temporal coherence. They both temporally define the waves that are to undergo interference. The retardation time associated with temporal coherence was defined in terms of the Michelson interferometer as

$$\tau = \frac{2d}{c},$$

where d is the difference in length of the two arms of the Michelson interferometer. The retardation time is redefined for spatial coherence in terms of the spacing of the slits in Young's two-slit experiment

$$\tau = \frac{r_2 - r_1}{c}.$$

All of the equations developed in temporal coherence can be utilized in spatial coherence theory by using this new retardation time.

In general, the spatial coherence function, $\gamma_{12}(\tau)$, is a complex function of the form

$$\gamma_{12}(\tau) = |\gamma_{12}(\tau)| \, e^{i\delta_{12}(\tau)},$$

where the real part of this relation is identical to Eq. (4.33). The intensity at point P produced by interference between the waves from slits 1 and 2 is given by

$$I_p = I_1 + I_2 = 2\sqrt{I_1 I_2} \, |\gamma_{12}(\tau)| \cos \delta_{12}(\tau), \tag{4.62}$$

which is identical to the results for the Michelson interferometer, except for $|\gamma_{12}(\tau)|$. It is obvious from this expression that the coherence function, $\gamma_{12}(\tau)$, reduces the fringe visibility when the degree of coherence is less than 1. The degree of coherence, $\gamma_{12}(\tau)$, is a function of $r_2 - r_1$ and thus describes how the visibility of the interference fringes vary as the observation point P is moved over the observation plane.

A Line Source

While the identical formalisms of spatial coherence and temporal coherence make it easy to develop coherence theory, it does not help in obtaining physical insight into the theory. We will expand our physical understanding of spatial coherence theory by performing a very simple experiment. We will examine Young's two-slit experiment, shown in

Figure 4.38; however, the experiment will be modified by the utilization of an extended source of length s, from ξ_1 to ξ_2 a distance R behind the slits. We will calculate the appearance of Young's fringes due to the source size and shape. The result will anticipate findings to be obtained in the discussion of diffraction—that spatial coherence limits the region, in space, over which Young's interference fringes can be observed.

We will assume that the source is a one-dimensional line source lying along the ξ-axis and extending from ξ_1 to ξ_2, where $\xi_2 - \xi_1 = s$ is the source's length. We have already derived the result if the source is limited to an element of size $d\xi$ located at position ξ above the optical axis. Allowing the source position to vary adds an additional complication to the derivation for Young's two slits. We must evaluate the phase change

$$\Delta\phi = \phi_2 - \phi_1,$$

which is no longer zero but

$$\Delta\phi = \frac{d_2 - d_1}{c}.$$

The optical path difference from the source to P, by way of the two slits, is

$$\Delta = n\left[(d_2 + r_2) - (d_1 + r_2)\right]. \tag{4.63}$$

(We will assume $n = 1$ in this discussion.) Earlier with Eq. (4.59), we found the optical path difference for the path from the two slits to the observation plane to be

$$c\tau = r_2 - r_1 = \frac{xh}{D}.$$

We now wish to perform an equivalent derivation for the optical path from the source to the slit plane

$$d_2 - d_1 = c\Delta\varphi,$$

using Figure 4.38.

The distances from the source to the two slits are given by

$$d_1^2 = R^2 + \left(\frac{h}{2} - \xi\right)^2 \qquad d_2^2 = R^2 + \left(\frac{h}{2} + \xi\right)^2$$

The difference between these two equations is

$$\begin{aligned} d_2^2 - d_1^2 &= R^2 + \frac{h^2}{4} + h\xi + \xi^2 - R^2 - \frac{h^2}{4} + h\xi - \xi^2 \\ &= 2h\xi, \end{aligned}$$

which can be written as the product between the sum and difference of the two distances

$$(d_2 - d_1)(d_2 + d_1) = 2h\xi.$$

Since the distance from the source to the slit plane, R, is large and the slit spacing is very small, $R \gg h$, we may write $d_1 \approx d_2 \approx R$. The sum of d_1 and d_2 can be approximated by $2R$

but we have to retain the difference. Using this approximation, the path difference from the source to the two slits is

$$d_2 - d_1 = \frac{h\xi}{R}. \tag{4.64}$$

From Eq. (4.12) we know that the intensity at P for a point source is

$$I_p = I_1 + I_2 + \sqrt[2]{I_1 I_2}\cos\delta,$$

where from Eq. (4.16)

$$\delta = k\Delta = \frac{kh\xi}{R} + \frac{khx}{D}. \tag{4.65}$$

We will assume that the total light reaching P from each pinhole, illuminated by $d\xi$, is the same

$$dI_1 = dI_2 = I_0 P(\xi)\,d\xi,$$

where $P(\xi)$ is the normalized spatial intensity distribution across the source. For this problem, the intensity distribution is unity over the length $\xi_2 - \xi_1 = s$:

$$P(\xi) = \begin{cases} 1 & \xi_2 \geq \xi \geq \xi_1 \\ 0 & all\ other\ \xi \end{cases}. \tag{4.66}$$

The intensity at point P from the source element, $d\xi$, is

$$dI_p = 2\left[1 + \cos\delta\right] I_0 P(\xi)\,d\xi. \tag{4.67}$$

To obtain the total intensity at point P we must integrate Eq. (4.67) over the source

$$I_p = 2I_0 \int P(\xi)\left[1 + \cos\delta\right] d\xi.$$

By analogy with Eq. (4.9), the degree of spatial coherence must be

$$\gamma = \int P(\xi)\cos\delta\,d\xi$$
$$= \int_{\xi_1}^{\xi_2} P(\xi)\cos\left[\frac{kh\xi}{R} + \frac{khx}{D}\right] d\xi. \tag{4.68}$$

Using a trigonometric identity, we can write

$$\gamma = \cos\left(\frac{khx}{D}\right)\int P(\xi)\cos\left(\frac{kh\xi}{R}\right)d\xi - \sin\left(\frac{khx}{D}\right)\int P(\xi)\sin\left(\frac{kh\xi}{R}\right)d\xi. \tag{4.69}$$

The two terms of Eq. (4.69) are the cosine and sine transforms of $P(\xi)$. As was the case for temporal coherence, where the degree of coherence is the Fourier transform of the spectral distribution of the light source, here the spatial coherence is the Fourier transform of the spatial distribution of the light source.

The fact that the spatial coherence is the Fourier transform of the spatial distribution of the light source is a manifestation of the *van Cittert–Zernike Theorem*. More precisely the van Cittert–Zernike theorem states that:

> The complex degree of coherence is equal to the diffraction pattern of an aperture of the same size and shape as the source and illuminated by a spherical wave whose amplitude distribution is proportional to the intensity distribution across the source. [8]

This means that incoherent sources viewed at large distances have some measure of spatial coherence. For example, we can measure the spatial coherence of a binary star to determine the separation of the two stars.[26]

If the source is centered on the optical axis of the experiment (the z-axis), then, because $\sin(kh\xi/R)$ is odd, the second integral of Eq. (4.69) is zero and the integral becomes

$$\gamma = s\frac{\sin\left(khs/2R\right)}{khs/2R}\cos\left(\frac{khx}{D}\right). \tag{4.70}$$

The intensity distribution on our observation plane has two different intensity distributions. The cosine term results from the interference of the light from the slit pair. The sinc function that forms the amplitude of the cosine function is due to interference of light from different parts of the source [9, 10].[27] A computer simulation of the interference pattern calculated using Eq. (4.70) is shown in Figure 4.40. In Figure 4.40a the source size was held constant and the pinhole spacing, h, was varied, as one might do in the measurement of the spatial coherence of a source. In Figure 4.40b, the pinhole spacing was held constant and the source size was varied, demonstrating the importance of source size on the observation of interference fringes.

4.7.1 Spatial coherence length

The operational definition of the longitudinal coherence length equated it to the difference in length between the two arms of the Michelson interferometer when the fringe visibility became 0. A *transverse coherence length*, ℓ_t, can be defined in an analogous way, in terms of Young's two-slit experiment. It represents a distance, in a plane normal to the direction of propagation, over which the phases at two points remain correlated and equals the slit separation for which the interference fringes disappear.

The spacing between the bright bands in the plane containing P of our two-slit geometry is from Eq. (4.61),

$$x_1 - x_2 = \frac{\lambda D}{h}.$$

The magnitude of the fringe spacing is independent of the source position, ξ. However, if the source position, ξ, is changed, the position of the fringes moves, as can be seen by taking the variation of (4.65) at constant phase, δ,

---

[26] A demonstration of the van Cittert–Zernike theorem using water waves can be observed using waves generated by swimming ducks.

[27] We will see this equation again when we develop the theory of diffraction.

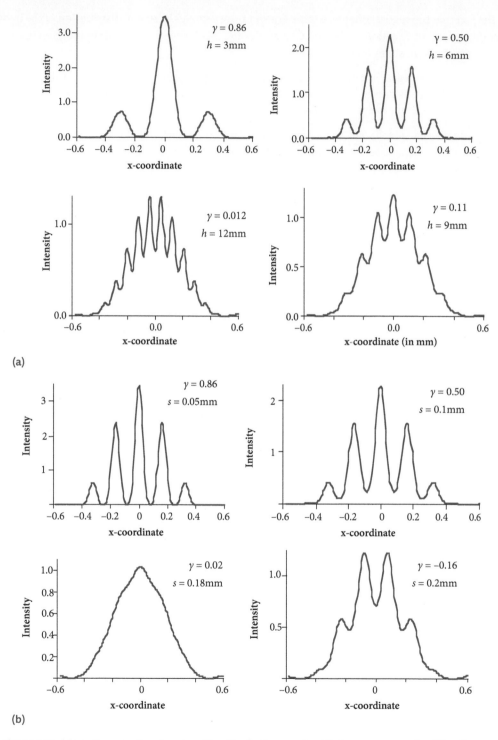

Figure 4.40 (a) Interference pattern from two slits with a fixed source size. A finite source was used and the slit spacing, h, was varied over a factor of 4. The degree of coherence is sinc(x) and is the cause of the fringe visibility dropping to 0 and then increasing. The overall shape of the pattern is due to diffraction. (b) The interference pattern for two slits with a fixed slit spacing. The source size was varied by a factor of 4.

$$\delta\left[\delta\right] = 0 = \delta\left[k\Delta\right] = \delta\left[\frac{kh\xi}{R} + \frac{khx}{D}\right]$$

$$0 = \frac{kh\delta\xi}{R} + \frac{kh\delta x}{D} \qquad (4.71)$$

$$\frac{\delta\xi}{R} = -\frac{\delta x}{D},$$

where δx is the distance a band moves when the source is moved a distance $\delta\xi$.[28] If there are two sets of fringes created by two points on a source located at ξ_1 and ξ_2, then the fringe visibility is destroyed when a bright band due to the light from ξ_1 falls on a dark band produced by ξ_2. The intensity on the observation screen does not vanish but the fringes do. This cancellation of fringe visibility occurs when $\delta x = \lambda D/2h$, which corresponds to a source displacement of

$$\delta\xi = -R\frac{\lambda}{2h}. \qquad (4.72)$$

This relationship allows the calculation of the maximum source size that can produce interference fringes at the observation point, a distance R away, when Young's slit spacing is equal to h. A source larger than this value will not produce a fringe pattern in Young's experiment but will rather produce a uniform illumination on the observation plane. The maximum source size that can produce interference may be described as an angular width, $\Delta\theta$, using the fact that $\delta\xi << R$:

$$\frac{\delta\xi}{R} = \tan\frac{\Delta\theta}{2} \approx \sin\frac{\Delta\theta}{2} \approx \frac{\Delta\theta}{2}$$

The value of h, calculated using Eq. (4.72), is equal to the transverse coherent length, λ_t, because that slit spacing is associated with a fringe visibility of 0. The source size associated with the transverse coherence length is called the *coherent source size*

$$\Delta\theta = \frac{\lambda}{h} = \frac{\lambda}{l_t}. \qquad (4.73)$$

This is an important parameter in microscopy. To get some idea of the magnitude of l_t, assume the source is $\xi = 1.2$ mm wide and is $R = 2$ m from the slits; then

$$\Delta\theta = \frac{\delta\xi}{R} = \frac{1.2}{2\times10^3} = 6\times10^{-4} \text{ radians.}$$

If $\lambda = 600$ nm, then the slits cannot be separated by more than 1 mm if interference is to be observed using this source, i.e., $l_t = 1$mm. The Sun produces an ℓ_t on the order of 1/20 mm. This coherence length is smaller than the resolution of the eye but in a microscope it can produce coherent effects.

4.8 Stellar Interferometer

Michelson made use of the concepts of spatial coherence and coherent source size to measure the angular dimensions associated with stellar objects. His stellar interferometer, Figure 4.41b,

---

[28] The minus sign means that the fringes move up if the source is moved down.

looks a great deal like Young's two-slit experiment but the scale is a good deal larger—the two input mirrors of his device were over 6 m apart.

With the geometry of Figure 4.41a, Eq. (4.65) becomes

$$\delta = kh_1 \Delta\theta + kh_2 \frac{x}{D}.$$

If we assume that the light intensities gathered by the two input mirrors are equal, then, at the fringe plane of Figure 4.41a, the intensity pattern would be given by

$$\begin{aligned} I_P &= 2I_0 \left(1 + \cos\delta\right) \\ &= 2I_0 \left[1 + \cos\left(kh_1\Delta\theta\right)\cos\left(kh_2\frac{x}{D}\right) - \sin\left(kh_1\Delta\theta\right)\sin\left(kh_2\frac{x}{D}\right)\right]. \end{aligned}$$

The angular extent of the source, $\Delta\theta$, is assumed small compared to the dimensions of the experiment, allowing the intensity distribution to be approximated by

$$I_P \approx 2I_0 \left[2 - kh_1\Delta\theta \sin\left(kh_2\frac{x}{D}\right)\right].$$

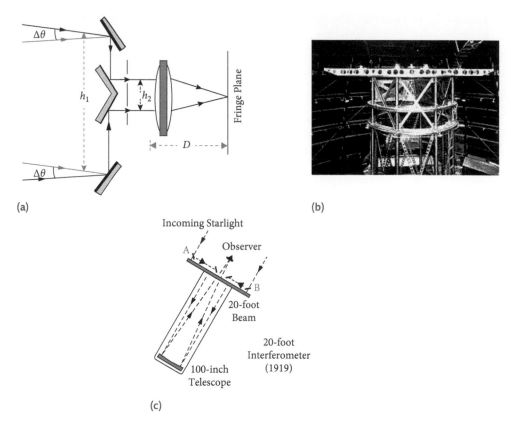

(a)

(b)

(c)

Figure 4.41 Michelson stellar interferometer. (a) Schematic representation of the interferometer. The distance h_1 is adjusted until the fringes on the fringe plane disappear. The separation, h_2, of the two apertures at the front of the lens establishes the fringe spacing to match the detector array in the fringe plane. (b) Image of the 20-ft Michelson interferometer mounted on the 100-in. Hooker telescope on Mount Wilson, near Pasadena California. By George Ellery Hale—The New Heavens, Public Domain, *https://commons.wikimedia.org/w/index. php?curid=1223713*. (c) A line drawing of the interferometer. This instrument is now on display at the Rose Center of the American Museum of Natural History in New York.

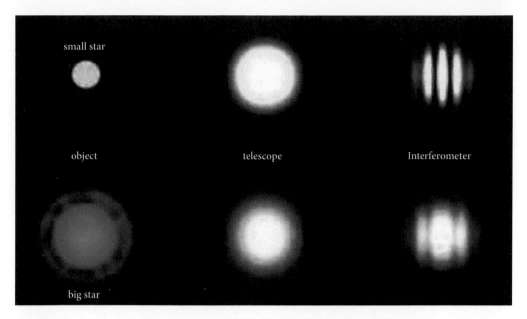

Figure 4.42 The power of a stellar interferometer over a conventional telescope to distinguish between a large and a small star (simulation).
Open source from European Southern Observatory: ESO PR Photo 10d/01 (18 March 2001) (https://cdn.eso.org/images/screen/eso0111d.jpg).

The fringe spacing at the fringe plane of Figure 4.41a is determined by the separation, h_2, of the two slits in the screen at the front of the lens and is calculated using Eq. (4.61). The two input mirrors sample the incoming wavefront at two points a distance h_1 apart and determine the contrast of the fringes. The observed fringe patterns would have an appearance similar to those in Figure 4.42.

Michelson proved the interferometric technique in 1891 by measuring the diameters of Jupiter's moons that have an angular extent of 1 s and are thus resolvable with conventional imaging techniques. Having proven the interferometric technique using a 12-in. interferometer, Michelson in 1919 applied the technique to the measurement of the separation of binary stars and the angular diameters of single stars with a 20-ft interferometer. One of the largest stars measured, in 1921 [11], was Betelguese with an angular diameter of 0.047 s. The spatial coherence length, λ_t, of this star is on the order of several meters.

At $\lambda = 600$ nm fringes produced by single wavelength from Betelgeuse disappeared for $h_1 = 2.63$ cm. This yields an angular diameter of Betelgeuse of 0.047 arc-sec:

$$\Delta\theta = \lambda/l_t = \frac{6 \times 10^{-7}}{2.63}$$

$$= \begin{cases} 2.28 \times 10^{-7} \text{ rad} \\ 1.31 \times 10^{-5} \text{ deg.} \\ 0.047 \text{ s} \end{cases}$$

A comparison of the images seen by the astronomer with an interferometer and a conventional telescope is seen in Figure 4.42.

4.9 **Intensity Interferometry**

The Michelson stellar interferometer is very sensitive to vibrations and to fluctuations in the relative phase of the two wave samples, due to propagation through the atmosphere. These disturbances prevent the use of a visible wave interferometer with a value of h_1 larger than about 5 or 6 m. Another technique developed in 1961 by *R. Hanbury Brown* and *Richard Q. Twiss* [12], which involves measurement of intensity from two detectors spaced a distance, *h*, apart, is not as sensitive to vibrations or atmospheric distortions. Dimensions of *h* associated with this type of interferometer exceed 300 m.

The theory behind the technique is based upon the assumption that the fluctuations in the outputs of two detectors must be correlated if the amplitudes of the two waves are correlated. At least a lower bound for the degree of coherence can be obtained by measuring the intensity correlation. We will give here a classical interpretation of the observed effect.

If we could see the instantaneous intensity distribution of the light, say, from the Sun or the stars, we would see a pattern such as that shown in Figure 4.43 labeled coherent. The dark and bright patterns are called speckle and are produced by instantaneous interference between light waves that have traveled different optical paths. The size of a single speckle is proportional to the wavelength and inversely proportional to the angular size of the source. The patterns fluctuate on a femtosecond time scale so we have had to simulate the pattern by using a coherent laser beam and opal glass to produce a speckle pattern that is stationary.

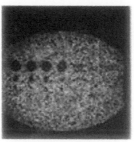

Coherent

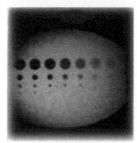

Averaged

Figure 4.43 Simulation of instantaneous intensity distribution of a star illumination produced by using coherent laser illumination is labeled "Coherent." What would be observed by temporally averaging the intensity distribution is labeled "Averaged." The Averaged image was produced by making 120 exposures of independent samples of coherent images. The simulation is required because there is no sensor fast enough to record the instantaneous intensity.

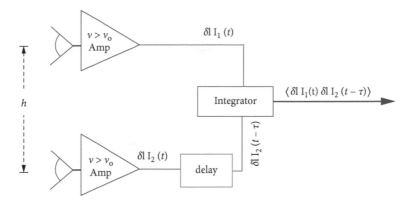

Figure 4.44 Hanbury Brown and Twiss interferometer for radio astronomy.

To simulate the temporal average we generated 128 statistically independent speckle patterns and averaged them by adding exposures of the independent patterns. The result of adding the statistically independent patterns is shown in the image label Averaged.[29]

In effect, Hanbury Brown and Twiss set up an experiment to measure the size of a bright speckle. To carry out the coherence measurement, the Michelson stellar interferometer of Figure 4.41 is replaced by two detectors spaced a distance h apart (Figure 4.44). Amplifiers, following the detectors, remove the time-invariant components, $<I_1>$ and $<I_2>$ by high-pass filtering. The signal from one detector is delayed, electronically, τ seconds with respect to the other signal and an electronic integrator calculates the correlation of the two signals by performing a time average of the product of the delayed output of one detector with the prompt output of the second detector.

To prove the technique met the theoretical requirements at visible wavelengths, Hanbury Brown and Twiss measured $|\gamma_{12}(0,\eta)|^2$ for a mercury arc lamp [13].

The difference between the intensity interferometer and the Michelson interferometer is that the Hanbury Brown intensity interferometer measures the square of the modulus of the complex degree of coherence while the Michelson interferometer also measures the phase.

The Hanbury Brown experiments introduced a great deal of controversy into the optics community, during the late 1950s and early 1960s. The effect can be explained by treating the electromagnetic radiation as a classical wave. From a quantum viewpoint, the basic assumption that the two detectors receive a correlated signal suggests that photons must arrive at the detectors in pairs. Since the stars are thermal sources, this seemed absurd to

---

[29] The test pattern consisting of a matrix of disks shown in the two images is not important in our current discussion but will be addressed in Chapter 11. Their function was to measure the effect of speckle noise on an imaging system involving both a human observer and a conventional imaging device. The diameter of the disks in each row is identical but the diameters decrease by a factor of 2 as we move down from row to row. The disks in each column have the same contrast but the contrast decreases from the left to right. (A. Rose developed this test pattern for the evaluation of television systems based on the human vision system.)

many physicists. If photons are emitted at random at the source, how could they arrive in pairs at the detectors?

The apparent conflict is a manifestation of the wave–particle duality of light. When we make certain type measurements on light, such an interference measurements, light appears to be a wave. For other types of measurements, such as the low light level images, the particle nature of light is evident.

The intensity interferometry seems to emphasize the particle nature of light because it recorded the simultaneous arrival of pairs of photons. The result of the measurement, however, seems to require an interpretation based on the wave nature of light. The two detectors appear to sample portions of a wavefront associated with a wave that originated at a point on the source and diverges toward the detector as a spherical wave with an ever-increasing diameter phase front. The solution to the conflict was to allow the original emitted photon to disappear and new photons to reform during propagation. The wave function associated with the photons would predict periodic increases in the probability of detection in a manner similar to the disappearance and reformation of a classical wave group.

Many people were reluctant to accept the concept of a photon that did not retain its identity throughout the propagation of light, from emission to absorption. Hanbury Brown and Twiss conducted a number of experiments that verified their approach and overcame the objections of other scientists. Purcell showed that a quantitative explanation of the observed correlation could be made based on the quantum statistical behavior of photons. Photons have a spin of 1 and thus are bosons. Bosons like to bunch together so it is more probable to see pairs of photons. R. J. Glauber later developed the quantum theory of optical coherence which gave a firm quantum theoretical basis for the experimental results of Hanbury Brown [14]. He shared the Nobel Prize in 2005 for this work.

4.10 Problem Set 4

(1) If we coat glass ($n = 1.5$) with another material ($n = 2.0$), what thicknesses gives maximum reflection? Minimum reflection? Assume that $\lambda = 500$ nm.

(2) A metal ring is dipped into a soap solution ($n = 1.34$) and held in a vertical plane so that a wedge-shaped film forms under the influence of gravity. At near-normal illumination with blue–green light ($\lambda = 488$ nm) from an argon laser, one can see 12 fringes per centimeter. Determine the wedge angle of the soap film.

(3) Eight interference fringes are spread over 2 cm on a screen 100 cm away from a double slit with 0.2-mm separation. What is λ?

(4) Use Figure 4.41 to develop a theoretical explanation of Lloyd's experiment.

(5) A solid Fabry–Perot interferometer is made of an uncoated 2-cm-thick slab of material ($n = 4.5$). Relying only on the reflectivities of the air–material interface, what is the fringe contrast and the resolving power at $\lambda = 500$ nm? A glass slab such as this, often with reflective coatings, is called a *solid etalon* and is used as a wavelength-selective filter.

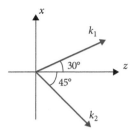

Figure 4.45 Propagation vectors of waves to be added together as described in Problem 7.

Figure 4.46 Fringes for Problem 8.

(6) Assume the mirrors of a Michelson interferometer are not perfectly aligned. We see as an output a 3×3 cm illuminated field containing 24 vertical bright fringes. What is the angle the two mirrors make with respect to each other? Assume $\lambda = 500$ nm

(7) Assume two plane waves of the same wavelength, $\lambda = 632.8$ nm., are polarized with their electric fields parallel to the y-axis. The propagation vectors for the two waves are shown in Figure 4.45.
Calculate the separation of intensity maxima along the x- and z-axes.

(8) *The fringes shown in Figure 4.46 were produced using a wavelength of 589 nm. The fringes are due to an air gap between two flat glass plates created by placing a scrap of tissue between the plates, along one edge. What is the angle of the air gap?

(9) What is the line width in nanometers and in hertz of light from a laser with a 10-km coherence length? The mean wavelength is 632.8 nm.

(10) A 1-mm-diameter pinhole 2 m from a double slit is illuminated by $\lambda = 589$ nm. What is the maximum slit spacing for which interference fringes are visible?

(11) A Hg^{198} low-pressure isotope lamp emits a line at $\lambda = 546.078$ nm. with a bandwidth of $\Delta \nu = 1000\ MH_z$. What is the coherence length and coherence time?

(12)* If a monochromatic light wave with $\lambda = 488$ nm is chopped at a frequency of 40 MHz, what is the bandwidth (in nm) of the light pulses?

(13) We stated that the transverse coherence length of the Sun was 1/20 mm. Assuming $\lambda \approx 600$ nm, what is the apparent angular diameter of the Sun?

(14) Michelson's stellar interferometer had a slit spacing of 6 m. At $\lambda = 500$ nm, what is the smallest angular diameter it could measure?

(15) A multilayer bandpass filter has a center wavelength of 632.8 nm and a bandwidth of 0.1 nm. What is the coherence length of the light that is transmitted by this filter?

 REFERENCES

1. ATLAS Collaboration, Evidence for light-by-light scattering in heavy-ion collisions with the ATLAS detector at the LHC. *Nature Physics* **13**: 852–8 (2017).

2. Young, T., Experiments and calculations relative to physical optics.*Philosophical Transactions of the Royal Society of London* **94**: 1–16 (1803).

3. Smith, W. J., How flat is flat. *Optical Spectra* (April 1974, May 1974, August 1978).

4. Hayes, J., Interference, in *Encylopedia of Optical Engineering*, edited by R. G. Driggers, pp. 945–65. Boca Raton, FL: CRC Press, 2003.

5. Blahnik, V., and B. Voelker, About the reduction of reflections for camera lenses: How T*-coating made glass invisible. Carl Zeiss, AG (2016). Available at https://pixinfo.com/wp-content/uploads/2016/04/en_About-the-reduction-of-reflections-of-camera-lenses.pdf.

6. Brexinski, M. E., *Optical Coherence Tomography*. San Diego: Academic Press, 2006.

7. Radhakrishnan, S., et al., Real-time optical coherence tomography of the anterior segment at 1310 nm.*Arch, Ophthalmol.* **119**(8): 1179–85 (2001).

8. Born, M., and E. Wolf, *Principles of Optics*, 7th edn, p. 952. Cambridge, UK: Cambridge University Press, 1999.

9. Gbur, G., You could learn a lot from a ducky: The van Cittert–Zernike theorem, in *Skulls in the Stars*. WordPress.com, 2010.

10. Knox, W.H., M. Alonso, and E. Wolf, Spatial coherence from ducks. *Physics Today* **63**(3): 11 (2010).

11. Michelson, A. A. and F. G. Pease, *Astrophys. J.* **53**: 249–59 (1921).

12. Brown, R. H., and R. Q. Twiss, A new type of iterferometer for use in radio astronomy.*Phil. Mag.* **45**: 663 (1954).

13. Brown, R. H. and R. Q. Twiss, Interferometry of the intensity fluctuations in light, II. An experimental test of the theroy for partially coherent light. *Proc. Roy. Soc.* **A243**: 291 (1957).

14. Klauder, J. R., and E. C. G. Sudarshan, *Fundamentals of Quantum Optics*. New York: Dover, 2006.

5 Polarizers

5.1 Detection of Polarization

In our discussion of interference, we discovered early that only parallel-polarized light waves interfere. We need some way to control polarization to optimize our ability to utilize interference.

Lots of animals and insects (for example, bees) not only can see polarized light but also apply that capability for such tasks as navigation. The human eye is at best weakly sensitive to polarization. Some humans describe an image called *Haidinger's brush* seen in the presence of polarized light. They describe a diffuse, elongated, yellowish, bowtie-shaped object with a similarly shaped blue bowtie crossing at 90° to the yellow object. In Figure 5.1 a drawing constructed from this description is shown. Only a drawing can be displayed because the pattern is generated in the macular pigment of the observer's eye.[1] To actually measure polarization we need an optical device.

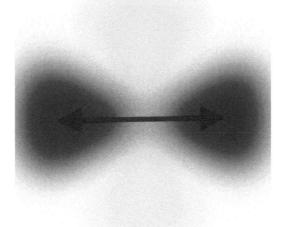

Figure 5.1 Artistic illustration of Haidinger's brush. The actual image occurs in the macula of the eye and cannot be recorded. The direction of the linear polarized light producing the brush is indicated by the red double arrow which is not part of the observed Haidinger's brush.

---

[1] This pigment is located in the center of the fovea. It handles excessive blue light levels, and lack of the pigment is associated with Macular Degeneration.

Modern Optics Simplified. B. D. Guenther. © B. D. Guenther 2020.
Published in 2020 by Oxford University Press. DOI: 10.1093/oso/9780198842859.001.0001

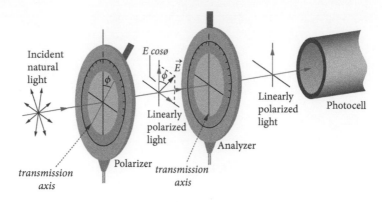

Figure 5.2 Polarizer–analyzer pair.

Another reason for wishing to control polarization is the fact that polarized light is ubiquitous in nature. Polarized sunlight scattered down to us from the sky and reflections from surfaces modify the polarization of the reflected light, as we can infer from curves like Figure 3.11. As an example of the performance enhancement gained by controlling polarization, we need only turn to photography. By placing polarizing filters on both the illumination and camera, complete control of reflections in a scene can be obtained.

We need a device that has different attenuation properties relative to the angle the device makes with the polarization vector. Malus' law, Eq. (5.1), gives the performance of a pair of such identical devices, called the polarizer and analyzer. The angle ϕ is measured between the polarization axis of the polarizer and that of the analyzer. The device pair is sometimes called a *polarimeter*; see Figure 5.2.

$$I = A^2\cos^2\phi. \tag{5.1}$$

When $\phi = \pi/2$, the polarizers are said to be *crossed* and no light propagates through the polarizer–analyzer pair. When $\phi = 0$ the polarizers are said to be *parallel* and the maximum amount of light is transmitted by the pair.

The polarizer prepares a linearly polarized beam of light from a beam of unpolarized light.[2] The transmission of the unpolarized light through the polarizer can be calculated using Eq. (5.1). Since the average value of $\cos^2\phi = 1/2$,

$$I/I_0 = 1/2.$$

We can place a sample we would like to study in this beam and use the analyzer to determine the effects of the sample on polarized light. The device shown in Figure 5.2 can be realized with a cellphone and a couple of low-cost plastic polarizers.

If we take the analyzer out of the system in Figure 5.2 we can measure a quantity called the *degree of linear polarization*, P, of the incident light. When we measure polarized light we determine not only the orientation of its electric field but also the degree of linear polarization, P,

---

[2] Unpolarized light is a uniform mixture of linearly polarized light at every orientation.

$$P = \frac{I_\parallel - I_\perp}{I_\parallel + I_\perp}, \tag{5.2}$$

where the parallel subscript refers to the intensity of light transmitted by a linear polarizer with its axis oriented parallel to the electric field vector of the incident wave and the perpendicular subscript indicates the axis is oriented perpendicular to the electric field of the incident wave. You can make a similar definition for right and left circular polarized light.

Note that when $P = 50\%$, the ratio of the intensity transmitted by the two orientations, called the *extinction ratio*, is

$$\frac{I_\parallel}{I_\perp} = 3$$

The extinction ratio is the ratio of the maximum transmission of the polarizer to the minimum transmission. Examples of the extinction ratio for a few types of polarizers are listed in Table 5.1.

Table 5.1 Extinction Ratio of Polarizers

Material	Extinction ratio
Polaroid	10^{-3}–10^{-5}
Tourmaline	10^{-2}
Wire grid	10^{-2}–10^{-3}
Calcite	10^{-7}
4-plate reflection	
Plate index ($n = 2.46$)	10^{-2}
Plate index ($n = 4.0$)	10^{-4}

5.2 Polarization Control

To build a polarizer to control the polarization and its measurement, a number of different physical mechanisms have been employed.

5.2.1 Absorption

5.2.1.1 Crystals

There are naturally occurring crystalline materials that can perform the role of a polarizer. Crystals that exhibit a complex index of refraction that varies with the crystallographic direction are said to exhibit *dichroism*.[3] In this type of crystalline material, one component of polarization is absorbed to a greater extent than its transverse partner. Naturally occurring

---

[3] The minerals that exhibit three colors are said to have pleochroism.

Figure 5.3 Dichroic crystal zoisite showing color variation with direction.
(By Rob Lavinsky, iRocks.com – CC-BY-SA-3.0, CC BY-SA 3.0, https://
commons.wikimedia.org/w/index.php?curid=10467426)

dichroism was first observed by Jean Baptiste Biot (1774–1862) and high-quality dichroic crystals are shown in Figure 5.3.

In addition to the zoisite shown in Figure 5.3, examples of naturally occurring dichroic minerals are tourmaline and herapathite (iodosulfate of quinine) in the visible and pyrolytic graphite in the infrared. By present-day standards, dichroic materials make poor polarizers but some dichroic crystals have been studied as possible optical data storage materials and as laser hosts. These crystals also make interesting gemstones because the color of the crystals vary with the viewing angle under polarized light. The dichroic crystal shown in Figure 5.3 is an example of this type of expensive gemstone.

5.2.1.2 Wire grids

We can construct a dichroic polarizer by stringing a set of wires as shown in Figure 5.4. Heinrich Hertz invented this type of polarizer in 1888 to study the properties of microwaves. The wire grid polarizer in Figure 5.4 is designed for use at millimeter wavelengths.

One might think, when first reviewing Figure 5.5, that it has been drawn improperly. If this were a mechanical wave the transmitted displacement vector should be parallel and not perpendicular to the openings between the wires. From a mechanical viewpoint we would naturally assume that the arrow indicating the displacement should physically fit through the grid. Our intuition using a mechanical model is wrong; the grid attenuates E_V because E_V induces currents in the wires. The resistance of the wires dissipates the energy provided by the wave and most of what is not absorbed is reflected. The grid looks like a metal mirror to E_V and like a dielectric to E_H. E_H does not couple strongly to the metal structure because the available interaction length is small. The grid performs badly in the role of polarizer if:

Figure 5.4 Corner of actual wire grid polarizer.

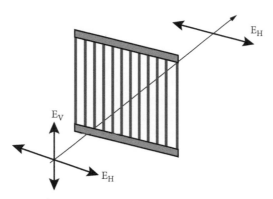

Figure 5.5 Operation of a wire grid polarizer.

- The wires are thick ($\mathbf{E}_H$ is then also attenuated);
- The spaces between the wires are large ($\mathbf{E}_v$ can leak through the grid because coupling to the wire is a function of the distance from the wire); or
- The wires have a high resistivity.

The wire grid polarizer has been shown to operate over a large range of angles of incidence, from 0° to 45°, and over a large wavelength region. The long wavelength limit of a polarizer is set by absorption in the supporting substrate if it is not free standing, as is the one in Figure 5.4. The short wavelength limit is established by the grid spacing and is equal to twice the grid spacing. Figure 5.6 displays the performance of a commercial wire grid polarizer.

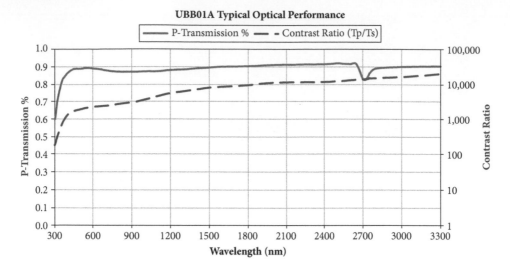

Figure 5.6 Performance of a Moxtek broadband polarizer.
Used with permission from Moxtek.

5.2.1.3 Polaroid sheet

"H-sheet" Polaroid is the most popular and familiar dichroic polarizer. Edwin Herbert Land invented this type of polarizer in 1928 while he was a 19-year-old undergraduate student. Stretching a sheet of polyvinyl plastic aligns the long-chain polymeric molecules of polyvinyl alcohol that make up the sheet. After stretching, the polyvinyl plastic is cemented to a rigid base such as cellulose acetate to prevent relaxation of the molecules to their original alignment. Iodine is then diffused into the polyvinyl alcohol layer where the iodine atoms affix themselves along the polymer molecule. The electrons from the iodine are free to move along the polymer chain, creating a conducting molecular wire. The Polaroid film performs as a molecular analog of the wire grid. The first version of Polaroid film (called J-sheet) was constructed using sheets of nitrocellulose with microscopic herapathite crystals (iodoquinine sulfate) imbedded in the plastic. The J-sheet was later replaced by the more achromatic H-sheet (polyvinyl alcohol with iodine). There is a third type called K-sheet with improved environmental properties.

A number of different grades of H-sheet are manufactured as a commodity item by a variety of companies. The visible transmission of polarized light for several grades of H-sheet, with transmission between 38 and 22%, is shown in Figures 5.7 and 5.8. To generate the curves, the transmission of linearly polarized light as a function of wavelength is measured. Figure 5.7 shows the spectral transmission properties of H-sheet Polaroid when the axis of the polarizer is parallel to the incident wave's direction of polarization. Figure 5.8 is similar to Figure 5.7 with the exception that the axis of the polarizer[4] has been rotated and is transverse to the polarization of the incident wave.

---

[4] The polarization axis defines the direction, parallel to the electric field of an incident linearly polarized wave, when the transmission through the polarizer is a maximum.

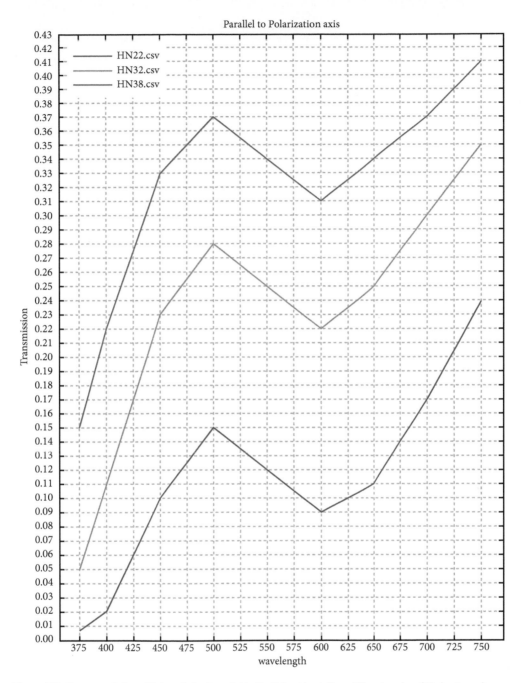

Figure 5.7 The transmission of light polarized parallel to the Polaroid axis. Two different grades of H-sheet are shown. These curves should be compared to those shown in Figure 5.8.

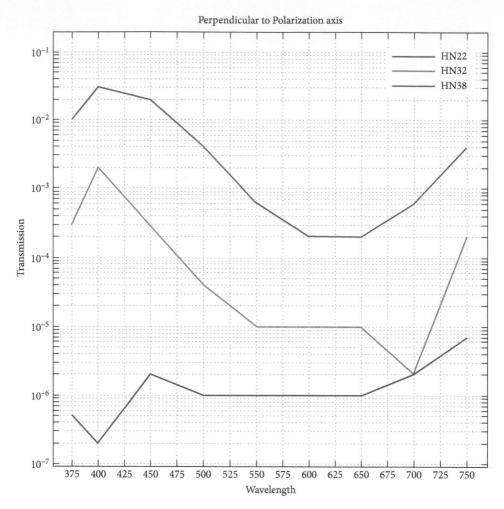

Figure 5.8 The transmission of H-sheet Polaroid with the light's plane of polarization normal to the axis of the Polaroid film.

5.2.2 **Reflection**

Rather than using absorption, we can use reflection to remove an undesired polarization component from the optical field of view. We saw earlier, using the boundary conditions associated with Maxwell's equations, that the reflection amplitude given by the Fresnel equations depends upon the polarization of the incident light wave. We also demonstrated that we could use reflection to turn unpolarized light into at least partially polarized light. Polarizers can be designed using this polarization dependent reflectivity but they have two drawbacks:

- The light beam is deflected through a large angle as multiple reflections occur and the polarizer becomes quite long when optics are added to redirect the deflected beam back parallel to the optical axis.

- The fraction of polarized light produced is often lower than desired and there is too much leakage of the unwanted polarization.

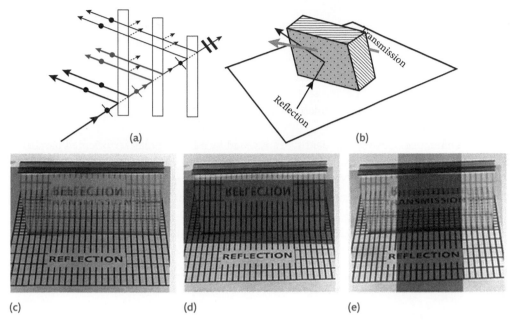

Figure 5.9 Polarization of light by a stack of glass plates. (a) This is a schematic display of the operation of a stack of glass plates that use Brewster's Angle to polarize light. To demonstration polarization by a stack of glass plates, the experimental arrangement shown in (b) was assembled. The word TRANSMISSION is positioned behind a stack of very thin glass plates and the word REFLECTION is placed in the foreground of the stack. In this configuration we see a mirror image of the word REFLECTION from the front surfaces of the glass plates and the word TRANSMISSION through the stack of plates. A photograph of the actual experiment is shown in (c). The illumination is a large fluorescent light. The images of the two images can be seen but because the light is unpolarized, there is no indication of the polarization properties of the two images. An H-sheet Polaroid polarizer with its axis of polarization oriented along its long dimension is placed in the foreground in (d) and (e). In (d) the polarizer has its axis oriented normal to the plane of incidence (as defined in Chapter 3). The word TRANSMISSION cannot be seen but the reflected word REFLECTION is visible. In (e) the polarizer is oriented with its axis in the plane of incidence. Here the word TRANSMISSION can be seen through the plates but the reflection of the word REFLECTION is not seen.

The utilization of the reflection polarizer, with the exception of lasers with Brewster angle windows, is limited to the infrared and ultraviolet where it is often the only option.

To demonstration polarization by reflection we use a stack of glass microscope cover slips, the experimental arrangement shown in Figure 5.9b. The word TRANSMISSION is positioned behind a stack of very thin glass plates (a stack of about 100 microscope cover slips) and the word REFLECTION is placed in the foreground of the stack. In this configuration we see a mirror image of the word REFLECTION from the front surfaces of the glass plates and the word TRANSMISSION through the stack of plates. A photograph of the actual experiment is shown in Figure 5.9c. The two images can be seen but there is no indication of the polarization properties of the two images. An H-sheet Polaroid polarizer with its axis of polarization oriented along its long dimension is placed in the foreground in Figures 5.9d and 5.9e. In Figure 5.9d the polarizer has its axis oriented normal to the plane of incidence. The word TRANSMISSION cannot be seen but the word REFLECTION, inverted by reflection, is visible. In Figure 5.9e the polarizer is oriented with its axis in the plane of incidence. Here the word TRANSMISSION can be seen through the plates but the reflected image of the word REFLECTION is invisible.

In Chapter 2 we introduced Stokes parameters as a way to analyze the effects of optical components on the polarization of a light wave. We are now in a position to calculate the Stokes parameters of an unpolarized light wave that is reflected an angle of 70° from a glass plate of index 1.5. (See the example box in section 3.4.1 for details on how to calculate the reflectivity of unpolarized light from a dielectric surface.) To find the amount of light reflected, we use from Chapter 2 the definitions of the Stokes parameters:

$s_0 \Rightarrow$ Total flux density

$s_1 \Rightarrow$ Difference between flux density transmitted by a linear polarizer oriented parallel to the x-axis and one oriented parallel to the y-axis. The x- and y-axes are usually selected to be parallel to the horizontal and vertical directions in the laboratory.

$s_2 \Rightarrow$ Difference between flux density transmitted by a linear polarizer oriented at 45° to the x-axis and one oriented at 135°.

$s_3 \Rightarrow$ Difference between flux density transmitted by a right-circular polarizer and that by a left-circular polarizer.

$$s_0 = E_P^2 + E_N^2 = (R_P + R_N) I$$

$$s_1 = (R_P - R_N) I$$

$$R_P = \frac{\tan^2 (\theta_i - \theta_t)}{\tan^2 (\theta_i + \theta_t)} = 0.04 \qquad R_N = \frac{\sin^2 (\theta_i - \theta_t)}{\sin^2 (\theta_i + \theta_t)} = 0.30$$

$$s_0 = (0.04 + 0.30) I = 0.34 I$$

$$s_1 = (0.04 - 0.30) I = -0.26 I$$

To find s_2 we must project E_P and E_N on lines at 45° and 135° to the horizontal coordinate in the laboratory.

Use the geometry in Figure 5.10 to get the project of E_P and E_N on a coordinate system rotated 45° to the original lab coordinate system.

The projections of the reflected components of the electric field are

$$E_{xP} = 0.2\sqrt{I_0} \cos 45° = 0.14\sqrt{I_0}$$

$$E_{xN} = 0.52\sqrt{I_0} \cos 45° = 0.39\sqrt{I_0}.$$

The sum of the two components is

$$E_x = (0.37 + 0.14) \sqrt{I_0} = 0.51\sqrt{I_0}$$

$$E_x^2 = 0.26 I.$$

The y-component of the field is

$$E_y = (0.37 - 0.14) \sqrt{I_0} = 0.23\sqrt{I_0}$$

$$E_y^2 = 0.05 I_0.$$

The Stokes component here is

$$s_2 = \left(E_x^2 - E_y^2 \right) = (0.28 - 0.06) I = 0.22 I.$$

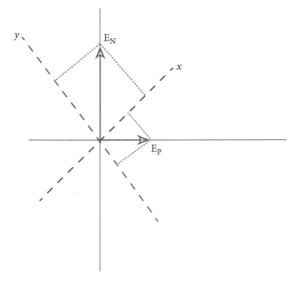

Figure 5.10 Projection of electric field onto lines at 45° and 135° from the horizontal to allow calculation of the Stokes variable s_2.

There is no phase shift occurring for the reflection so s_3 is zero. The Stokes vector is thus

$$\begin{pmatrix} 1 \\ -0.26 \\ 0.22 \\ 0 \end{pmatrix}$$

With this vector we can calculate the degree of polarization (Eq. (2.42)) of the partially polarized wave produced by reflection, $V = 0.693$.

5.2.3 Interference

Interference polarizers are used in laser systems when the incident radiation will strike the dielectric layer at an angle. A popular polarizing beam splitter design called the MacNeille cube design ensures that the internal angles match Brewster's angle so that the p-polarized wave is transmitted without reflection. The radiation is incident on the dielectric layer at an angle of 45°. The wave transmitted by the layer is polarized with its electric vector in the plane of incidence (p-type) while the reflected wave has its electric vector normal to the plane of incidence (s-type), as shown in Figure 5.11. Because interference is the basis of this optical device it can have strong wavelength dependence, as shown in Figure 5.12, unless a complex broadband dielectric stack design or a wire grid is used in place of the dielectric layer. The advantage of this type of polarizer is that it can have a large aperture, allowing it to be used in high-energy laser systems.

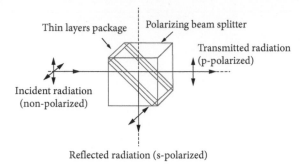

Figure 5.11 Polarizing beam splitter. s is the polarization perpendicular to the plane of incidence and p is the polarization in the plane of incidence.

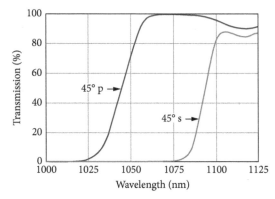

Figure 5.12 Transmission as a function of wavelength for the two polarizations in a polarizing beam splitter.

5.2.4 Birefringence

Another way to throw an undesired polarization component out of the optical field of view is through the use of birefringence. Before we describe the polarizer designs we need to discuss the concept of birefringence.

5.2.4.1 Birefringent properties

In crystals with cubic symmetry, the propagation properties of light are isotropic, there is a single index of refraction, and the crystal behaves optically like a noncrystalline material such as glass (at least to first order). All other classes of crystals are optically *anisotropic*. The index of refraction (and naturally the dielectric constant) depends on the direction of propagation of the light, relative to the crystal axes. Tensor calculus must be used to mathematically describe propagation of light in these materials. The polarization direction of the electromagnetic wave in the medium is denoted by the direction of the displacement, **D**. Dividing the wave into two component waves with orthogonal polarizations allows the calculation of the propagation behavior. The two waves, with orthogonal polarizations, $\mathbf{D}_1$ and $\mathbf{D}_2$, will have different propagation vectors and will exhibit double refraction.

Table 5.2 Uniaxial Crystals

Mineral	Index of refraction (Na-D)	
	n_o	n_e
Ice (H_2O)	1.309	1.313
Sellaite (MgF_2)	1.378	1.390
Quartz	1.5443	1.5534
Sodium Nitrate	1.5854	1.3369
Wurtzite (ZnS)	2.356	2.378
Rutile (TiO_2)	2.616	2.903
Cinnabar (HgS)	2.854	3.201
Calcite ($CaO \cdot Co_2$)	1.658	1.486
Tourmaline	1.669	1.638
Sapphire	1.7681	1.7599

There are special directions in an anisotropic crystal for which both polarizations have the same propagation velocity. The special direction is called an *optical axis*.[5] If two such directions exist, the crystals are called *biaxial* and if there is only one direction for which the two orthogonal polarizations have the same propagation velocity, the crystal is called *uniaxial*.

The physical origins of birefringence can be understood by assuming that the forces that bind the crystal are different along different coordinate directions. You can view this classically as atoms connected via springs with different spring constants. The springs represent the binding forces that hold the crystal together. If the binding forces are equal in two coordinate directions, then there will be two indices of refraction, i.e., a uniaxial crystal (Table 5.2). If there are three unequal binding strengths, you have a biaxial crystal with three indices of refraction (Table 5.3).

The uniaxial crystal calcite, shown in Figure 5.13, provides a useful model for understanding birefringence. A crystal of calcite is made up of a three-dimensional array of unit cells. The structure of a unit cell is shown in Figure 5.13. Each carbon atom in the crystal lies at the center of an imaginary equilateral triangle with oxygen atoms at each corner of the triangle. The direction, normal to the planes containing these triangles, is the optical axis of calcite. From the arrangement of atoms it seems reasonable to assume that the binding energy in the planes containing the oxygen atoms is different than the binding energy normal to these planes. A classical view with springs representing the bond gives us a harmonic oscillator with a spring constant proportional to the bonding energy. When light is incident on a crystal of calcite

---

[5] The optical axis of the crystal is a direction in the crystal and not a line even though we call it an axis and indicate it with a line.

Table 5.3 Biaxial Crystals

Mineral	Index of refraction		
	n_α	n_β	n_γ
Tridymite	1.469	1.47	1.473
Mica(muscovite)	1.5601	1.5936	1.5977
Turquoise	1.61	1.62	1.65
Topaz	1.619	1.62	1.627
Sulfur	1.95	2.043	2.240
Borax	1.447	1.47	1.472
Lanthanite	1.52	1.587	1.613
Stibnite (Sb_2S_3)	3.194	4.303	4.46

parallel to the optical axis, shown in Figure 5.13, the electric displacement, **D**, of the light wave lies in the same plane as the oxygen atoms. The electric field of the light wave causes an amplitude displacement in the imaginary harmonic oscillator. The electric polarization induced by the electric field is the same for all polarization directions in this case and the resulting index of refraction would be independent of polarization direction. When a light wave is incident on the calcite crystal normal to the optical axis, the electric polarization induced by the light wave's electric field is different, resulting in a second index of refraction.

A light wave propagating in an arbitrary direction through this crystal can be decomposed into two waves, with orthogonal polarizations, propagating at different velocities. Any wave whose electric displacement lies in the plane of the oxygen atoms has a propagation velocity that is isotropic and the wave is called an *ordinary wave* with an index of refraction n_o. Any other wave, with its electric displacement at an angle to the plane containing the oxygen atoms, has an anisotropic propagation velocity. This wave is called an *extraordinary wave* (see Figure 5.13) and its index of refraction is n_e.

The electrons can move quite easily in the planes containing the oxygen atoms and for our classical model, the spring constant of the carbon–oxygen bond is small relative to that observed in the direction normal to the planes. In the direction normal to the oxygen planes, the electrons experience strong binding forces. For this reason, the propagation velocity in the direction of the optical axis (where the index is n_o) is less than the propagation velocity normal to that direction (where the index is n_e). The index of refraction associated with the ordinary wave is thus greater than the index associated with the extraordinary wave, $n_o > n_e$. Crystals with this property are called *negative* uniaxial crystals:

$$(n_e - n_o) < 0$$

The polarization direction for the ordinary wave of a negative uniaxial crystal is called the *slow axis* (the light's propagation velocity is less than all other directions). The polarization direction (the direction of **D**), for the wave that has the higher propagation velocity, is called the *fast axis*.

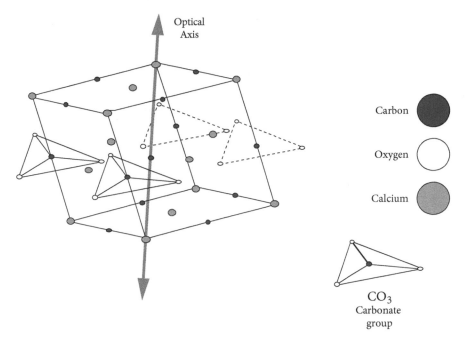

Figure 5.13 The unit cell of calcite. Only four of the carbonate groups that make up the unit cell are shown in their entirety; the other groups are represented by only the carbon atoms at their center.

For a *positive* uniaxial crystal ($n_e - n_o > 0$) the definitions are reversed and the fast wave is the ordinary wave.

5.2.4.2 Propagation of light in a birefringent crystal

In general, when an unpolarized beam of light is incident normal to the surface of a plane parallel plate of an uniaxial crystal, two beams will emerge from the back side of the crystal, as shown in Figure 5.14. This is called double refraction. Snell's law works for the ordinary wave using n_o but not the extraordinary wave. If the angle the incident wave makes with the optical axis of a uniaxial crystal is θ, then the refracted angle would be calculated using an effective index given by

$$n = 1/\sqrt{\cos^2 \theta/n_o^2 + \sin^2\theta/n_e^2}.$$
(5.3)

Obtaining this relationship and others concerning anisotropic materials requires the use of tensors along with Maxwell's equations to generate what are called Fresnel equations of the wave normal. We will not cover Fresnel equations in this text [1].

The ordinary wave, labeled O in Figure 5.14, is polarized with its displacement vector, **D**, normal to the plane containing the optical axis. The extraordinary wave labeled E is polarized with its displacement vector in the plane containing the optical axis. We will limit our discussion of birefringent polarizers to the use of uniaxial crystals.

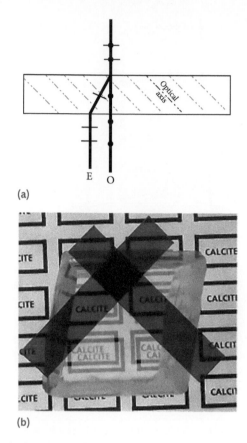

(a)

(b)

Figure 5.14 (a) Uniaxial crystal calcite showing the failure of the extraordinary wave to obey Snell's law. The lines in the crystal indicate the direction called the optical axis. (b) The birefringence shown in (a) results in a double image. The polarizations of the two images are orthogonal. The polarizers are oriented orthogonally and have maximum transmission along their long axis.

5.2.5 **Birefringent polarizer types**

5.2.5.1 Calcite

The simplest design for a birefringent polarizer uses the naturally occurring birefringent crystal, calcite, along with a set of "stops" positioned to remove either of the two beams; see Figure 5.15. Because the separation of the two waves is small (only 6.2° in calcite), this technique can only be used with very narrow beams. Because of the limitation on optical beam size, other designs have been developed with the objective to increase the product of the polarizer's angular aperture (semi-field angle) and circular cross-sectional area of the polarizer.

5.2.5.2 Wollaston and Rochon polarizers

The Wollaston and Rochon polarizers increase the separation of the two beams, over that obtainable with a single crystal, by using two single crystals, usually quartz, cut and polished into two prisms; see Figure 5.16. The Rochon polarizer has an entrance prism with its optical axis oriented perpendicular to the incident face of the polarizer, in the plane of Figure 5.16,

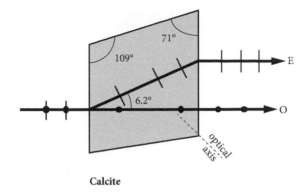

Calcite

Figure 5.15 Separation of the extraordinary (the upper beam) and the ordinary (lower beam) waves in calcite. The dotted line labeled optical axis indicates the direction of the optical axis. The rhombohedron shown is the naturally occurring form of a single crystal of calcite.

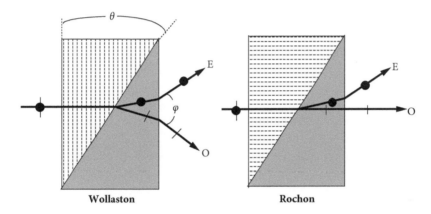

Wollaston   **Rochon**

Figure 5.16 Two polarizers designed to increase the separation of the extraordinary and ordinary waves. The entrance prism on both polarizers has its ordinary wave polarized out of the paper. The second prism has its ordinary wave polarized in the plane of the paper. The lines and dots used to shade the two prisms indicate the direction of the optical axis.

and a second, exit prism, glued to the first with its optical axis perpendicular to the first prism's axis, perpendicular to the plane of Figure 5.16.

The light wave whose polarization is perpendicular to the second prism's axis, and in the plane of Figure 5.16, sees no index discontinuity at the interface between the two prisms because the wave also has its polarization perpendicular to the optical axis of the first prism. For this reason, the ordinary wave is not deviated by the Rochon polarizer. The extraordinary wave of the second prism, with its polarization parallel to the optical axis, however, is an ordinary wave in the first prism. It sees a discontinuous change in the index of $(n_e - n_o)$ as it crosses the boundary between the two prisms. The prism angle θ, and the difference between the extraordinary and ordinary indices, determine the angle φ between the E and O beams. A rule

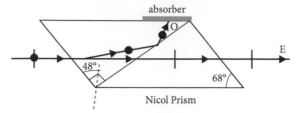

Figure 5.17 The geometry of a Nicol prism polarizer. The direction of the optical axis is indicated by the dotted line.

of thumb number for the deviation is 10°. The Wollaston polarizer produces twice the deviation as the Rochon polarizer (about 20°) by causing both the ordinary and extraordinary waves to see the same discontinuous change in the index. This occurs because the extraordinary wave and ordinary wave interchange roles in the two prisms. The advantage of the Rochon prism is that it keeps the ordinary wave on the optical axis.

5.2.5.3 Nicol prism

Another method of separating the two orthogonal polarizations in a birefringent crystal is to couple total internal reflection and birefringence to deflect one beam toward an absorber; see Figure 5.17. The Nicol prism, named after *William Nicol* (1768–1857), was the first of this type of polarizer. A calcite crystal is cut into two prisms with the optical axis neither parallel nor perpendicular to the entrance face (see the dotted line in Figure 5.17) and polished to obtain the angles shown in Figure 5.17. The cement used to assemble the two prisms is selected so that the ordinary wave experiences total internal reflection at the glue joint, but the extraordinary wave passes through the polarizer. Historically, the cement called Canada balsam (resin of the balsam fir tree with an index of 1.55) has been used but it limits uv operation. (The Nicol polarizer will operate only with rays arriving within a cone whose angle, called the acceptance angle, is 24°. The acceptance angle of this type of polarizer is determined by the difference between the critical angle of the ordinary and extraordinary rays.) The cement restricts the wavelength range of the polarizer so designs were developed that eliminated the cement to allow operation at shorter wavelengths.

5.2.5.4 Glan–Foucault, Glan–Thompson, Glan–Taylor

The Glan type of polarizer has the optical axis in the plane of the entrance face. It has replaced the Nicol polarizer in modern applications, and uses total reflection as in the Nicol to extract the unwanted polarization. Its advantages are that it replaces glue with air at the interface of total internal reflection, and the incident and exit faces are perpendicular to the light wave. The Glan–Foucault is a more modern polarizer design; see Figure 5.18. The acceptance angle of this polarizer is only 7° but it will operate over a wavelength range from 0.23 to 5.0 μm. The Glan–Taylor prism, a popular design, has an advantage over the Glan–Foucault because the angle of incidence at the gap is reasonably close to Brewster's angle, giving it better transmission. If the two prisms of the Glan–Foucault polarizer are cemented together, the name is changed to Glan–Thompson polarizer. This design modification increases the field of view to 30° but it cannot be used in the uv.

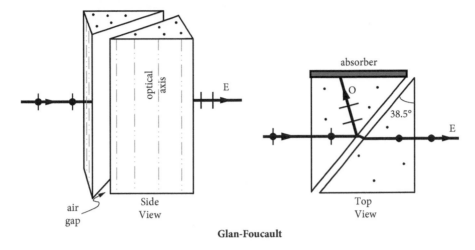

optical axis

air gap

Side View

Top View

absorber

O

38.5°

E

Glan-Foucault

Figure 5.18 Two views of a calcite Glan–Foucault polarizer showing its operation. The parallel lines indicate the direction of the optical axis oriented parallel to the prism sides.

5.2.5.5 Prism design

One probably wonders how the design of one of these prism polarizers is carried out. As an example, we will calculate the angle one must cut the prism used in a Glan–Foucault polarizer if the material used is calcite. The index of refraction for the birefringent calcite is

$$n_o = 1.658 \qquad n_e = 1.486$$

We have two angles to worry about: one associated with the ordinary beam, θ_o, which we want to be totally reflected (Eq. (3.43)) at the air gap and one with the extraordinary beam, θ_e, which we want to be less than the critical angle so that it is transmitted:

$$\sin \theta_{co} = 1/n_o = 1/1.658 \Rightarrow \theta_{co} = 37.1°$$
$$\sin \theta_{ce} = 1/n_e = 1/1.486 \Rightarrow \theta_{co} = 42.3°$$

We set the angle to exceed the critical angle for the ordinary wave; in Figure 5.18 it is set at 38.5°. If we try to use quartz as a polarizer the two angles are 40.36° and 40.07°. The required tolerance in prism angle makes quartz impossible to use.

One of the problems in constructing polarizers, using birefringent material, is finding or growing a birefringent crystal of a size and optical quality to be useful. The Feussner polarizer design uses isotropic prisms with a thin plate of birefringent material sandwiched between the prisms; see Figure 5.19. The two prisms in the Feussner polarizer are made of a glass with an index equal to the higher index of the birefringent material. If the birefringent material is calcite, then the ordinary ray is transmitted. The optical axis can be oriented normal to the slab (Feussner polarizer) or parallel to the entrance face (Bertrand type).

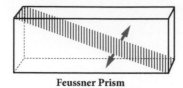

Figure 5.19 A Feussner polarizer. The optical axis of the birefringent material is normal to the slab of material as shown by the arrows [2].

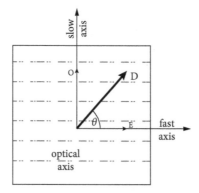

Figure 5.20 Retarder plate operation. A linearly polarized wave is incident on a negative uniaxial crystal cut so that its face is parallel to the optical axis.

5.2.6 Retarder

Retarders (also known as *waveplates*) modify the polarization of an incident wave by changing the relative phase of two orthogonally polarized components of the incident wave represented by Eq. (3.2),

$$\tilde{\mathbf{E}} = E_{0x}e^{i(\omega t - kz + \phi_1)}\hat{\mathbf{i}} + E_{0y}e^{i(\omega t - kz + \phi_2)}\hat{\mathbf{j}}$$

but they do not attenuate or cause a deviation of the beam. To discuss the operation of a retarder, we will rewrite this equation by replacing the unit vectors along the *x*- and *y*-axes by unit vectors identified with the fast $\hat{\mathbf{f}}$ and slow $\hat{\mathbf{s}}$ axes of the retarder plate:

$$\tilde{\mathbf{E}} = \left[E_{0x}\hat{\mathbf{f}} + E_{0y}e^{i(\phi_2 - \phi_1)}\hat{\mathbf{s}} \right] e^{i(\omega t - kz)}. \tag{5.4}$$

Either birefringence or reflection can be used to produce the desired phase change. We will limit our discussion to birefringent retarders.

To understand how birefringent retarders operate, and to learn how they are constructed, we will examine a uniaxial, birefringent plate, cut with the optic axis parallel to the plate's face. In Figure 5.20, we show a linearly polarized wave incident on such a crystal with the electric displacement, **D**, making an angle θ with the optical axis. We describe the polarized wave using Eq. (5.4). The different propagation velocities cause the component of polarization parallel to the slow axis (vertical in Figure 5.20) to lag the component parallel to the fast axis.

The birefringent plate modifies the polarization of the incident light because of the different propagation velocities, characterized by the two indices, n_0 and n_e. If the retarder plate has a thickness, d, the optical path difference between the two orthogonally polarized waves is

$$N\lambda = \pm (n_e - n_o)\, d. \tag{5.5}$$

N is called the *retardation* and is expressed in fractions of a wavelength. For example, $N = \frac{1}{4}$ corresponds to a quarter-wave retardation.

The phase difference, between the two orthogonally polarized waves, generated by propagating through the retarder plate, is simply 2π times the retardation,

$$\delta = \phi_2 - \phi_1 = 2\pi N = \pm \frac{2\pi d (n_e - n_o)}{\lambda}. \tag{5.6}$$

This phase is used to determine the polarization, after propagating through the retarder plate.

If the displacement, **D**, of the linearly polarized light is incident on the birefringent plate, with the direction of polarization parallel to either the fast or slow axis of the plate, the plate will not modify the polarization. In all other orientations, the plate will modify the polarization of the incident light because of the phase difference between the polarization components parallel to the fast and slow axes.

5.2.6.1 Quarter-wave plate

If the plate's thickness results in Eq. (5.6) having a value of $\delta = \pi/2$, a phase retardation, equivalent to a shift of one quarter of a sinusoidal wave, is generated between the fast and slow waves and the plate is called a *quarter-wave plate, $N = \frac{1}{4}$*. Mica or quartz is normally used for retarders because it is easy to produce plates of the desired thickness. In quartz, a plate 13.7 µm thick produces a quarter-wave phase retardation while in mica, the quarter-wave thickness is 22.3 µm at $\lambda = 500$ nm.

Mica is actually biaxial with the angle between the two optic axes varying between $0°$ and $42°$. The mica selected for retarders has a biaxial angle near $0°$. Mica can be cleaved into very thin plates, by virtue of its crystal structure and this ease overrides any complication introduced by the small departure from uniaxial behavior. The cleavage planes that allow mica to be formed into thin plates are not parallel to either optic axis. For the crystal orientation with the faces of the retarder determined by the cleavage planes, the ordinary and extraordinary indices are $n_0 = n_\gamma$ and $n_e = n_\beta$; see Table 5.3.

If linearly polarized light is incident on a quarter-wave plate such that $\theta = 45°$ in Figure 5.20, then the linear polarization is converted to circularly polarized light:

$$\tilde{\mathbf{E}} = e^{i(\omega t - kz)} \left[E_{0x}\hat{\mathbf{f}} + E_{0y}e^{i\pi/2}\hat{\mathbf{s}} \right] = E_0 \left[\hat{\mathbf{f}} + i\hat{\mathbf{s}} \right] e^{i(\omega t - kz)}$$

$$e^{i/\pi/2} = i$$

If circularly polarized light is incident on a quarter-wave plate, it will be converted to linearly polarized light.

By combining a polarizing beam splitter and a quarter-wave plate you can create an optical isolator that prevents back reflections into the preceding optical system.

Wave plates with a single birefringent material are called multiple order wave plates. They generate the desired retardation plus an integer. One advantage is that they can be designed to function at several wavelengths. The shortcoming is that they are more strongly affected by temperature changes or actual operating wavelength. By stacking several multiple order wave plates you can create a zero-order wave plate which will have an effective retardation equal to the difference between the two multiple order retardations. The zero-order wave plate is more stable than the multiple order device.

5.2.6.2 Half-wave plate

If the plate's thickness results in Eq. (5.6) having a value of $\delta = \pi$, then the plate is called a *half-wave plate*. The half-wave plate changes the orientation of the polarization but does not change its form. For example, suppose in Figure 5.20 that **D** represents the polarization of an incident plane wave,

$$
\begin{aligned}
\tilde{\mathbf{E}} &= D\left[\cos\theta\hat{\mathbf{f}} + \sin\theta e^{i\delta}\hat{\mathbf{s}}\right] e^{i(\omega t - kz)}\\
\delta &= \pi \Rightarrow e^{i\pi} = -1\\
\tilde{\mathbf{E}} &= D\left[\cos\theta\hat{\mathbf{f}} - \sin\theta\hat{\mathbf{s}}\right] e^{i(\omega t - kz)}
\end{aligned},
$$

the emerging wave will have **D** oriented at an angle of $-\theta$ with respect to the fast axis in Figure 5.20. The half-wave plate will have rotated the polarization through an angle of 2θ.

5.2.6.3 Compensator

Because the desired phase retardation produced by a wave plate occurs only at the design wavelength, the plate will be too thick for shorter wavelengths and too thin for longer wavelengths. It is useful to have a component with continuously variable phase retardation for use over a range of wavelengths. Such a device, called a *compensator*, would allow the measurement of an unknown retardation by comparison with the known retardation of the compensator.

One way to produce continuously varying phase retardation is to construct a wedge using a birefringent material—the upper wedge shown in Figure 5.21 would meet the requirement. The simple wedge is not too useful because it cannot produce zero phase retardation. If two wedges are used, with the slow axis of one normal to the slow axis of the other, then by moving one wedge relative to the other, a continuous variation of phase retardation can be produced. This design is called a *Babinet compensator*.

The *Soleil–Babinet compensator*, shown in Figure 5.21, produces the same phase retardation as the Babinet compensator but over a wide aperture. The aperture of the compensator is determined by the small wedge and plate at the bottom of Figure 5.21. When the small wedge is positioned as shown at the left of Figure 5.21, the two overlapping wedges, producing zero retardation, exactly cancel the phase shift produced by the rectangular plate. When the small

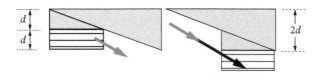

Figure 5.21 Soleit–Babinet compensator. On the left is the configuration of the compensator that produces no retardation. On the right the maximum retardation is produced. The compensator is designed to produce a uniform retardation across the aperture of the device, which is the width of the small sliding component.

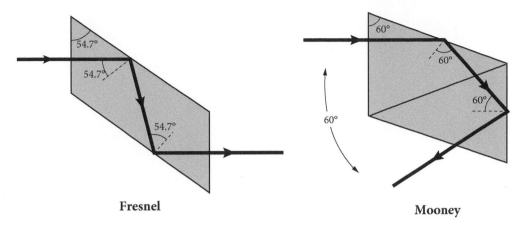

Fresnel

Mooney

Figure 5.22 Two types of rhombs.

wedge is positioned to the extreme right of the large wedge, as shown on the right of Figure 5.21, the two wedges produce a phase retardation equal to twice the negative of the phase retardation of the rectangular plate. This demonstrates that this design can produce as much as two waves of retardation.

5.2.6.4 Rhomb

The rhomb-type retarder is a popular substitute when the strong wavelength dependence of the birefringent, phase-shifter design is a problem. The rhombs derive their phase retardation from the phase change introduced by total internal reflection (see Eqs. (3.46) and (3.47), and Figure 3.22). Figure 5.22 shows two typical examples from a large number of rhomb designs. All of the designs require that the rhomb be constructed of homogeneous, isotropic material.

Optical Activity

François Arago observed, in 1811, that the direction of polarization was rotated when plane polarized light was propagated through quartz, parallel to the optical axis of quartz. *Jean Baptiste Biot*, in 1815, extended Arago's observations and discovered that the continuous rotation of the plane of polarization occurred in gases and liquids, as well as crystalline material. Materials that rotate the plane of polarization are said to exhibit *chirality*.[6] An object is chiral if it cannot be superimposed on its mirror image. For example, your hands and feet each exhibit chirality. In chemistry, anything that can exhibit a "handedness"[7] is said to be *chiral*. This characteristic is applied to atoms that differ only in the direction they rotate polarized light. Such atoms are said to differ in their *chirality*.

Biot observed that if one looked at the light source, through optically active materials, the materials could be grouped into two classes: those that rotated the polarization to the left—*levorotatory*—and those that rotated the polarization to the right—*dextrorotatory*.

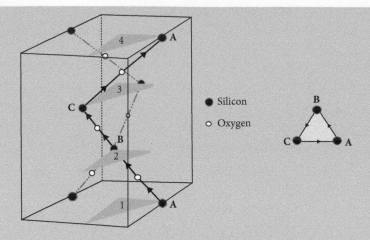

Figure 5.23 Crystal structure of quartz.

Fresnel developed the first theoretical description of optical activity by demonstrating, experimentally, that linearly polarized light can be decomposed into right- and left-circularly polarized waves. Optical rotation was then explained by assuming that the two circularly polarized components, of a linearly polarized wave, propagated at different velocities. Fresnel suggested that the observed optical activity might be due to "a helical arrangement of molecules of the medium."

Quartz appears as both right-handed and left handed. The right-hand form of quartz is shown in Figure 5.23. By following along the Si-O bond from silicon atom to silicon atom (A⇒B⇒C⇒A from bottom to top in Figure 5.23), a right-handed spiral is traced out—this path is shown in Figure 5.23 by the dark arrows.

If the silicon atoms are projected onto a plane, they form the vertices of an equilateral triangle, shown on the right of Figure 5.23. To aid in visualizing the helical path, a shaded version of the triangle is shown at each of four planes in the quartz unit cell.

Two classical models have been developed to explain optical activity. One model is based upon the interaction of the electromagnetic wave with charges confined to a helical path. This model approximates the behavior of charge carriers in quartz and in many optically active molecules.

A second classical model treats the interaction of two coupled, harmonic oscillators that lie in separate planes, with an electromagnetic wave. The two-oscillator model approximates the behavior of tetrahedral-bonded atoms, such as the molecule shown in Figure 5.24. The coupled oscillator model can be shown to be equivalent to the helical model but we will accept that fact without a mathematical proof.

The unique property possessed by optically active molecules is that a mirror image of the molecule exists. For example, the molecule and its mirror image, shown in Figure 5.24, are the simplest optically active molecule pair. They will lose their separate identities and their optical activities if we arbitrarily replace any two of the bonded atoms by an identical pair. For example, if the fluorine atom is replaced by a hydrogen, then the mirror image

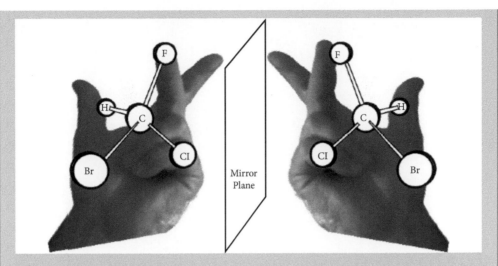

Figure 5.24 The molecular structure on the left is the simplest optically active molecule. The mirror image of this tetrahedral molecule, shown on the right, is a unique molecule that cannot be reproduced by rotation of the molecule on the left. If twins replace any two of the atoms on this tetrahedral-bonded molecule, then the molecule is no longer dissymmetric and it is not optically active.

does not exist because we can create an identical image by a simple rotation about the normal to the plane of the figure.

To follow a plane polarized wave, as it propagates through an optically active material, we assume that at the surface of the optically active material, $z = 0$, we have a vertically polarized beam. Using the Jones vector notation, Eq. (2.43),

$$\frac{E_0}{2} \begin{pmatrix} 0 \\ \sqrt{2} \end{pmatrix} e^{i\omega t},$$

we decompose the plane polarized wave, in the optically active material, into right- and left-circularly polarized components; see Table 2.5. The wave traveling in the positive z-direction is

$$\frac{E_0}{2\sqrt{2}} \begin{pmatrix} -i \\ 1 \end{pmatrix} e^{i(\omega t - kz)} + \frac{E_0}{2\sqrt{2}} \begin{pmatrix} i \\ 1 \end{pmatrix} e^{i(\omega t - kz)}. \tag{5.7}$$

After traveling through a thickness, d, of a crystal of index of refraction, n, the phase of light, relative to the initial wave, at the surface $z = 0$, is

$$nkd = \frac{2\pi}{\lambda} nd.$$

We assume that the right-circularly polarized wave sees an index of refraction n_+ and the left-circularly polarized waves sees n_-. Upon leaving the optically active material, the wave is

$$\frac{E_0}{2\sqrt{2}} \begin{pmatrix} -i \\ 1 \end{pmatrix} e^{i(\omega t - n_+ kz)} + \frac{E_0}{2\sqrt{2}} \begin{pmatrix} i \\ 1 \end{pmatrix} e^{i(\omega t - n_- kz)}. \tag{5.8}$$

This can be rewritten

$$\frac{E_0}{2\sqrt{2}} e^{i\left[\omega t-(n_+ +n_-)^{kd/2}\right]} \left\{ \begin{pmatrix} 1 \\ -i \end{pmatrix} e^{i(n_- -n_+)^{kd/2}} + \frac{E_0}{2\sqrt{2}} \begin{pmatrix} 1 \\ i \end{pmatrix} e^{i(n_+ -n_-)^{kd/2}} \right\}. \tag{5.9}$$

To interpret the physical meaning of Eq. (5.9), we simplify the equation by defining

$$\psi = \frac{kd}{2}(n_+ + n_-) \qquad \theta = \frac{kd}{2}(n_- - n_+).$$

With these definitions we can write Eq. (5.9) as

$$\frac{E_0}{\sqrt{2}} e^{i(\omega t-\psi)} \left\{ \begin{pmatrix} 0 \\ 1 \end{pmatrix} \left(\frac{e^{i\theta} + e^{-i\theta}}{2}\right) + \begin{pmatrix} 1 \\ 0 \end{pmatrix} \left(\frac{e^{i\theta} - e^{-i\theta}}{2i}\right) \right\},$$

$$\frac{E}{\sqrt{2}} e^{i(\omega t-\psi)} \left\{ \begin{pmatrix} 0 \\ 1 \end{pmatrix} \cos\theta + \begin{pmatrix} 1 \\ 0 \end{pmatrix} \sin\theta \right\}. \tag{5.10}$$

Equation (5.10) describes a plane polarized wave with the polarization direction at an angle

$$\theta = \pi d/\lambda (n_- - n_+). \tag{5.11}$$

The incident, plane polarized wave has had its plane of polarization rotated through an angle, θ, proportional to the length of path traversed in the substance. If $(n_- > n_+)$, then the right-circularly polarized wave travels faster than the left and the material is called dextrorotatory. For the opposite case, the material is called levorotatory.

For most crystals, $(n_- - n_+)$ is on the order of 10^{-4} or less; see Table 5.4. For quartz, the index difference is 7.1×10^{-5} at 589.3 nm; thus, a 1-mm-thick plate of quartz cut

Table 5.4 Specific Rotation of Solids

Material	Rotation, β (deg/mm)
HgS	+32.5
Lead Hyposulfate	+5.5
Potassium Hyposulfate	+8.4
Quartz	+21.684
NaBrO$_3$	+2.8
NaClO$_3$	+3.13

normal to the optical axis rotates a linearly polarized wave through an angle of 21.7°. The angle through which the plane of polarization is rotated for a 1-mm-thick crystal is defined as the *specific rotation (or rotatory power*, Figure 5.25), $\beta = \dfrac{\theta}{1\ mm}$. We see from Eq. (5.11) that the specific rotation of any material has an explicit wavelength dependence of $1/\lambda$ (Figure 5.25). For quartz, the index difference, $(n_- - n_+)$, also varies inversely with wavelength, so that the specific rotation of quartz should vary as $1/\lambda^2$.

As Biot discovered, fluids also exhibit optical activity (Table 5.5). A fluid is isotropic in the sense that the molecules are randomly oriented throughout. The observed optical activity is attributed to molecules that have no center of symmetry or plane of symmetry. The molecules are randomly oriented in a liquid or gas but still, they produce optical activity because the preferred direction of rotation associated with each molecule is independent of orientation. To aid in visualizing this property, think of a box of screws; they remain right-handed, no matter what orientation they have in the box.

If an optically active compound is dissolved in an inactive solvent, then the rotation is nearly proportional to the amount of the compound dissolved; for this reason, the specific rotation is defined for 1 gram of solute in one cubic centimeter of solution. The specific rotation of a liquid is smaller than for a crystal so it is usually defined for a 10-cm path length rather than a 1-mm path length (Table 5.6). If the concentration of the solution is

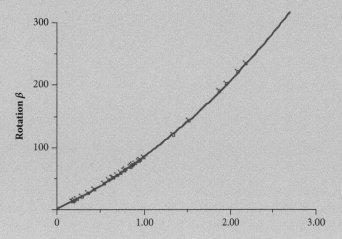

Figure 5.25 Frequency dependence of the specific rotation of quartz. Note that it has a quadratic wavelength dependence.

Table 5.5 Specific Rotation of Liquids

Material	Rotation, β (deg/dm)
Amyl alcohol	−5.7
Camphor	+70.33
Menthol	−49.7
Nicotine	−162
Turpentine	−37

Table 5.6 Specific Rotation of Solutions

Substance Solvent Rotation, β	Solvent	deg/dm
Camphor	alcohol	+54.4
Camphor	benzene	+56
Camphor	ether	+57
Galactose (milk sugar)	water	+83.9
d-glucose (dextrose)	water	+52.5
l-glucose	water	−51.4
Lactose	water	+52.4
Maltose	water	+138.48
Nicotine	water	−77
Nicotine	benzene	−164
Sucrose	water	+66.412

m grams/cc, then the rotation of the plane of polarization, θ, produced by a solution of an optically active material is

$$\theta = \beta \frac{md}{10}.$$

For a pure fluid, m is replaced by the density of the fluid.

One of the major industrial applications of optical activity is in the measurement of the concentration of inverted sugar. The inverted sugar is a result of converting table sugar (sucrose) by hydrolysis to glucose and fructose. Sucrose exhibits right-handed rotation and the mixture of processed sugar is inverted when the optical rotation becomes left-handed. Tracking of this process is of importance as a sweetener in canning, confectionery, and candy production, as a raw material in beer and wine production, as a preservative in some food processing, and in quality control in the ethanol industry. Other industrial applications are found in medicine, cosmetics, and the pharmaceutical industries. As an example of the impact of chirality in drug effects consider Penicillamine: the left-hand version is used to treat primary chronic arthritis while the right-hand version has no therapeutic effect but is instead highly toxic.

[6] Optical activity is sometimes known as *natural rotation* to distinguish it from magnetic rotation of the plane of polarization discovered by Michael Faraday and called the *Faraday effect*. The Faraday rotation occurs when the light propagates parallel to an applied magnetic field.

[7] We are referring to the exhibiting of right-handed or left-handed appearance.

5.3 Problem Set 5

(1) Calculate the maximum index of refraction that the cement in a Nicol prism can have.

(2) Discuss the reasoning behind the statement that wire polarizers should have a low resistivity for optimum performance

(3) Prove that when $P = 50\%$ $I_\parallel = 3I_\perp$. Derive the ratio when $P = 60\%$.

(4) Two polarizing sheets have their polarizing directions parallel, so that the intensity of the transmitted light is maximized (with value I_{max}). Through what angle, θ, must either sheet be turned, if the transmitted light intensity is to drop to ½?

(5) If crossed polarizers block all light, why does putting a third polarizer at 45° between them result in some transmission of light?

(6) What should be the orientation of a polarizer used in your sunglasses?

(7)* What is the effect of stacking two linear half-wave retarders in series with the fast axis of one at 0° and the second at $\theta = 45°$? Use the Mueller calculus to obtain your result.

(8)* A 1-mm-diameter beam of light of wavelength $\lambda = 589.3$ nm, travels through a 1-cm-thick calcite crystal, at an angle of 32° with respect to the optical axis. How far apart will the extraordinary and ordinary beams be upon exit from the crystal?

(9) What is the minimum thickness of a quartz quarter-wave plate to be used at 589 nm?

(10) A solution of camphor in alcohol, in a 20-cm-long cylinder, rotates linearly polarized light 33°. The specific rotation of camphor is 54°. What is the concentration of camphor in the solution?

(11) Describe the polarization of a wave, initially unpolarized, after passing through a linear polarizer and then a half-wave plate. What happens if we reverse the order of the two components?

(12) Light is incident on eight glass plates of index $n = 1.75$, at Brewster's angle. What is the percent polarization of the transmitted light?

(13) Why do most polarizers using birefringent materials contain two prisms?

(14) A 10-cm-long cylinder contains a 20% solution of maltose. If the observed rotation of linearly polarized light, after passing through the solution, is 28.8°, what is the specific rotation of maltose?

REFERENCES

1. Guenther, B. D., *Modern Optics*, 2nd edn. Oxford: Oxford Univeristy Press, 2015.

2. Sleeman, P. R., Dr. Feussner's New Polarizing Prism, in *Scientific American Supplement* (1884).

6 Geometrical Optics

6.1 Early Optics

In the development of the theory of optics, we have not taken a historical path but rather assumed our current understanding of light as electromagnetic waves was the proper view. The historical approach to optical theory prior to the twentieth century was that optics was confined to phenomena within the visible light range and was divided into two topical areas: geometric optics and physical optics. Physical optics assumes light is a wave, the approach to optics selected in this book.

Geometric optics is based on the laws of reflection and refraction, topics we have already covered. There are no new physics theories but instead geometrical optics is concerned with the applications of these two simple laws to the design and construction of optical elements such as mirrors, prisms, lenses, and fibers using the geometrical concept of light rays and ignoring the wave nature of light. The mathematical treatment is simple and the equations are not too complicated. For those reasons, geometrical optics is a staple of many introductory physics courses at the high school and college levels. The actual design and optimization of the elements making up an optical system is not as simple as presented in those courses. Before about 1960 the optical designer spent a large amount of time tracing out the path of optical rays using tables of logarithms and trigonometric functions to perform accurate calculations. The optical designer spent much of his training devoted to learning numerous techniques that would reduce the number of ray traces that were needed and techniques for organizing the data in an easily accessible way. After 1960, the development of the computer changed the approach to lens design, leading to the realization of useful designs based on computerized ray tracing and system optimization.

The ray concept resulted in a simple model for the propagation of light but provided no knowledge of the actual physical properties of light. In the publication *Catoptrics* (280 BC), Euclid is attributed with having recognized that light travels in straight lines and that its propagation can be predicted using simple geometrical constructions of lines of zero thickness, called rays, to represent the paths taken by light. Euclid formulated the law of reflection in *Catoptrics,* and Hero of Alexandria (born in AD 20) wrote a similarly titled manuscript that included a derivation of the law of reflection by assuming the principle of the shortest path.[1] We identify the principle of the shortest path as the "Fermat principle," in recognition of Fermat's

---

[1] An interesting aside that gives some indication of the creativity of Hero is the fact that he invented and built the first vending machine, which provided the customer with holy water for a coin.

Modern Optics Simplified. B. D. Guenther. © B. D. Guenther 2020.
Published in 2020 by Oxford University Press. DOI: 10.1093/oso/9780198842859.001.0001

mathematics contribution of the principle of least action, which has been applied not only to optics but to quantum theory.

The lack of a mathematical formalism for trigonometry held back the understanding of refraction because the concept of an angle measurement was not understood until between the ninth and tenth centuries. Instead, measures were based on the length of a cord.[2] Many of the current trigonometric laws and formulas were expressed geometrically in terms of cords, and tables of cords were prepared by a number of early Greek, Indian, and Islamic mathematicians. Tables of sines and cosines did not appear until the ninth century. The lack of a mathematical foundation delayed formulation of an accurate law of refraction until independently derived by Willebrod Snellius and René Descartes.

Actually, the Persian scientist Ibn Sahl was first to describe the law of refraction in 984 but failed to receive any credit because portions of his manuscript were accidently separated and deposited in libraries located in Damascus and Tehran. It was not until 1990 that Roshdi Rashed discovered the separated segments and recognized that they should be reassembled into a single manuscript. Figure 6.1 shows a copy of one of the figures from the manuscript giving a geometric presentation of the law of refraction. This redrawing of the important segment of the original figure applies a modern interpretation of the cord-based presentation of Sahl.

It is interesting to note that Ibn Sahl was able to develop an accurate design of an aspheric lens based on his newly discovered law of refraction. A drawing of the design is contained in the same manuscript as the presentation of the law of refraction [1] in Figure 6.1.

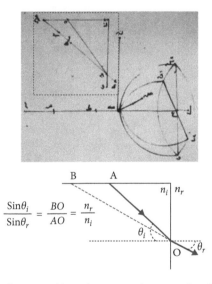

Figure 6.1 Ibn Sahl diagrams for refraction and for a plano-convex lens at top from "On the Burning Instruments" written in AD 984. The lower drawing is a modern representation of Snell's law plot highlighted with a dotted box in the upper figure.

By Ibn Sahl (Abu Saʿd al-ʿAlaʾ ibn Sahl) (c.940–1000) (Milli MS 867, fol. 7r, Milli Library, Tehran) [Public domain], via Wikimedia Commons.

---

[2] A cord is the length of a line that subtends an arc of a circle with a specified radius of r. The definition using modern trigonometry is cord $(\theta) = r\,[2\sin(\theta/2)]$.

6.2 **Lens Design and Matrix Algebra**

The design of optical components was at first guided by the use of geometrical constructions, such as shown in Figure 6.1. It was pointed out by Thomas Young in 1801 that geometrical methods did not have the necessary accuracy, so he developed algebraic equations to go with the geometrical constructions. We will derive the equations needed for ray tracing with a goal of using the predictions of ray tracing to determine the optical system's performance before it is constructed.

Ideally the optical system would collect light from an object point and focus it to an image point. In practice the theory we use is an oversimplification and the actual image point is blurred. The lens designer creates a merit function and attempts to minimize the merit function by adjusting the lens material, shape, and position. Before the computer, this was a labor-intensive process and every attempt was made to minimize the number of rays that needed to be traced [2]. Now modern lens design programs generate an optical system with desired performance criteria by varying selected lens parameters and tracing many rays [3]. A typical algorithm would numerically calculate the derivatives of the desired merit function to find the direction to move in design space to realize a smaller merit function. This process finds a local minimum; however, to find a global minimum, experience and "tricks" have to be called upon.

Typical steps taken by the designer are the following:

1. *Pre-design*: A pencil and paper calculation of the optical system needed to accomplish the desired performance is made. The reader will learn in this chapter some of the tools used to perform this step.

2. *Initial conditions*: The initial conditions are provided as the starting point for the optimization software and are generated by reviewing patents, books, past design projects, and the results of the pre-design. Changes in the initial conditions can result in major changes in the optimization results.

3. *Starting performance*: Calculations of the baseline performance of the system based on the initial conditions selected are made.

4. *Optimization*: The design software will search for an optimum design guided by the injection of changes to parameters by the lens designer.

5. *Final*: It is often necessary to return to step 2 and run through the design steps again to realize a usable design that can be fabricated.

Figure 6.2 shows the output of a typical lens design program [4], showing a two-lens optical system optimized by the software to produce minimum spherical aberration. A simple two-element, air-spaced lens system as shown in the figure has nine variables (four radii of curvature, two thicknesses, one airspace thickness, and two glass types), requiring finding a minimum in 9-dimensional space. Cost and environmental performance also play a role in the optimization but we will ignore these issues.

The ray-tracing formulas used by the lens designers are developed by applying the laws of refraction and ray propagation, as obtained from Fermat's principle, at progressive surfaces[3]

---

[3] A surface is anywhere along the optical axis that you want to know the ray height and can include the object, image, refracting surfaces, and dummy surfaces.

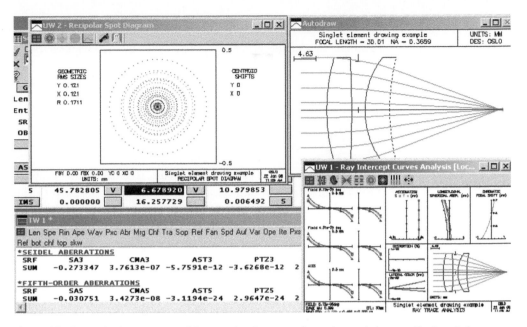

Figure 6.2 The graphical presentation of the optical performance of a two-lens optical system. The lens design program, Oslo [4], was asked to minimize both spherical aberration and coma.

through the optical system. We will place the resulting equations in matrix form to produce a generalized formula for the propagation of light through any optical system, from an optical fiber to a complex zoom lens. Lens designers use the ray-tracing formula[4] rather than matrices and keep everything in order by using a spreadsheet [5, 6].

In this section, we will, in a stepwise fashion, follow a ray through a thick lens with two refractive surfaces. These steps will provide us a set of simple equations that can be repeatedly applied to trace a ray through a complicated optical system. The equations derived make it possible to code a general computer ray-tracing program[5] and allow steps to be taken to optimize the optical design. For tracing a few rays through an optical system, a spreadsheet can be used, which is available for download from some websites [5, 7]. To trace just a few rays it is simpler to use a matrix formulation called the ray transfer matrix analysis, because all of the lenses in the system can be combined into a single transfer matrix between the object and image. It is possible to formulate an exact matrix involving sine, cosine, and tangent functions [8]; however, we will not do so. Our goal is to create a sketch of the general process used within a modern optical design program but not to study the procedure [3, 9]. We will restrict our discussion to the following approximation that allows the sine and tangent functions to be replaced by their angle in radians.

---

[4] We need to alert the reader to the fact that real lens designers represent the vertical dimension by y and the angle by u and call the ray trace an *ynu*-ray trace.

[5] The term, ray tracing, is also associated with computer rendering of images. We will not cover this topic. Note that the major effort of image rendering is to develop methods that trade optical accuracy for rendering speed through the use of illusions and shortcuts.

6.2.1 Paraxial approximation

The sine, cosine, and tangent of a small angle, γ, can be approximated by the Taylor series

$$\sin \gamma = \gamma - \frac{\gamma^3}{3!} + \frac{\gamma^5}{5!} - \frac{\gamma^7}{7!} + \cdots,$$

$$\cos \gamma = 1 - \frac{\gamma^2}{2!} + \frac{\gamma^4}{4!} - \frac{\gamma^6}{6!} + \cdots,$$

$$\tan \gamma = \gamma + \frac{\gamma^3}{3} + \frac{2\gamma^5}{15} + \frac{17\gamma^7}{315} + \cdots.$$

The paraxial approximation assumes that γ, measured in radians, is small so that only the first term of each expansion is needed. The numerical error is shown in Table 6.1. This approximation removes the need for the use of trigonometric functions as well as removing the need for second-order differential equations in following the propagation of light waves. The paraxial approximation is also called the first-order theory; retention of higher order terms lead to the third-, fifth-, seventh-, etc., order theories and their accuracies are graphically indicated in Figure 6.3.

6.2.2 Sign convention

Before we begin, we will establish a sign convention[6] based on the Cartesian coordinate system shown in Figure 6.4. This system is the foundation of analytic geometry and is used throughout the sciences and engineering.[7] The optical surface under consideration is assumed to intersect the optical axis at the point V, called the vertex of the optical surface and the origin of the x,z-coordinate system is positioned at this vertex. The z-axis lies along the axis of rotational symmetry of the optical components, called the optical axis.[8] The plane defined by x and z is called the *tangential plane*. The plane perpendicular, the y,z-plane, is called the *sagittal plane*. All distances along z to the right of V and all distances in the x-direction, above the optical

Table 6.1 Errors Introduced by Paraxial Approximation

γ in degrees	γ in radians	$\tan \gamma$	$\sin \gamma$	$\cos \gamma$
1	0.0175	0.0175	0.0175	0.9998
5	0.0873	0.0875	0.0872	0.9962
10	0.175	0.176	0.174	0.9848
15	0.262	0.268	0.259	0.9659
20	0.349	0.364	0.342	0.9397

[6] The sign convention used here is tied to the Cartesian coordinate system. It is easy to remember and works well with multiple lenses. There is no "best" sign convention and a large number of other conventions are used in geometrical optics. Moving from one convention to another can be a source of confusion so no other convention will be described here.

[7] The Cartesian sign convention was selected due to its almost universal use in mathematics.

[8] Most lens designers work in the y,z-plane but the notation has no effect on the fundamentals discussed here.

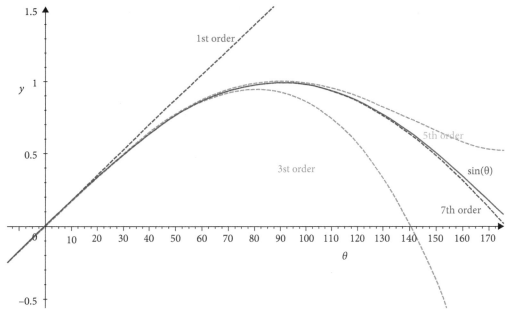

Figure 6.3 Graphic representation of approximation of sine function compared to actual function.

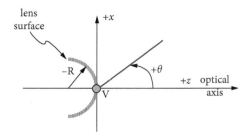

Figure 6.4 Coordinate system that establishes the sign convention used for geometrical optics. Distances to the right of V and above the z-axis are positive. Angles measured counterclockwise from the positive z-axis are positive. The optical surface under consideration intersects the z-axis at the position V and is rotationally symmetric about the z-axis. The z-axis is defined as the optical axis. The radius of curvature of the optical surface is defined as being negative if the center of curvature is to the left of V.

axis (the z-axis), are positive, as are angles measured counterclockwise from the positive z-axis. (Normally the sign of the incident, reflected, and refracted angles need not be specified but assume the sign dictated by the ray-tracing procedure.) We assume that the optical surfaces will all have spherical curvature with a radius of curvature, R, measured from the surface to the center of curvature and thus is negative if the center of curvature of the surface is to the left of the vertex, V. The index of refraction to the left of the optical surface has a lower subscript than the index to the right of the surface; i.e., the indices are ordered n_1, n_2, n_3, etc. Subscripts are assigned to parameters to correspond with the index subscript.

6.2.3 Ray tracing

We will analyze a general lens by considering each surface independently. This will require that a new coordinate system be constructed at each surface. Figure 6.5 contains the first optical surface to be considered, with the previously defined coordinate system constructed at its vertex, labeled V in Figure 6.5. Light will travel from the object point O on the left of V in Figure 6.5 toward the right, hitting the image point I after refraction at the optical surface a distance x_1 above the optical axis, labeled P. Any ray, such as the one just defined, that has as its origin the object–optical axis intersection and is confined to the plane containing the optical axis and the object is called a meridional (or axial) ray. Note because of the sign convention we have selected, the distance traveled from O to P is a positive number as would be the case in any Cartesian coordinate system.

The first surface has a positive radius of curvature R_1, centered on point C of Figure 6.5, to the right of V. The radius of curvature through the point where the light ray intersects the optical surface makes an angle ϕ with the optical axis

$$\sin \phi = \frac{x_1}{R_1}.$$

We will use the paraxial approximation so that $\sin \phi \approx \phi$. In the paraxial theory, all rays leaving the object point arrive at the image point as required by Fermat's principle; however, in the geometrical theory this is not the case for a spherical refracting surface. The failure of all the light rays from an object to converge to a single image point, after passing through an optical system, is called optical aberration. To treat the aberrations mathematically, the paraxial theory is extended to third, fifth, seventh, etc., order by including additional terms in the expansion of the sine function [8, 10].[9] We will do a detailed ray trace in Chapter 7 to demonstrate the occurrence of aberration.

Snell's law is applied using the angle, shown in Figure 6.5, between the incident ray and the normal to the curved surface, the line "radius," at the point, P, where the ray intersects the optical surface. Using the first-order approximation

$$n_1 \theta_i = n_2 \theta_t.$$

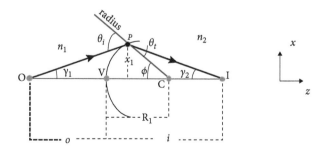

Figure 6.5 Geometry used for the development of the lens design equations.

---

[9] A treatment of aberrations consistent with the present analysis can be found in the book by Nussbaum where computer routines are provided for calculating the various aberrations of an optical system.

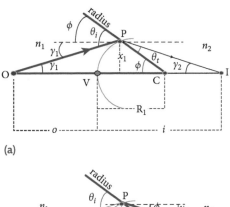

(a)

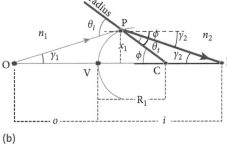

(b)

Figure 6.6 (a) Triangle constructed using the incident ray from the object O, the optical axis from point O to point I, and the radius of curvature of the optical surface. (b) Triangle constructed using the transmitted ray traveling toward the image point, I, the optical axis, and the radius of curvature of the optical surface.

the angle of incidence, θ_i and the angle of transmission, θ_t, can be written in terms of known angles using triangles from Figure 6.5 that are redrawn in Figure 6.6 with their boundaries highlighted for clarity.

6.2.3.1 Refraction matrix

Figure 6.6a is the triangle used to find the angle of incidence at point P, where ϕ is measured from the horizontal to the radius (clockwise) so the measured value is negative:

$$\theta_i = \gamma_1 - \phi.$$

Figure 6.6b is the triangle used to find the angle of transmission at point P, where again ϕ is measured from the horizontal to the radius,

$$\theta_t = -\phi - (-\gamma_2)$$

$$\theta_t = \gamma_2 - \phi.$$

Note from Figure 6.5 that γ_1 is the angle the ray leaving the object O makes with the optical axis and γ_2, the negative in this sign convention, is the angle the ray, approaching the image, I, makes with the optical axis. From Figure 6.5, we have

$$\sin \phi = -\frac{x_1}{R_1}.$$

The paraxial approximation allows us to write

$$\phi \approx -\frac{x_1}{R_1}.$$

Snell's law can now be written[10]

$$n_1\left(\gamma_1 + \frac{x_1}{R_1}\right) = n_2\left(\gamma_2 + \frac{x_1}{R_1}\right).$$

Solving this equation for γ_2 yields

$$\gamma_2 = \frac{n_1}{n_2}\gamma_1 + \frac{x_1(n_1 - n_2)}{n_2 R_1}. \tag{6.1a}$$

The quantity

$$\Phi_1 = -\frac{n_1 - n_2}{R_1} \tag{6.1b}$$

is called the *power* of the surface, with units of diopters when R_1 is in meters.[11] If the lens under consideration is located in air, then $n_1 = 1$ and the power for this surface is a positive number when the radius is positive,

$$\Phi_1 = \frac{n_2 - 1}{R_1} > 0.$$

There is no change in the ray's height above the optical axis, x_1, across the surface so that

$$x_1 = x_1'', \tag{6.2}$$

where the double prime denotes that we are now across the boundary into the material with index n_2. Each ray is characterized by its slope and its distance from the optical axis. These can be combined into a vector allowing Eqs. (6.1a) and (6.2) to be written in matrix form

$$\begin{pmatrix} 1 & 0 \\ \dfrac{n_1 - n_2}{n_2 R_1} & \dfrac{n_1}{n_2} \end{pmatrix} \begin{pmatrix} x_1 \\ \gamma_1 \end{pmatrix} = \begin{pmatrix} x_1'' \\ \gamma_2 \end{pmatrix}.$$

A matrix that describes a ray crossing an output reference plane, given the ray is leaving an input reference plane, is called the *ABCD matrix* or the *ray transfer matrix*. The reference plane we are using here is located at the lens vertices V_1. The matrix, multiplying the ray vector, describes refraction across the interface and is called the *refraction matrix*:

$$R_1 = \begin{pmatrix} 1 & 0 \\ \dfrac{n_1 - n_2}{n_2 R_1} & \dfrac{n_1}{n_2} \end{pmatrix}. \tag{6.3}$$

This matrix defines the ray's new direction after it has undergone refraction at surface 1.

---

[10] Many lens design texts combine the index of the medium with the ray slope, i.e., $\gamma_2 n_2$ or in the customary notation $u n_2$. This is only important if you explore other optical design books.

[11] The introduction of power and the replacement of R by the curvature $C = 1/R$ were used before computers to reduce the calculation burden made by requiring division, and custom has retained this notation.

6.2.3.2 Transfer matrix

We now must consider the translation from the front surface at V_1 to the back surface at V_2 within the lens medium of index n_2. We will denote in Figure 6.7 the entry point into the lens as point A and the exit point from the lens as point B. We will retain the same coordinate system, with its origin at V_1 for this part of the ray-tracing problem but the problem does not change if you move to the new origin.

Referring to Figure 6.7, we can write

$$x_2'' - x_1'' = \Delta x = d \tan \gamma_2 \approx d\gamma_2,$$

where d is a positive number because the light ray is traveling in the positive z-direction, from left to right $(d = (z_2 - z_1))$. As can be seen in Figure 6.7, because of rectilinear propagation, the angle, γ_2, does not change from A to B but the height above the optical axis changes from $x_1'' \to x_2''$:

$$x_2'' = x_1'' + d\gamma_2 \tag{6.4a}$$
$$\gamma_2 = \gamma_2 \tag{6.4b}$$

These equations can be rewritten in matrix form

$$\begin{pmatrix} 1 & d \\ 0 & 1 \end{pmatrix} \begin{pmatrix} x_1'' \\ \gamma_2 \end{pmatrix} = \begin{pmatrix} x_2'' \\ \gamma_2 \end{pmatrix}.$$

The matrix that describes the propagation from one surface (here V_1) to another (V_2) is called the *transfer matrix*,

$$\mathbf{T} = \begin{pmatrix} 1 & d \\ 0 & 1 \end{pmatrix}. \tag{6.5}$$

To follow the ray across the surface at B, we must construct another refraction matrix using the same procedure as used to construct Eq. (6.3). The origin of the coordinate system is moved to the vertex of the second surface, V_2, in Figure 6.7, and the following matrix equation is constructed:

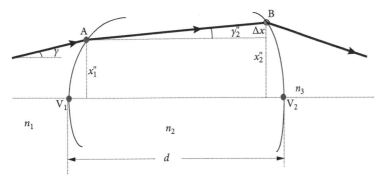

Figure 6.7 Geometric construction for propagation of a light ray in the lens. The first surface has a radius of curvature of R_1 and the second surface has a radius of curvature of R_2, a negative quantity once we move the coordinate system to V_2.

$$\begin{pmatrix} 1 & 0 \\ \dfrac{n_2 - n_3}{n_3 R_2} & \dfrac{n_2}{n_3} \end{pmatrix} \begin{pmatrix} x_2'' \\ \gamma_2 \end{pmatrix} = \begin{pmatrix} x_2' \\ \gamma_3 \end{pmatrix}.$$

The refraction matrix for the second surface is, therefore,

$$\mathbf{R}_2 = \begin{pmatrix} 1 & 0 \\ \dfrac{n_2 - n_3}{n_3 R_2} & \dfrac{n_2}{n_3} \end{pmatrix}. \tag{6.6}$$

The radius of curvature of this second surface is negative according to the sign convention we have established. If the lens is in air, then $n_3 = 1$ and the power for this surface is

$$\Phi_2 = \frac{n_3 - n_2}{R_2} = \frac{1 - n_2}{-R_2} > 0.$$

The power of the lens is the sum of the powers of the two surfaces: $\Phi = \Phi_1 + \Phi_2$.

6.2.3.3 Ray transfer (ABCD) matrix

The product of the three matrices Eqs. (6.3), (6.5), and (6.6) is called the ray transfer matrix for the lens

$$\mathbf{S} = \mathbf{R}_2 \mathbf{T} \mathbf{R}_1,$$

or the ABCD Matrix,

$$\begin{pmatrix} 1 & 0 \\ \dfrac{n_2 - n_3}{n_3 R_2} & \dfrac{n_2}{n_3} \end{pmatrix} \begin{pmatrix} 1 & d \\ 0 & 1 \end{pmatrix} \begin{pmatrix} 1 & 0 \\ \dfrac{n_1 - n_2}{n_2 R_1} & \dfrac{n_1}{n_2} \end{pmatrix} = \begin{pmatrix} A & B \\ C & D \end{pmatrix}. \tag{6.7}$$

The elements of the system matrix are calculated by simply multiplying row times column (RC) to generate each of the elements for the product[12]:

$$\underbrace{\begin{pmatrix} 1 & 0 \\ \dfrac{n_2 - n_3}{n_3 R_2} & \dfrac{n_2}{n_3} \end{pmatrix} \cdot \underbrace{\begin{pmatrix} 1 & d \\ 0 & 1 \end{pmatrix} \cdot \begin{pmatrix} 1 & 0 \\ \dfrac{n_1 - n_2}{n_2 R_1} & \dfrac{n_1}{n_2} \end{pmatrix}}_{\text{1st matrix multiply}}}_{\text{2nd matrix multiply}}$$

$$A = 1 + \frac{d(n_1 - n_2)}{n_2 R_1} \qquad B = \frac{dn_1}{n_2}$$

$$C = \frac{1}{n_3}\left[\frac{n_2 - n_3}{R_2} + \frac{n_1 - n_2}{R_1} + \frac{d(n_1 - n_2)(n_2 - n_3)}{n_2 R_1 R_2} \right] \qquad D = \frac{n_1}{n_3}\left[\frac{1 + d(n_2 - n_3)}{n_2 R_2} \right].$$

These terms are the elements of an ABCD matrix for a thick lens. If the optical element is thin enough to allow us to assume $d \to 0$ and if we assume the optical system will be used in air, we have $n_1 = n_3 = 1$. The system matrix Eq. (6.7) for the thin lens becomes

$$\mathbf{S} = \begin{pmatrix} 1 & 0 \\ [n_2 - 1]\left[\dfrac{1}{R_2} - \dfrac{1}{R_1} \right] & 1 \end{pmatrix}. \tag{6.8}$$

---

[12] $\begin{pmatrix} a & b \\ c & d \end{pmatrix} \circ \begin{pmatrix} A & B \\ C & D \end{pmatrix} = \begin{pmatrix} aA + bC & aB + bD \\ cA + dC & cB + dD \end{pmatrix}$

Note that the C element of the system matrix Eq. (6.8) (i.e., S_{21}) is simply the power of the lens:

$$\Phi = \Phi_1 + \Phi_2.$$

The ratio of the height of the image to that of the object is called the *lateral magnification* and is defined as $\beta = x_3/x_1$. Because the elements of the ABCD matrix are $A = 1$ and $B = 0$ when we assume $d = 0$, the values of the x-coordinate at the vertices V_1 and V_2 are unchanged by the system matrix (Eq. (6.5) has no effect on the height of the light ray) and $\beta = 1$:

$$\left(\begin{matrix} 1 & 0 \\ [n_2 - 1]\left[\dfrac{1}{R_2} - \dfrac{1}{R_1} \right] & 1 \end{matrix} \right) \begin{pmatrix} x_1 \\ \gamma_1 \end{pmatrix} = \begin{pmatrix} x_3 \\ \gamma_3 \end{pmatrix} \Rightarrow x_1 = x_3.$$

This is equivalent to saying that the front and back vertices of the thin lens define planes of unit magnification. The planes of unit magnification are important reference planes for both thick ($d \neq 0$) and thin ($d = 0$) lenses. We will introduce a technique for locating the planes of unit magnification for a thick lens later.

6.2.3.4 Object image matrix

The matrix for the general imaging condition can be shown to be equivalent to the more familiar thin lens equation by calculating the imaging of point O at point I in Figure 6.8. To obtain the equivalence we first apply the transfer matrix from the object point, O, in Figure 6.8, to the lens, then apply the lens' system matrix, and finally use a second transfer matrix to reach the image point, I:

$$\begin{pmatrix} 1 & s' \\ 0 & 1 \end{pmatrix} \left(\begin{matrix} 1 & 0 \\ (n_2 - 1)\left(\dfrac{1}{R_2} - \dfrac{1}{R_1} \right) & 1 \end{matrix} \right) \begin{pmatrix} 1 & -s \\ 0 & 1 \end{pmatrix} \begin{pmatrix} x_1 \\ \gamma_1 \end{pmatrix} = \begin{pmatrix} x_3 \\ \gamma_3 \end{pmatrix}.$$

A negative s is used in the propagation matrix for the region to the left of the lens because of the sign convention; s is a negative quantity, measured in the negative direction, and the negative

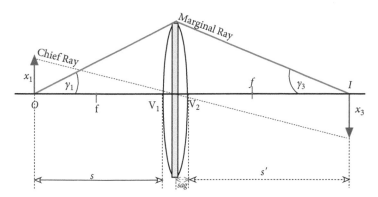

Figure 6.8 Thin lens imaging system. We have chosen the marginal ray, that is, the ray that makes the maximal angle with the optical axis as it leaves the base of the object and passes through the maximum opening—called the aperture; see discussion following Figure 6.12. The chief ray is that ray that leaves the top of the object and crosses the optical axis at the center of the aperture. The sag (abbreviation of sagittal) is the thickness of material required to accommodate a lens surface of given radius of curvature with a given aperture. It is an indication of the error introduced by assuming a thin lens.

sign ensures that the propagation matrix contains a positive quantity for propagation from left to right.[13]

$$\begin{pmatrix} 1 + s'(n_2 - 1)\left(\dfrac{1}{R_2} - \dfrac{1}{R_1}\right) & s' - s - ss'(n_2 - 1)\left(\dfrac{1}{R_2} - \dfrac{1}{R_1}\right) \\[2ex] (n_2 - 1)\left(\dfrac{1}{R_2} - \dfrac{1}{R_1}\right) & 1 - s(n_2 - 1)\left(\dfrac{1}{R_2} - \dfrac{1}{R_1}\right) \end{pmatrix} \begin{pmatrix} x_1 \\ \gamma_1 \end{pmatrix} = \begin{pmatrix} x_3 \\ \gamma_3 \end{pmatrix}$$

This is called the object–image matrix, **O**.

6.2.3.5 Thin lens equation

If x_3 is an image of x_1, then all rays, regardless of their value of γ, must arrive at x_3 according to Fermat's principle. This means x_3 is independent of γ and the object–image matrix element, O_{12}, must be zero. This requirement, called the *imaging requirement*, leads to

$$s' - s - ss'(n_2 - 1)\left(\frac{1}{R_2} - \frac{1}{R_1}\right) = 0,$$

which can be rewritten as the *thin lens equation*

$$\frac{1}{s'} - \frac{1}{s} = (n_2 - 1)\left(\frac{1}{R_1} - \frac{1}{R_2}\right). \tag{6.9}$$

If we allow the index of refraction in the object and image space to differ, then Eq. (6.9) is modified to read

$$\frac{n_3}{s'} - \frac{n_1}{s} = \frac{n_2 - n_1}{R_1} - \frac{n_3 - n_2}{R_2}.$$

The elements of the object–image matrix can be identified with some simple parameters of the thin lens. From our definition of the lateral magnification of the imaging system, we find that the element O_{11} of the object–image matrix is equal to the lateral magnification

$$\begin{aligned} \beta = \frac{x_3}{x_1} &= O_{11} \\ &= 1 + s'(n_2 - 1)\left(\frac{1}{R_2} - \frac{1}{R_1}\right). \end{aligned} \tag{6.10}$$

We can complete the identification of the elements of the object–image matrix by defining the concept of *angular magnification* that we find to be equal to the element O_{22}:

$$\beta_\gamma = \frac{\gamma_3}{\gamma_1} = O_{22} = 1 - s(n_2 - 1)\left(\frac{1}{R_2} - \frac{1}{R_1}\right). \tag{6.11}$$

6.2.3.5.1 *Lens maker's equation*

When the indices are the same in object and image space the object–image matrix now simplifies to

$$\mathbf{O} = \begin{pmatrix} \beta & 0 \\ \Phi & \beta_\gamma \end{pmatrix}.$$

---

[13] If we used Cartesian coordinates for each point and subtracted to get distances, the signs of the coordinates would not be an issue. We are here using distances in the equation because they are less cluttered but they can be confusing on first reading.

We can easily show that the power Φ is related to the focal point of a lens. When we put the object at infinity, the image point, located a distance, f_2, from the lens vertex, V_2, is defined as the focal point of the lens, F_2 (see Figure 6.9):

$$\frac{1}{s'} = \frac{1}{f_2} = (n_2 - 1)\left(\frac{1}{R_1} - \frac{1}{R_2}\right). \tag{6.12}$$

This relationship for the focal length is called the *lens maker's equation*. We discover that the power of the lens is

$$\Phi = -1/f_2$$

and the object–image matrix becomes

$$\mathbf{O} = \begin{pmatrix} \beta & 0 \\ -\dfrac{1}{f_2} & \dfrac{1}{\beta} \end{pmatrix}. \tag{6.13}$$

The ABCD system matrices for some common optical elements and media are shown in Figure 6.10. One optical component matrix has not yet been derived, i.e., the reflection matrix. This matrix can be obtained from the refraction matrix by setting $n_2 = -n$. A negative index

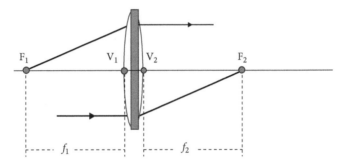

Figure 6.9 Definition of the focal points and focal planes of a lens.

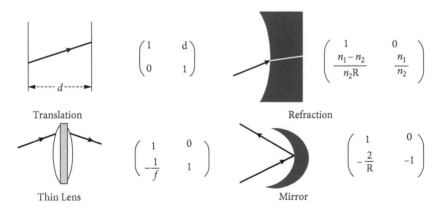

Translation	Refraction
$\begin{pmatrix} 1 & d \\ 0 & 1 \end{pmatrix}$	$\begin{pmatrix} 1 & 0 \\ \dfrac{n_1 - n_2}{n_2 R} & \dfrac{n_1}{n_2} \end{pmatrix}$

Thin Lens	Mirror
$\begin{pmatrix} 1 & 0 \\ -\dfrac{1}{f} & 1 \end{pmatrix}$	$\begin{pmatrix} 1 & 0 \\ -\dfrac{2}{R} & -1 \end{pmatrix}$

Figure 6.10 The ABCD system matrix for translation in a uniform medium and refraction at an interface. Also shown are the ABCD matrices for a thin lens and a mirror. (We have not carried out the analysis of a ray trace for a mirror in the text. The mirror equation has the same form as the thin lens equation, Eq. (6.13), with a focal length given by $f = -R/2$.)

of refraction, -n, is interpreted physically as associated with a ray traveling in the opposite direction from the ray associated with the positive index, n.[14]

6.2.3.5.2 Gaussian formalism

If we measure a distance equal to the focal distance, f_2, from V_1 at the front of the lens, we locate a second focal point, F_1. If the index of refraction is the same on both sides of the lens, then the focal length, i.e., the distance from each focal point to its vertex, is $-f_1 = f_2 = f$. Using the definition of the focal length given by Eq. (6.12) we can rewrite Eq. (6.9) as

$$\frac{1}{s'} - \frac{1}{s} = \frac{1}{f}. \tag{6.14}$$

This is called the Gaussian form of the thin lens equation. If f_2 is positive, the lens is called a positive lens; it will focus a plane wave to a point located a distance f_2 from the lens vertex. If f_2 is negative, the lens is called a negative lens; it will cause a plane wave to diverge, as if it originated from a point located a distance $-f_2$ from the lens vertex.

It is necessary to point out the negative sign on the left in Eq. (6.14) is a result of our Cartesian sign convention. The use of the sign convention used in most introductory physics textbooks (also called the "real is positive" sign convention) will result in a plus sign replacing the negative sign. Our sign convention reduces the difficulty of handling complex optical systems and also provides a useful view of the operation of a lens. The light leaving the object point at s is a spherical wave and will have a wavefront curvature, the vergence, given by $1/s$.[15] The lens can be thought of as simply adding curvature (power) to the object's wave vergence. The curvature provided by the lens is given by Eq. (6.12)—the reciprocal of the focal length of the lens. The operation of the lens on the wavefront curvature is easily understood by rewriting Eq. (6.14) as a vergence equation with all distance in meters:

$$\underbrace{\frac{1}{s'}}_{\substack{\text{output} \\ \text{curvature}}} = \underbrace{\frac{1}{s}}_{\substack{\text{input} \\ \text{curvature}}} + \underbrace{\frac{1}{f}}_{\substack{\text{added} \\ \text{curvature}}} .$$

Before we go any further, let's apply the thin lens equation to a simple problem. We want to locate an image of our object a distance T from the object. We could say we want to "throw the object a distance T." Using Figure 6.8, the throw distance is given by

$$T = -s + s'.$$

The Cartesian sign convention makes the distance s a negative number. Using Eq. (6.14) we can solve for s by replacing s':

---

[14] If this modification is used on the equation for the power of a surface, we find that a mirror with a positive power must have a negative radius of curvature, which is in agreement with the knowledge that a concave mirror will focus sunlight.

[15] Power and vergence, $V = n/r$, are encountered in ophthalmic optics. The curvature of wavefront produced by an object or that will create an image is described by V using the same dimensions as the diopter of a lens.

$$\frac{1}{s'} = \frac{1}{s} + \frac{1}{f}$$

$$s' = \frac{sf}{s+f}$$

$$T = -s + \frac{sf}{s+f}.$$

To find s we simply solve the resulting quadratic equation:

$$s^2 + Ts + Tf = 0$$

$$s = 1/2\left[-T \pm \sqrt{T^2 - 4Tf}\right].$$

The minimum value T is found by setting the result under the radial equal to 0, resulting in $T_{\min} = 4f$.

Optical Invariant

The determinants of the object–image matrix is equal to 1 if the index of refraction is the same in both the object and image space. The determinant is equal to the product of the elements of the ABCD object–image matrix:

$$\|\mathbf{O}\| = AD - BC = \left(O_{11}{\cdot}O_{22} - O_{12}{\cdot}O_{21}\right) = 1.$$

For the imaging condition, we must have $B = 0$, requiring that

$$\|\mathbf{O}\| = AD = 1.$$

Because the determinate is unity, the product of the lateral and angular magnifications is

$$\|\mathbf{O}\| = \beta\beta_\gamma = \frac{x_3}{x_1}\frac{\gamma_3}{\gamma_1} = 1,$$

which can be rewritten as a statement of the optical invariant of an optical system:

$$x_3\gamma_3 = x_1\gamma_1.$$

A more general form of this invariant can be obtained by recalling that when the index of refraction differs in the object and image space, then the object–image determinant is

$$\|\mathbf{O}\| = \beta\beta_\gamma = \frac{n_1}{n_3}.$$

The *optical invariant* is then given by

$$\mathcal{H} = \mathbf{n}_1 x_1 \gamma_1 = \mathbf{n}_3 x_3 \gamma_3. \tag{6.15}$$

This relationship is also called the Lagrange invariant or Smith–Helmholtz equation. The invariant was derived using ray parameters from object and image space. The relationship holds for any two conjugate planes[16] of the optical system. The Lagrange invariant is a measure of the information-handling capacity of the optical system.

[16] A pair of conjugate planes of an optical system contain an object–image pair.

6.2.4 **Cardinal points**

6.2.4.1 Focal points: Newtonian formalism

The pair of focal points, F_1 and F_2, is the first of three pairs of points called the *cardinal points* of the lens. If someone handed you a complex lens, you would not be able to use any of the equations we have derived because the only references we have are the vertex of a thin lens. The two focal points can be used to establish a new set of reference planes as replacements for the vertices of the lens. The object and image distances are then measured with respect to the focal planes. The lens equation, Eq. (6.14), is modified to reflect the new reference points by substituting

$$Z = s - f_1 \quad Z' = s' - f_2,$$

yielding the *Newtonian form of the thin lens equation*

$$ZZ' = -f_1 f_2. \tag{6.16}$$

6.2.4.2 Principal points

The thin lens results in a very simple ABCD system matrix. It is possible to retain this simplicity for very complex optical systems by defining a pair of cardinal points, called the principal points, H_1 and H_2, to replace the vertices, V_1 and V_2, as reference points for the lens and also in place of the focal points used in the Newtonian formalism; see Figure 6.11 for the generic location of the cardinal points. These new principal points define planes of unit magnification, just as the vertices did, but they may not lie anywhere near the outer surfaces of the optical system; see Figure 6.12. With these new reference points, the ABCD matrix generated for any complex lens assumes the simple form of the thin lens shown in Figure 6.10 with the vertices replaced by H_1 and H_2. These reference points are said to be conjugate points because the points are related by the optical system as an object–image pair.

To create the new ABCD system matrix for a complex lens system, we begin by using a standard ray trace to produce the system ABCD matrix for the thick lens, Eq. (6.7). The system matrix is multiplied by two transfer matrices: first multiply the system's ABCD matrix by the

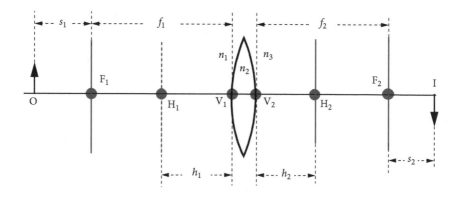

Figure 6.11 Principal (cardinal) points for a lens.

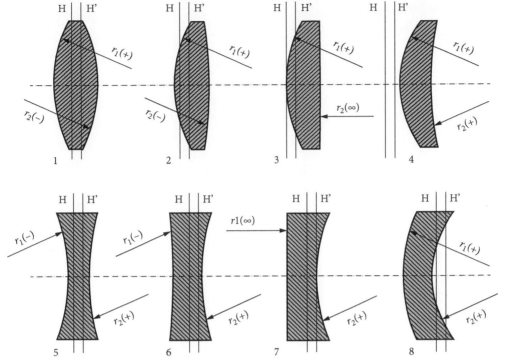

Figure 6.12 Various lens shapes, and the location of the principal planes. By Tamasflex (CC BY-SA 3.0 (https://creativecommons.org/licenses/by-sa/3.0) or GFDL (http://www.gnu.org/copyleft/fdl.html)), from Wikimedia Commons.

transfer matrix that describes the ray propagation from the objects to H_1, then multiply this new matrix by the transfer matrix describing the propagation from the principle plane H_2 to the image:

$$\mathbf{M} = \begin{pmatrix} 1 & h_2 \\ 0 & 1 \end{pmatrix} \begin{pmatrix} A & B \\ C & D \end{pmatrix} \begin{pmatrix} 1 & -h_1 \\ 0 & 1 \end{pmatrix}. \tag{6.17}$$

Here, as with the thin lens object–image matrix, the propagation matrix for the region to the left of the lens contains a negative distance because of the sign convention selected for geometric optics:

$$\mathbf{M} = \begin{pmatrix} A + h_2 C & B - h_1 A + h_2 (D - h_1 C) \\ C & D - h_2 C \end{pmatrix}. \tag{6.18}$$

The principal planes, H_1 and H_2, located at h_1 and h_2, are the equivalent of the thin lens vertices; thus, they are planes of unit magnification. Because the principal planes are planes of unit magnification, the (11) element of the thick lens matrix is

$$\mathbf{M}_{11} = 1.$$

We can therefore write

$$A + Ch_2 = 1$$

$$h_2 = \frac{1 - A}{C}. \tag{6.19}$$

Because the two planes are defined by conjugate points, the imaging condition must hold between them. This means that

$$\mathbf{M}_{12} = 0.$$

Since the determinate of the matrix is

$$\|\mathbf{M}\| = \frac{n_1}{n_3},$$

we must have

$$\mathbf{M}_{22} = \frac{n_1}{n_3} = D - Ch_1$$

$$h_1 = \frac{D - n_1/n_3}{C}. \tag{6.20}$$

When $n_3 = n_1$, then

$$h_1 = \frac{D - 1}{C}.$$

We now have defined the location of the principal points with respect to the lens' vertices. For these equations to be meaningful, we must have that $C \neq 0$ (it occurs in the denominator of h_1 and h_2). In Eq. (6.13) we associated C with $-1/f$; here, C is equal to the reciprocal of the effective focal length, $f_{eff} = f_2$, measured from the principal plane, H_2, i.e.,

$$C = -\frac{1}{f_{eff}}. \tag{6.21}$$

There is a special optical system for which $C = 0$; *the afocal system*. This optical system produces an image at infinity of an object located at infinity. Because an astronomical telescope performs this optical operation, an afocal system is often called a telescopic system. Because the object and the image are both located at infinity, it is more useful to specify the size of the object by the angle it subtends. We, therefore, use angular magnification, Eq. (6.11), for the telescopic system, as a measure of its performance:

$$\beta_\gamma = \frac{\gamma_3}{\gamma_1} = D.$$

6.2.4.3   Nodal points

The third and final set of cardinal points are called the nodal points, N_1 and N_2, and the pair of planes, normal to the optical axis, that contain these points are called nodal planes. The nodal points are defined as those points for which

$$\gamma_3 = \gamma_1,$$

Or, equivalently, the points of unit angular magnification

$$\beta_\gamma = 1.$$

The image of an object will not move if the imaging lens is rotated a small amount about an axis passing through a nodal point perpendicular to the optical axis. Thus, it is quite easy to locate the nodal points of a lens experimentally. If the index of refraction on each side of an optical system is the same, $n_1 = n_3$, then the nodal points coincide with the principal points and we need only deal with four cardinal points.

6.2.5 Aperture stop and pupils

The finite size of an optical system limits the bundle of axial rays leaving the object point O on the optical axis of Figure 6.8. The obstruction that limits the maximum angle, γ_M, of a ray from the object to the image is called the *aperture stop*. The rays that leave the object at angles nearly equal to γ_M are called *marginal rays*. In Figure 6.13 we show a marginal ray leaving the base of the object at the angle γ_M and just skimming the aperture stop labeled A (colored dark blue in Figure 6.13). The ray from the top of the object, passing through the center of the aperture at A, is called the *chief ray* (also called the *principal ray*; colored light blue in Figure 6.13).

To locate the aperture stop is quite easy. In conventional ray tracing, we trace a single ray from plane to plane through the system, as illustrated in Figure 6.14. We then calculate the ratio of aperture radius, r_j, to ray height, x_j:

$$\frac{r_j}{x_j}.$$

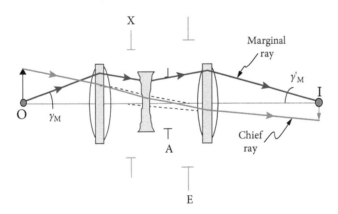

Figure 6.13 A complex lens system (Cooke triplet) containing a stop, A, that acts as the aperture stop for the lens. The rays that just skim pass the aperture stop are called marginal rays. Any ray that passes through the center of the aperture stop is called a chief ray. The image of the aperture stop as seen from the object is called the entrance pupil, labeled E in the figure. The image of the aperture stop as seen from the image is called the exit pupil, labeled X. Because the entrance and exit pupils are conjugates of the aperture stop, the chief ray will also pass through the center of these pupils, or, as in this figure, they will appear to do so as indicated by the dotted lines.

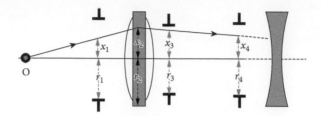

Figure 6.14 Geometry for finding the aperture stop of an optical system. The stops contained in the lens system are the black aperture limits defined in the figure.

The smallest ratio indicates the aperture stop. With the matrix technique, we create an ABCD matrix from the object O, on the optical axis, to each stop in the optical system, as illustrated in Figure 6.14. Because the object height at the starting point is zero, the matrix equation for each ray height would then be

$$\begin{pmatrix} x_j \\ \gamma_j \end{pmatrix} = \begin{pmatrix} A & B \\ C & D \end{pmatrix}\begin{pmatrix} 0 \\ \gamma \end{pmatrix}.$$

From this equation the value of x_j in each aperture is given by

$$x_j = B\gamma_j.$$

If we set each x_j equal to the radius of the stop at that point, r_j, then the object angle for the marginal ray of that stop is

$$\gamma_j = \frac{r_j}{B}.$$

Thus, to find the aperture stop we simply calculate the matrix element B for the optical elements to the left of each stop. The stop with the smallest ratio of the stop radius to matrix element B is the aperture stop.

The maximum angle γ'_M, equal to the angle for the ray just passing through the aperture stop, determines the numerical aperture of the lens:

$$\text{N.A.} = n_3 \sin\gamma'_M, \tag{6.22}$$

and the f/# of the lens

$$f/\# = \frac{1}{2\sin\gamma'_M}. \tag{6.23}$$

The equation normally used to define and calculate the $f/\#$ is a special case of Eq. (6.23). If a plane wave fully illuminates the aperture of the optical system, then γ'_M is given by

$$\gamma' = \frac{r_M}{\mathbf{f}_2},$$

where $\mathbf{f}_2$ is the effective focal length and r_M is the aperture radius. Applying the paraxial approximation allows us to write the normally encountered definition of the $f/\#$

$$f/\# = \frac{\mathbf{f}_2}{D},$$

where D is the diameter of the aperture stop.

The image of the aperture stop formed by the optical system to the stop's left, i.e., the image of the aperture stop as seen from the object position, is called the *entrance pupil*, and is labeled E in Figure 6.13. The image of the aperture stop formed by the optical system to its right, i.e., the image of the aperture stop as seen from the image position, is called the *exit pupil*, and is labeled X in Figure 6.13.

As was noted earlier, the ray passing through the center of the aperture stop is called the chief ray.[17] Because the entrance and exit pupils are conjugates (images) of the aperture stop, the chief ray passes through the centers of these pupils also. In Figure 6.13, the entrance and exit pupils are virtual images; thus, a chief ray appears to originate at the centers of these pupils.

Other stops exist in an optical system; they may be placed there to control aberrations, to act as lens mounts, or to reduce scattered light (*glare stops*). The stop that limits the maximum oblique chief ray that can propagate through the optical system is called the *field stop*. This stop limits the angular size of the image.

If a ray bundle is created by the chief ray and the marginal ray for an object point on the optical axis, we find that a ray bundle of this size cannot propagate through the system for object points located off the optical axis. Another way to state this fact is that oblique ray bundles will not fill the aperture stop. This means that the flux density that can pass from the object to the image is a function of the distance off the optical axis. This effect is called a *vignette*. Vignette is not always a defect. Sometimes lens designers "hide" aberrations in the low-light region produced by a vignette. Also vignette may be used to emphasize part of the image or for artistic effect as in Figure 6.15.

It will be helpful in understanding our previous discussion by examining the Cooke triplet lens design shown in Figure 6.16 with the thin lens equation and the matrix formalism. To find the focal length of the triplet lens system we use an object located at infinity in front of lens L_1 in the thin lens equation:

$$\frac{1}{s'} - \frac{1}{s} = \frac{1}{s'} - \frac{1}{-\infty} = \frac{1}{f_1} = \frac{1}{0.2}$$
$$s' = 0.2m.$$

The image position, s', is the object for lens L_2, i.e., a simple shift of the origin of our optical coordinate system to V_2:

$$s_2 = s'_1 - 0.06 = 0.14.$$

The image produced by lens L_2 is then

$$s'_2 = \frac{s_2 f_2}{s_2 + f_2} = \frac{0.14\,(-0.065)}{0.14 - 0.065} = -0.113.$$

For the third lens, L_3, another coordinate shift is made from V_2 to V_3:

$$s_3 = -0.113 - 0.06 = -0.173$$
$$s'_3 = \frac{-0.173(0.1)}{0.1 - 0.173} = 0.237.$$

---

[17] Sometimes the term *chief ray* is used and other times the term *principal ray* is used. When principal ray is used, the term chief ray is reserved for the center ray of a ray bundle. Most of the time the two rays are identical.

Figure 6.15 Akbar's Tomb—an example of vignette for artistic enhancement. By Anupamg—own work, CC BY-SA 3.0, https://commons.wikimedia.org/w/index.php?curid=28343185 from Wikimedia Commons.

This is the focal length of the triplet system. Now we create the system ABCD matrix for the triplet in Figure 6.16:

$$
\begin{pmatrix} 1 & 0 \\ -\dfrac{1}{0.1} & 1 \end{pmatrix} \begin{pmatrix} 1 & 0.06 \\ 0 & 1 \end{pmatrix} \begin{pmatrix} 1 & 0 \\ \dfrac{1}{0.0625} & 1 \end{pmatrix} \begin{pmatrix} 1 & 0.06 \\ 0 & 1 \end{pmatrix} \begin{pmatrix} 1 & 0 \\ -\dfrac{1}{0.2} & 1 \end{pmatrix}
$$
$$
= \begin{pmatrix} 1.072 & 0.178 \\ -4.52 & 0.18 \end{pmatrix}.
$$
(6.24)

$$\frac{1}{s'} - \frac{1}{s} = \frac{1}{s'} - \frac{1}{-\infty} = \frac{1}{f_1} = \frac{1}{0.2}$$

$$s' = 0.2 \text{ m}$$

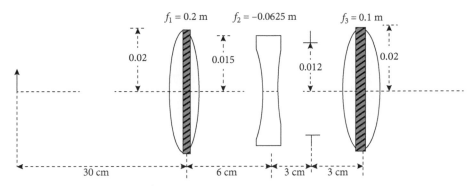

Figure 6.16 Cooke triplet.

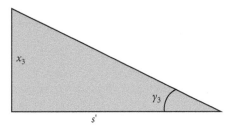

Figure 6.17 Trig determination of image position.

The matrix equation for a ray trace is then

$$\begin{pmatrix} 1.072 & 0.178 \\ -4.52 & 0.18 \end{pmatrix} \begin{pmatrix} x_1 \\ \gamma_1 \end{pmatrix} = \begin{pmatrix} x_3 \\ \gamma_3 \end{pmatrix}.$$

As we assumed with the thin lens equation, the object is at infinity so that $\gamma_1 = 0$. We will assume that the ray height is 1 m to make our calculation easy. The ray matrix vector is then

$$\begin{pmatrix} x_3 \\ \gamma_3 \end{pmatrix} = \begin{pmatrix} 1.072 \\ -4.52 \end{pmatrix}.$$

To get the image position from this vector we have to use a simple trig relation. The ray height x_3 is measured at the surface of lens L_3 and the ray angle is downward toward the optical axis; Figure 6.17. The ray will intercept the optical axis at s':

$$s' = \frac{x_3}{\gamma_3} = \frac{1.072}{4.52} = 0.237.$$

The radii of each of the lenses and the aperture between lens L_2 and L_3, r_j, are shown in Figure 6.16 as is a test object located 30 cm in front of L_1. We follow a ray from the object through lens L_1 and obtain the ray's height

$$\begin{pmatrix} 1 & 0 \\ -5 & 1 \end{pmatrix}\begin{pmatrix} 1 & 0.3 \\ 0 & 1 \end{pmatrix} = \begin{pmatrix} 1 & 0.3 \\ -5 & -0.5 \end{pmatrix} \Rightarrow B = 0.3$$

$$\gamma_1 = \frac{r_1}{B} = \frac{0.02}{0.3} = 0.067.$$

The angle given is the maximum ray angle that can occur at L_1. There is no change in x as we pass through the lens. From L_1 to L_2 we have

$$\begin{pmatrix} 1 & 0.06 \\ 0 & 1 \end{pmatrix}\begin{pmatrix} 1 & 0.3 \\ -5 & -0.5 \end{pmatrix} = \begin{pmatrix} 0.7 & 0.27 \\ -5 & -.05 \end{pmatrix} \Rightarrow B = 0.27$$

$$\gamma_2 = \frac{0.015}{0.27} = 0.056$$

and again no change in x as we pass through L_2. Refract through L_2

$$\begin{pmatrix} 1 & 0 \\ 16 & 1 \end{pmatrix}\begin{pmatrix} 0.7 & 0.27 \\ -5 & -0.5 \end{pmatrix} = \begin{pmatrix} 0.7 & 0.27 \\ 6.2 & 3.82 \end{pmatrix}$$

and travel to the aperture where no refraction occurs

$$\begin{pmatrix} 1 & 0.03 \\ 0 & 1 \end{pmatrix}\begin{pmatrix} 0.7 & -.27 \\ 6.2 & 3,82 \end{pmatrix} = \begin{pmatrix} 0.886 & 0.385 \\ 6.2 & 3.82 \end{pmatrix} \Rightarrow B = 0.385$$

$$\gamma_4 = \frac{0.012}{0.385} - 0.031.$$

Finally, we go from the aperture to L_3 which is a simple translation

$$\begin{pmatrix} 1 & 0.03 \\ 0 & 1 \end{pmatrix}\begin{pmatrix} 0.886 & 0.385 \\ 6.2 & 3.82 \end{pmatrix} = \begin{pmatrix} 1.07 & 0.5 \\ 6.2 & 3.82 \end{pmatrix}$$

$$\gamma_3 = \frac{0.02}{0.5} = 0.04.$$

We see that the smallest γ is γ_4 at the aperture between L_2 and L_3; therefore, it is the system aperture stop.

6.3 Problem Set 6

(1) Prove that the reflected ray from a plane mirror turns through an angle 2θ when the mirror rotates through θ.

(2) What length plane mirror must you purchase for you to see your full height when it is mounted in a vertical position? Do not assume that you have eyes on top of your head.

(3) A fish appears to be 2 m below the surface of a pond ($n = 1.33$). What is its actual depth?

(4) Given the Cooke triplet specified in Figure 6.18, find the (a) aperture stop, (b) the exit pupil, and (c) the entrance pupil.

(5) Calculate the object–image matrix for the lens system shown in Figure 6.19, where the lens on the left has a focal length f_1 and the one on the right has a focal length f_2. Prove that conjugate planes occur for $s = -f_1$ and $s' = f_2$. This latter matrix is sometimes called the focal plane matrix.

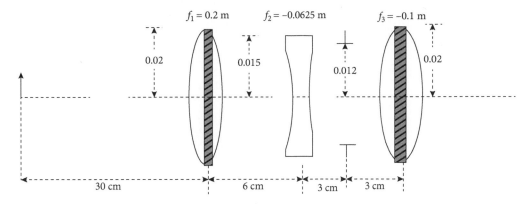

Figure 6.18 Cooke triplet.

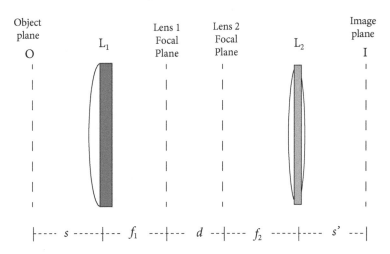

Figure 6.19 General two-lens optical system.

(6) Using the object–image matrix from Problem 5, (a) calculate the positions of the cardinal points when $s = -f_1$ and $s' = f_2$. (b) What is the effective focal length of this optical system if the two lenses are in contact with each other?

(7) Using the results of Problem 5, assume $f_1 = 2f_2$ and sketch out the optical system, including the positions of the principal planes for: (a) a Huygens eyepiece: $d = -1.5f_2$, (b) an astronomical telescope $d = 0$, and (c) a compound microscope $d = 2f_2$. The condition of $d < 0$ occurs when the distance between the two lenses is less than $f_1 + f_2$.

(8) A lens of focal length +30 mm is in front of a lens of focal length -60 mm. The lenses are separated by 160 mm. An object is located 35 mm in front of the first lens. Find the final image location, characterize it, and get the overall magnification.

(9) The human eye is approximately spherical with a diameter of around 25 mm. At the back of the eye is the retina while at the front is a compound lens comprising the cornea (of fixed focal length) and the lens itself (with variable focal length), which can

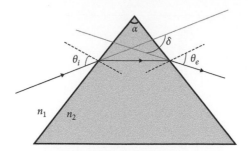

Figure 6.20 Prism with apex angle equal to α.

be approximated as a thin lens at a distance of 17 mm from the retina. The eye is capable of focusing on objects at distances varying from the near point (about 25 cm) to infinity. Estimate the range of lens powers that can be achieved by the lens and cornea in focusing between the near and far points.

(10) Trace a ray through the prism shown in Figure 6.20. The angle between the initial and final rays is called the *angle of deviation*, denoted by δ in Figure 6.20. Derive an equation in terms of the apex angle, α, in Figure 6.20, for the angle of deviation. Assume the prism has an index of refraction n_2 and is in a medium of index n_1. What is the minimum angle of deviation? Prove that the angle of minimum deviation occurs when $\theta_i = \theta_e$.

 REFERENCES

1. Khan, S., Medieval Islamic achievements in optics.*Il Nuovo Saggiatore***31**(1–2): 36–45 (2015).

2. Conrady, A. E., *Applied Optics and Optical Design*, new edn. New York: Dover, 1992.

3. Smith, W. J., *Modern Lens Design*. New York: McGraw-Hill Professional Engineering, 2004.

4. Oslo. 2008. Free limited version of optical design program for educational perposes]. Available from http://www.lambdares.com/contact_us/oslo/.

5. Stroschine, D., *OptomechSpreadsheet_001.xlsx* S. Sparrold, Editor. 2011: Optics Realm.

6. Berger, C. G.https://opticsthewebsite.com/raytrace.aspx. opticsthewebsite 2018; Available from https://opticsthewebsite.com/Default.aspx.

7. Sparrold, S., Chief and marginal rays, YNU, in *OpticsRealm*. Edmund Optics, 2015. Available from http://www.opticsrealm.com/home.

8. Nussbaum, A., and R. A. Phillips, *Contemporary Optics for Scientists and Engineers*. Englewood Cliffs, NJ: Prentice-Hall, 1976.

9. Geary, J. M., *Introduction to Lens Design*. Richmond, VA: Willmann-Bell, 2002.

10. Nussbaum, A., *Optical System Design*, 1st edn. Englewood Cliffs, NJ: Prentice-Hall, 1997.

7 Aberrations

A perfect optical system, from a geometric optics point of view, is one which redirects all light rays from an object point through a single conjugate (image) point as implied by the thin lens equation we introduced in the previous chapter:

$$\underbrace{\frac{1}{s'}}_{} = \underbrace{\frac{1}{s}}_{} + \underbrace{\frac{1}{f}}_{}$$

output curvature   input curvature   added curvature

For a real image, if we view the geometrical ray as the normal to a wavefront, then the wavefront associated with each ray leaving the optical system must be spherical, centered upon the image point. Fermat's principal states that the optical path length of each ray will be identical and the travel time a minimum.

If the perfect optical system were a positive lens, then a plane wave incident upon the lens would be focused to a point. If we inspected the light distribution in planes on either side of the focal plane, based on geometric optics, we would expect to see uniformly illuminated disks of light, sections through cones arranged symmetrically about the focus. We would expect that the size of the patterns would decrease in diameter as we approached focus and increased in diameter as we moved away from focus. Experimentally produced light distributions for a perfect lens are shown in Figure 7.1.

As we already know, light is a wave and geometric optics is only an approximate theory. We expect to see a departure from the predictions of geometric optics when dimensions are on the order of a wavelength or when the light is very coherent. In Figure 7.2 the departure from the predictions of geometric optics around the lens focus is obvious. The patterns you see in the light distributions displayed along the z-axis were collected by a microscope objective that produced magnified images of the light distribution transverse to the z-axis. It is easy to see that the uniform distributions predicted by geometric optics are not present. The actual patterns generated are produced by diffraction. The lens producing these images is said to be a diffraction limited lens.

The light distributions are quite complicated with the light intensity on the optical axis passing through zero periodically. The key item to note is that there are identical distributions in planes symmetrically positioned about the focal plane. These light distributions are not aberrations but rather are due to diffractive effects and a critical analysis of them can be found in the literature [1]. We will discuss diffractive effects later.

Modern Optics Simplified. B. D. Guenther. © B. D. Guenther 2020.
Published in 2020 by Oxford University Press. DOI: 10.1093/oso/9780198842859.001.0001

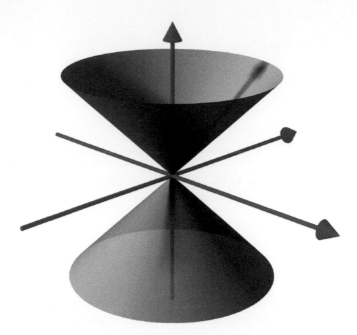

Figure 7.1 Cone representing the prediction of the outer edge of light focused by a lens according to geometrical optics.

By Lars H. Rohwedder (User:RokerHRO) (own work) (public domain), via Wikimedia Commons.

7.1 **Ray Trace of Single Refractive Surface**

To discuss the appearance of aberrations we will examine the simple refractive surface shown in Figure 7.3. The lens modeled here is often found in the photo sensor of a sunlight-activated street or yard light. The lens found in a yard light usually has a lower index of refraction than we use in this example and therefore has a different length than the $2R$ value used here.

We saw in Table 6.1 that we could use Snell's law, in the paraxial approximation up to angles as large as $20°$ with less than 1% error. We will assume the dot-dash line in Figure 7.3 is a paraxial ray coming from infinity with θ_i no larger than $20°$ so that ρ for this ray is no larger than about 30% of the radius of the refractive surface,

$$\rho = R \sin\theta_i = (0.342)R.$$

We want the dotted (paraxial) ray to focus on the z-axis a distance $2R$ behind the vertex of the curved surface. This is where our detector will be. Rather than adjust the image distance, we have adjusted the index of the lens to produce the desired image at the distance $2R$. The index required for this to happen can be obtained by using the matrix formalism we have derived to ray trace this lens. First, refraction is across the curved surface

$$\mathbf{R} = \begin{pmatrix} 1 & 0 \\ \dfrac{n_1 - n_2}{n_2 R_1} & \dfrac{n_1}{n_2} \end{pmatrix} \tag{7.1}$$

Then translation to the detector at the back surface of the lens is

$$\mathbf{T} = \begin{pmatrix} 1 & d \\ 0 & 1 \end{pmatrix}. \tag{7.2}$$

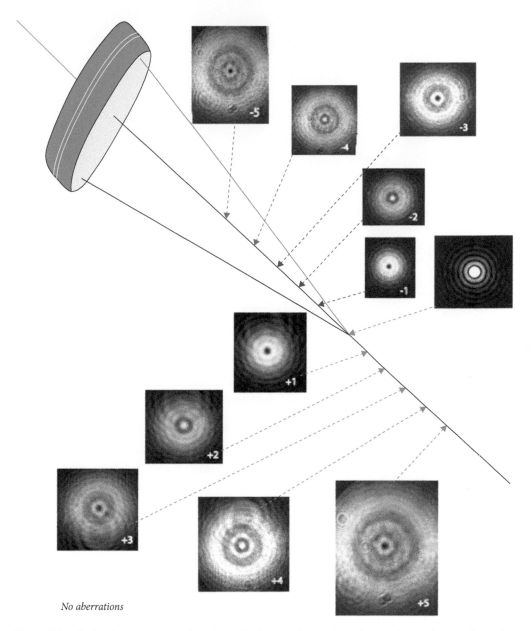

No aberrations

Figure 7.2 Light distributions found in planes normal to the optical axis and equally spaced along the optical axis of a diffraction limited lens illuminated by a plane wave from a coherent light source. The planes were space about 1 mm apart and the focal length of the lens was about 500 mm. Note that the images are symmetric about the focus. The small bull's-eyes seen on the various images are due to interference between waves scattered by dust particles on the lens surfaces.

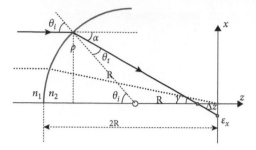

Figure 7.3 Geometry for finite lens trace of simple lens with only one refractive surface.

The ray transfer for the lens is then

$$\mathbf{TR}\begin{pmatrix} x_1 \\ \gamma_1 \end{pmatrix} = \begin{pmatrix} 1 & d \\ 0 & 1 \end{pmatrix} \begin{pmatrix} 1 & 0 \\ \dfrac{n_1 - n_2}{n_2 R_1} & \dfrac{n_1}{n_2} \end{pmatrix} \begin{pmatrix} x_1 \\ \gamma_1 \end{pmatrix} = \begin{pmatrix} x_2 \\ \gamma_2 \end{pmatrix}.$$

Multiplying by only the refraction matrix we get from the refraction matrix

$$n_1 \left(\gamma + \frac{x_1}{R} \right) = n_2 \left(\gamma'' + \frac{x_1}{R} \right)$$

$$n_2 \gamma'' = n_1 \gamma - x_1 \frac{n_2 - n_1}{R}$$

There is no change in x_1 across the boundary, $x_1 = x_1''$. As the light propagates to the back surface, the angle does not change but the height decreases to

$$x_2' = x'' + 2R\gamma'' = x_1 + 2R\gamma''.$$

We have assumed that the paraxial (dot) ray is associated with a plane wave from infinity so that $\gamma = 0$. The paraxial ray strikes the z-axis a distance $2R$ behind the vertex of the curved surface where $x_2' = 0$. The index required for this to happen can be obtained from these equations:

$$x_1 = -2R\gamma''$$

$$n_2 \gamma'' = -\frac{x_1 (n_2 - n_1)}{R}$$

$$n_2 = 2n_1.$$

If $n_1 = 1.0$ (the lens is located in air), then $n = 2$.

The angle the solid refracted ray in Figure 7.3 makes with the optical axis can be calculated by using trigonometry. The maximum aperture for the solid ray is $\rho = R \sin \theta_i$. The distance, L, to the z-intercept of that ray is given by the law of sines (Figure 7.4),

$$\gamma'' = \alpha = \theta_i - \theta_t.$$

$$\tan \gamma'' = \frac{R}{L} \approx \sin \gamma''$$

allows us to calculate the z-axis intercept,

$$\Delta z = \frac{R \sin \theta_i}{2 \sin (\theta_i - \theta_t)} - R.$$

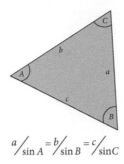

$$a\big/_{\sin A} = b\big/_{\sin B} = c\big/_{\sin C}$$

Figure 7.4 Law of sines applied to Figure 7.3.

$$\ell_1 n_1 + \ell_2 n_2 = \text{constant}$$

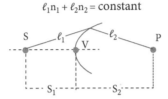

Figure 7.5 Ellipsoidal surface required for imaging by Fermat's principal.

This is called longitudinal spherical aberration. We see that the optical system will have a focus that varies with the height of the ray in the aperture, $R \sin \theta_i$. This variation is illustrated at the lower right in Figure 7.6, where the paths of a fan of rays through the lens are generated by a commercial lens design program. When spherical aberration causes the focal length to decrease with increasing zone radius, as it does in Figure 7.7, the aberration is said to be undercorrected. When the marginal rays have a larger focal length than the paraxial rays the aberration is said to be overcorrected and are usually associated with diverging lenses.

Why does our simplified math associated with paraxial theory fail us and lead to this aberration? The answer is simple. The optical surface we are considering is spherical in shape; however, Fermat's principle requires that the surface be Cartesian; see Figure 7.5. The time to travel from S to P over all optical paths should be

$$t_1 + t_2 = \text{constant}.$$

In terms of the parameters in Figure 7.5

$$\frac{n_1 \ell_1}{c} + \frac{n_2 \ell_2}{c} = \text{constant}.$$

For an object at infinity, the required Cartesian surface is an ellipsoid with an eccentricity of ½. Why don't we use this Cartesian surface? The answer is simple: it cost more to manufacture than a spherical surface and cost is one of the major optimization parameters. Because our surface is not the surface required by Fermat's principal, we should expect to encounter aberrations. And we do. Figure 7.6 displays an evaluation of our single surface lens in a commercial lens design program.

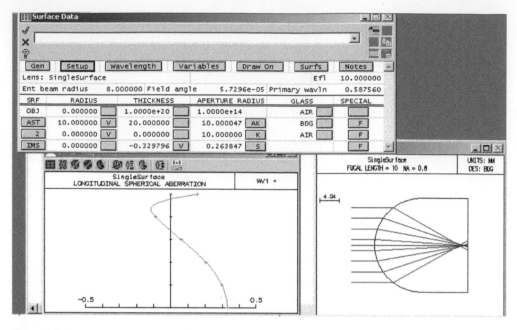

Figure 7.6 The graphical presentation of the optical performance of the single surface lens of Figure 7.3. The lens design program, Oslo [2], was used to display the spherical aberration.

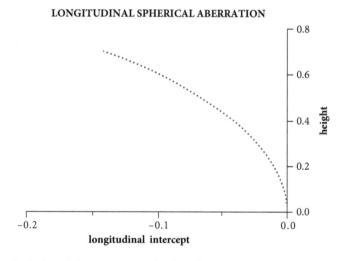

Figure 7.7 Longitudinal spherical aberration of a single spherical refracting surface. This is under-corrected spherical aberration. Longitudinal spherical aberration varies as the square of the ray height, as is suggested by the plot.

We can plot the displacement of the ray intercept from the paraxial intercept along the z-direction, as a function of the ray height; see Figure 7.7. This quadratic behavior[1] agrees with the longitudinal spherical aberration calculated by the lens design program.

---

[1] Note that the funny curve reversal at the top of the longitudinal spherical aberration plot of Figure 7.6 is due to the ray missing the lens at large apertures in the automated program.

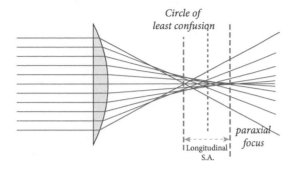

Figure 7.8 A fan of rays through a simple lens with spherical aberration. The envelope of the rays shows the existence of a region with a higher density of rays. This is the defocus position called "circle of least confusion."
(By Mglg—made by Mglg, uploaded to English Wikipedia, public domain, https://commons.wikimedia.org/w/index.php?curid=3655426).

A magnification of the fan of rays traced through a lens shown on the lower right of Figure 7.6 is redrawn in Figure 7.8. The fan of rays suggests that if we moved the focal plane toward the lens from the paraxial plane, then we could improve the image by putting more energy near the optical axis. This operation is called defocus and the resulting light distribution is called the circle of least confusion. Defocus is used by lens designers to minimize the effects of aberrations.

The set of images for an optical system with spherical aberration taken at positions identical to those shown in Figure 7.2, for a perfect optical system, are shown in Figure 7.9. In Figure 7.9 the best focus appears to be at about plane -1 where the focus would be brightest. This defocus[2] does not remove the aberration; in fact, spherical aberration cannot be completely eliminated from a single spherical lens.

The characterization of spherical aberration by the longitudinal displacement of the ray is not as useful to the lens designer as is the use of the image plane intercept, ε_x, because it is difficult to construct an optimization procedure with the longitudinal displacement.[3] Longitudinal is normally replaced by transverse spherical aberrations, which is displayed in Figure 7.3 as ε_x.

There are actually two basic ways to characterize the aberrations that are compatible with optimization codes and theoretical explanations of aberrations.

7.1.1 Wavefront aberration coefficients

These coefficients examine a wavefront, normal to the individual rays, produced in the optical system. This approach for characterizing aberrations was introduced by William Hamilton while he was in his early twenties. A wavefront is constructed by generating the normal to

[2] Defocus is considered one of the third-order Seidel aberrations and suggests that aberrations can be played off against one another to improve the overall performance.

[3] The difficulty with the longitudinal intercept is that it has a large range of possible values $(-\infty, \infty)$. This makes it hard to specify improvement criteria or to check for convergence. Skew rays may never intercept the optical axis. This means separate code would have to be written to handle skew rays. With transverse intercept the range of values can be restricted and a single code can handle skew and meridional rays by making measurements with respect to the chief rays.

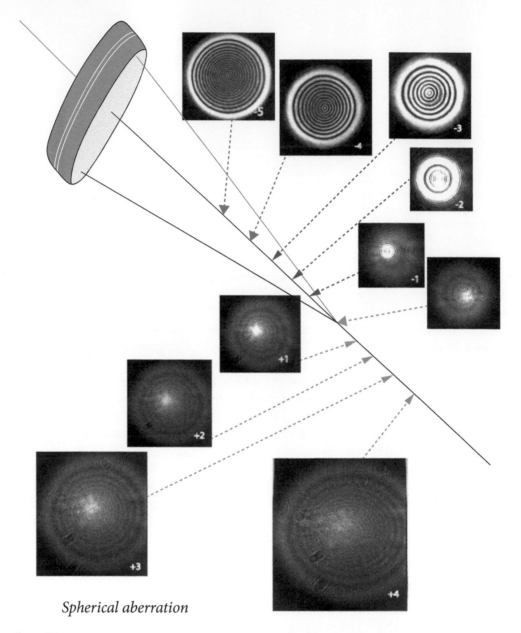

Spherical aberration

Figure 7.9 Images recorded in planes spaced equally on either side of the paraxial focal plane of an optical system with under-corrected spherical aberration is illuminated by a plane wave. The planes selected are identical to the planes recorded in Figure 7.2. Note that the light distributions observed are not symmetric about the focus.

each traced ray and is a sphere for a point image generated by a perfect optical system. The spherical surface for a perfect system, predicted by the paraxial theory, is called the *reference sphere*. The aberration of an optical system is described in terms of the departure of the geometric wavefront from a spherical shape centered on the paraxial image; see Figure 7.10. The

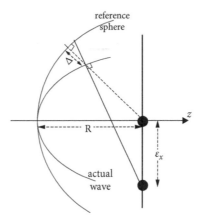

Figure 7.10 Reference sphere and actual wavefront are shown in the exit pupil of a lens. Also the optical path difference and the ray intercept in the image plane are shown.

wavefront produced by the optical system is constructed by generating the normal to each ray generated by geometrical optics. *Zernike circle polynomials* are used to fit the surface produced by the ray trace, allowing the characterization of the departure of the geometrical wavefront from the reference sphere. The Zernike polynomials are used because of their mathematical properties:

- All their derivatives are continuous.
- They provide an efficient represent of common optical errors (e.g., coma, spherical aberration).
- They form a complete orthonormal set over the unit circle.

If we calculate the distance between the reference sphere and the actual wavefront, in the exit pupil, we generate the wave aberration function, Δ. This function defines the optical path difference between the two surfaces and is measured along the radius of the reference sphere. Because the optical system is axially symmetric, Δ must not change when we simultaneously rotate the vectors ρ and $\mathbf{h}$ shown in Figure 7.13. This means that we should be able to describe the functional dependence of Δ in terms of scalar quantities. The simplest independent scalar relationships involving the variables are

$$\rho^2, \quad \rho \cdot \mathbf{h} = \rho h \cos\theta, \quad h^2$$

We expand Δ as a power series in these scalar variables to obtain a description of the aberrations in the optical system. We will not worry about how we generate this expansion but simply identify the aberrations associated with each term so the power dependence on exit pupil parameters can be seen:

$$\Delta\left(h^2,\rho^2,\rho h\cos\theta\right) = \Delta_{000}$$

$$+\ \Delta_{020}\rho^2 + \Delta_{111}h\rho\cos\theta$$

$$+\ \underbrace{\Delta_{040}\rho^4}_{\text{Spherical}} + \underbrace{\Delta_{131}h\rho^3\cos\theta}_{\text{Coma}} + \underbrace{\Delta_{222}h^2\rho^2\cos^2\theta}_{\text{Astigmatism}}$$

$$+\ \underbrace{\Delta_{220}h^2\rho^2}_{\text{Field Curvature}} + \underbrace{\Delta_{311}h^3\rho\cos\theta}_{\text{Distortion}}. \tag{7.3}$$

There are eight terms. The second and third are linear terms associated with magnification and defocus but do not affect image quality. The remaining five, called the third order aberration coefficients, affect the image quality.

The functions are quite complicated. The Zernike polynomials are a poor choice for describing some types of data but have found use in modeling the eye. Here is one for spherical aberration using Zernike polynomials.

Plots like the one shown in Figure 7.11 are used in wavefront-guided LASIK[4] to analyze the patient's cornea both before and after surgery and to guide the laser ablation during surgery. The use of these curves has led to a significant improvement on outcomes especially for night vision. This sophisticated approach attempts to correct not only third- but higher order aberrations. Some people are not eligible for this type of surgery and complications produced by changes

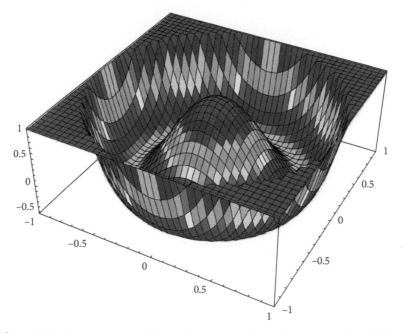

Figure 7.11 A three-dimensional deformation of the wavefront using Zernikes polynomials [3]. Permission granted by James C. Wyant, University of Arizona, Created with Wolfram webMathematica (http://www.wolfram.com).

---

[4] LASIK stands for laser in-situ keratomileusis.

during healing can prevent everyone from obtaining super vision but patient satisfaction rates as high as 96% have been reported.

Before computers when ray tracing was a labor-intensive process it was convenient to have a single number to characterize aberration. One technique used to generate such a number was to set ρ and **h** equal to their maximum values, then Δ (see Figure 7.10) would equal the optical path difference contribution from all of the aberrations. This number is called the *optical path difference (OPD)* and is usually quoted in wavelength.

7.1.1.1 Optical path difference

An optical system with an OPD of ¼ wave is considered a perfect system from a geometrical optics point of view. This criterion is often called the *Rayleigh criterion*.[5] A system that meets the Rayleigh criterion has its performance controlled by the wave properties of light and is called a diffraction limited system like the one whose performance is shown in Figure 7.2. In such systems it is necessary to use an additional criterion to specify its performance.

As displayed in Figure 7.2, a diffraction limited system produces an image of a point that consist of a small disk surrounded by a set of rings. The ring pattern at the image plane in Figure 7.2 is known as the Airy pattern. (We will give a mathematically description of this pattern when we discuss diffraction.) The fraction of energy contained in the central disk is used to characterize the performance of a nearly perfect, diffraction limited system. The characterization is made through use of the *Strehl ratio*, which is the ratio of light intensity at the peak of the diffraction pattern of an aberrated image relative to the peak in a perfect image. A perfect lens, with an OPD of 0, should contain 84% of the energy within the central disk of the Airy pattern at paraxial focus as indicated in Figure 7.12.

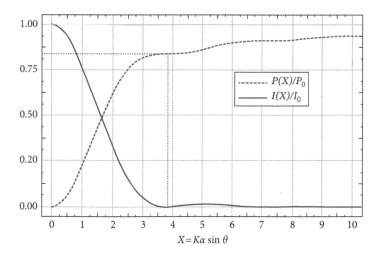

Figure 7.12 A mathematical representation of an Airy pattern. The energy contained in the central disk is indicated by the dotted line. The solid line is the normalized intensity in the focal plane.
(By inductive load (own work) (public domain), via Wikimedia Commons.)

---

[5] This is an approximate measure of imaging capability proposed by Lord Rayleigh in 1879. It is not the same thing as the Rayleigh limit to be discussed later in the theory of diffraction.

Table 7.1 Strehl Ratio

OPD	Fraction of the Rayleigh criteria	Intensity	Strehl ratio
0	Perfect lens	84%	1.0
$\lambda/16$	¼	83	0.988
$\lambda/8$	½	80	0.952
$\lambda/4$	1	68	0.8095

The origin of this number will be clear once we discuss diffraction. It is conventional to assumed a diffraction limited lens has a Strehl ratio > 0.80. Table 7.1 compares several measures of lens perfection.

7.1.2 Transverse ray coefficients

The displacement of the ray intercept from the paraxial image point, i.e., the chief ray intercept shown in Figure 7.3, is a second method used to characterize aberrations. A set of rays is followed through the optical system, as we did in Figure 7.3. In the general case illustrated in Figure 7.13 the entrance pupil of the system is divided into an array of points, as was the case for wavefront analysis. The rays are traced from the object to the entrance pupil, through the system to the exit pupil, and finally from the exit pupil to the image plane.

In Figure 7.13 the various coordinate planes used in the following discussion are defined. The vector **h** defines the object in the object plane and $\overline{\mathbf{h}}$ is the image in the image position in the image plane calculated using the paraxial thin lens equation. In Figure 7.13 we see that the ray starts at the tip of **h**, $(h, 0)$ and is traced to the exit pupil where it crosses the point (ρ, θ) and continues to the image where the point of the tip of $\overline{\mathbf{h}}$ is found. In addition to the tip of $\overline{\mathbf{h}}$, a second point, the actual image point, shown in the image plane is determined by finite ray tracing as was done in Figure 7.2. The coordinate position of this second point, $(\varepsilon_x, \varepsilon_y)$, is a measure of the aberration. To produce a general purpose equation containing aberration information, we fit an equation that will describe the image plane intersection coordinates, $(\varepsilon_x, \varepsilon_y)$ in terms of h, ρ, and θ in a similar way to that of wavefront analysis.

As we did for the wavefront aberration representation, we can expand ε_x and ε_y as a power series in h, ρ, and θ. The third-order terms affecting image quality are

$$\varepsilon_x = \underbrace{\sigma_1 \rho^3 \cos\theta}_{\text{Spherical}} + \underbrace{\sigma_2 \rho^2 h (2 + \cos 2\theta)}_{\text{Coma}} + \underbrace{3\sigma_3 \rho h^2 \cos\theta}_{\text{Astigmatism}} + \underbrace{\sigma_4 \rho h^2 \cos\theta}_{\text{Field curvature}} + \underbrace{\sigma_5 h^3}_{\text{Distortion}}$$

$$\varepsilon_y = \underbrace{\sigma_1 \rho^3 \sin\theta}_{\text{Spherical}} + \underbrace{\sigma_2 \rho^2 h \sin 2\theta}_{\text{Coma}} + \underbrace{\sigma_3 \rho h^2 \sin\theta}_{\text{Astigmatism}} + \underbrace{\sigma_4 \rho h^2 \sin\theta}_{\text{Field curvature}}. \tag{7.4}$$

For the simple example of a single refracting surface in Figure 7.3, the image plane intercept, ε_x, is a measure of the transverse spherical aberration and can be calculated by using the simple trigonometric relationship suggested by Figure 7.3,

$$\varepsilon_x = \Delta z \tan \gamma'' = R \left[\frac{\sin \theta_i}{2 \cos (\theta_i - \theta_t)} - \tan (\theta_i - \theta_t) \right].$$

to generate the *ray intercept plot*, Figure 7.14.

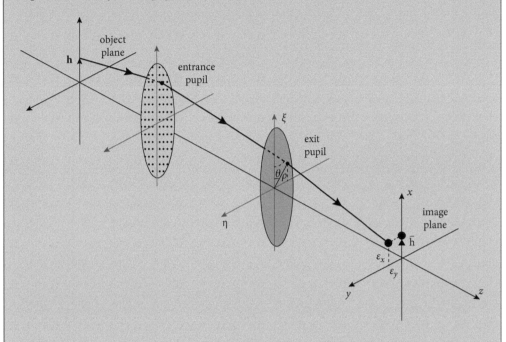

Figure 7.13 Coordinate system for aberration theory. The lenses in this optical system are not shown.

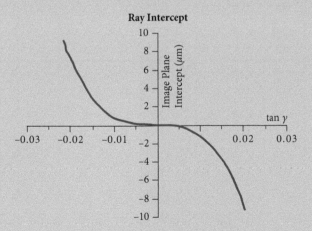

Figure 7.14 Ray intercept plot for a single spherical refracting surface. The horizontal coordinate is normalized to the radius of the aperture of the lens. Note the curve exhibits cubic behavior with respect to the ray height ($\rho \propto \tan \gamma$), as is clear in Eq. (7.4) for spherical aberration.

7.2 **Third-Order Aberrations**

The third-order aberration coefficients can all be calculated from paraxial ray trace data by tracing two rays: a marginal and a chief (also called principal) ray [4]; see Figure 6.12. In modern computer-based lens design these coefficients are calculated by the software using these two rays and a number of different graphical representation are provided to allow the evaluation of the lens performance. It is difficult to independently analyze a single aberration with a simple optical system but in this qualitative discussion we will assume that only the aberration under discussion is present; all other aberrations will be assumed to be zero. In addition to the presentations we have already discussed we will also use spot diagrams. To produce a spot diagram, a number of rays are traced through the optical system to the image plane. The intersection of these rays with the image plane is represented by small circles or squares. Normally, the wave properties of light and the intensity of the rays are not used in the creation of spot diagrams. Experimentally produced spot diagrams are images of the focus of a plane wave. It naturally involves the wave properties of light and its utility is based on linear, time-invariant, system theory that we will discuss in the chapter on imaging.

We have used experimentally determined spot images to display some of the aberrations we are discussing. The impulse response (our spot image) is generated by illuminating the optical system with a uniform plane wave and recording the energy distribution in the focal plane. Actual images of a focal spot in an optical system with various amounts of spherical aberration are shown in Figure 7.15.

7.2.1 **Spherical aberration**

We have selected only one aberration to explore in detail, spherical aberration. We will, however, generate representations of all of the aberrations to demonstrate their effects on images. Spherical aberration can be considered alone because it is the only term independent of h. On axis, it is the only aberration that is not identically zero; see Eq. (7.4). Spherical aberration

Spherical Aberration

Increasing Spherical Aberration

Figure 7.15 Spot diagrams of a lens as the observation point is adjusted to best focus. The aberration shown is spherical aberration.

is also symmetric about the optical axis, so we can perform all ray tracing in the meridional (x,z)-plane.

The ray intercept curve in Figure 7.14 is a cubic function of the radial position in the exit pupil, ρ, as predicted by Eq. (7.4). We find that as the aperture of the optical system increases so does the spherical aberration and more and more energy appears outside the paraxial focus. In Figure 7.15 we have recorded the image of a point source in the paraxial focal plane for an optical system with increasing amounts of spherical aberration present. We increased the spherical aberration by increasing the aperture of the optical system. At first the optical system is diffraction limited, diffraction dominates, and as the aperture size increases, the Airy disk decreases in size as predicted by diffraction theory (we will explore this property later in out discussion of diffraction). Quickly, however, the spherical aberration dominates and we see a spreading of energy in the paraxial focal plane. The image of an on-axis object point is circular in shape but with no internal structure. For the human eye the transition from diffraction limit to spherical aberration occurs for an eye aperture of about 2.5 mm.

We observe in Figure 7.8 that rays through the outer zones focus nearer the lens than the paraxial rays in the inner zones. Spherical aberrations became a major issue upon the launch of the Hubble Space Telescope (April 24, 1990). Once in orbit it was discovered that Hubble did not meet its intended level of performance. The performance shown in Figure 7.16 was not much better than what could be produced by ground-based telescopes. The problem was identified as spherical aberrations.

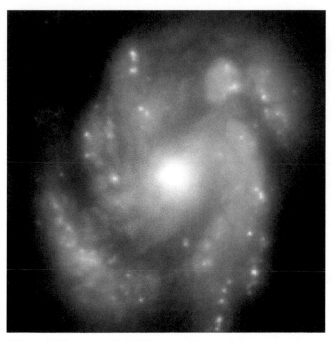

Figure 7.16 Image of Galaxy M100 taken by the Hubble Space Telescope before spherical aberrations were corrected. (News release ID: STScI-1994-01. Release date: January 13, 1994. Credit: NASA, STScI.)

1

Figure 7.17 The same galaxy as in Figure 7.16 made by the Hubble Space Telescope after spherical aberration correction.
(News release ID: STScI-1994-01. Release date: January 13, 1994. Credit: NASA, STScI.)

What had happened was that a washer had been placed in a testing jig as a temporary spacer during shaping of the mirror surface but was not removed before final testing. The result was an error in the shape of the mirror that was not identified before launch. The outer edges were too flat by as much as 2 μm. Optical designers were able to save the telescope by adding secondary mirrors that removed the spherical aberrations; see Figure 7.17.

7.2.2 Coma

The first off-axis aberration we wish to discuss is called coma. The image of an off-axis spot in the presence of coma contain circular light distributions, displaced with respect to each other, forming a comet shape; thus, the name, coma (Figure 7.19). The aberration is determined by measuring the distance between the ray intercept of the chief ray, the head of the comet, and a zonal ray in the comet tail. Physically, the origin of the coma spot is due to the magnification of the object rays passing through marginal zones, having different values from rays passing through zones near the optical axis.

To describe the formation of the comet, it is necessary to follow several skew rays. We will not treat skew rays but indicate their complexity. To treat the skew rays, each ray is decomposed into a sagittal component and a tangential component. The *tangential plane* (another name for the meridional plane) is defined as any plane containing the object point and the optical axis of the system; see Figure 7.18. Throughout the book all optical drawings have been of the

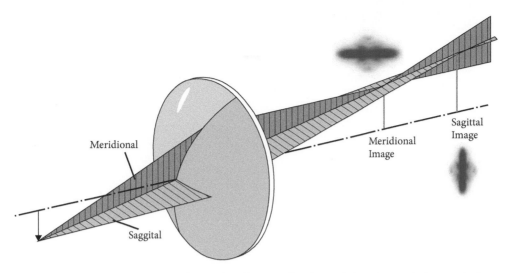

Figure 7.18 Graphical representation of the meridional (tangential) and sagittal planes in an optical system. In the presentation here, the x,z-plane is the meridional plane. The images (they are real images and thus are representative of wave behavior) labeled Meridional Image and Sagittal Image will be discussed in covering astigmatic aberrations.

x,z-plane, the tangential plane. In the exit pupil of Figure 7.13 it is defined as the ξ,z-plane. All rays confined to this plane are called tangential rays.

The plane at right angles to the tangential plane and containing the chief ray is called the *sagittal plane*, in Figure 7.18. The sagittal plane changes at each surface but at the pupils and the aperture stop, the sagittal plane is also the η,z-plane (or y,z) in Figure 7.13. This is because the chief ray crosses the optical axis at these positions. It should be pointed out that the chief ray lies in both the tangential and sagittal planes—it is the only ray that does so.

Because of the change in magnification with ray height, rays passing through a marginal zone are focused at a different height above the optical axis than the chief ray, labeled O in Figure 7.19. To generate the pattern formed by this aberration assume that the image height, h, and the zone radius, ρ, are fixed. The fixed zone radius defines a large circle in the exit pupil shown at the left of Figure 7.19. The value of h defines the position of O in the image plane and scales the size of the comet. Table 7.2 uses Eq. (7.4) to determine the ray intercepts in the sagittal and tangential planes for rays passing through the points labeled A,B,C,D,A.

From the results in Table 7.2 we see that the coma is given by

$$C_T = 3\sigma_2^2\rho^2 h \quad C_s = \sigma_2^2\rho^2 h. \tag{7.5}$$

About 55% of the light falls in the region between O and C in Figure 7.19 so that the sagittal coma, C_S, is a useful measure of the image quality. Because of the quadratic dependence of C_S on ρ, Eq. (7.5), we would expect the ray intercept curve to be parabolic in shape. Figure 7.20 shows, on the left, a ray intercept curve for a Cooke triplet lens. Coma was the dominant aberration in this lens design so the general shape of the curve is parabolic.

In Figure 7.21 a set of images of a point source produced by an optical system with all off-axis aberrations other than coma minimized. At the top is a image from a diffraction limited optical system, the Airy pattern. The images were made with increasing amounts of coma. Unlike the

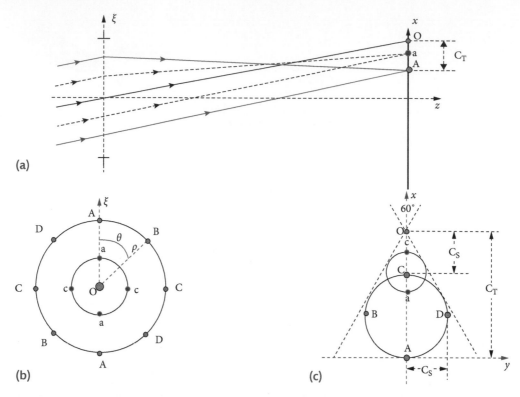

Figure 7.19 Coma is due to aperture zones having different magnification. (a) Off-axis rays are focused at different heights above the optical axis with the height proportional to the square of the zone radius; i.e., each aperture zone has a different magnification. (b) We select rays from two different zones in the exit pupil. (c) At the image plane the tangential rays, A and a, focus at a distance C_T from the principal ray O while the sagittal rays, C and c, focus a distance C_S from the principal ray. The envelope of the circular images created by rays passing through individual zones form a 60° wedge, which means that $C_T = 3C_S$.

Table 7.2 Ray Intercepts for Coma

Label	θ	Tangential	Sagittal
A	0°	$3\sigma_2^2\rho^2 h$	0
B	45°	$2\sigma_2^2\rho^2 h$	$\sigma_2^2\rho^2 h$
C	90°	$\sigma_2^2\rho^2 h$	0
D	135°	$2\sigma_2^2\rho^2 h$	$-\sigma_2^2\rho^2 h$
A	180°	$3\sigma_2^2\rho^2 h$	0

spot diagrams, these images represent the wave performance of the optical system as we stated in our discussion of spherical aberration. The fact that the fourth spot from the top is inverted from the others is due to an error in reproduction and not associated with the actual optical system. These spots can be used to estimate the imaging performance of the optical system.

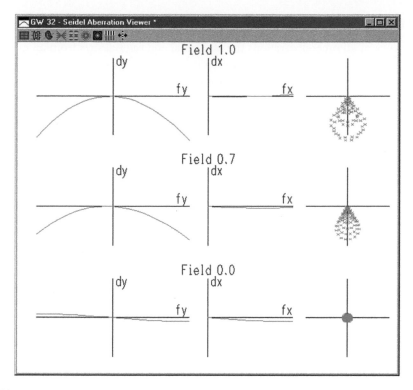

Figure 7.20 A ray intercept curve of a lens whose dominate aberration is coma. Spot diagrams show amount of coma.

Figure 7.21 Images of a point source produced by an optical system with increasing amounts of coma. The top image is the diffraction limited spot, the Airy pattern. The fourth figure from the top was accidently inverted in production.

7.2.3 **Astigmatism**

In the absence of all other aberrations the tangential and sagittal rays from an off-axis point do not focus in the same plane. This aberration is due to the lens exhibiting a different power[6] in the tangential plane and in the sagittal plane. If we look at the wavefront expansion term (see Eq. (7.3)) we see that there is no aberration in the sagittal plane, but there is a quadratic distortion of the wavefront in the tangential plane. As far as the off-axis wave is concerned, the lens is tilted and the curvature of the lens is different in the two planes. The additional curvature brings the tangential component of the wave to a focus earlier than predicted by paraxial theory. A spherical lens with astigmatism behaves, for off-axis rays, as if a cylindrical lens were in contact with the spherical lens. Figure 7.22 displays the focal properties of such a lens.

The rays in the tangential plane come to a focus in the tangential image plane, called the tangential field. The tangential rays continue on from the focus to form a line image of the point source, called the sagittal focal line or the sagittal field. Before reaching their focal plane, the sagittal rays form a line image, called the tangential focal line. Figure 7.18 clarifies this description by showing that the sagittal focal line, which defines the sagittal field, lies in the tangential plane and that the tangential focal line, which defines the tangential field, lies in the sagittal plane.

In an optical system with astigmatism two-line images of a point source are produced. Astigmatic images of a point source, recorded in the paraxial plane, can be seen in Figure 7.22. Astigmatism of the optical system increases as you move down the images from top (diffraction limited) to bottom.

A graphic presentation using Figure 7.23 can help identify these two image planes. Because of the axial symmetry of the optical systems, we know that the tangential plane represents an infinite number of planes about the optical axis over the angular range

Figure 7.22 Images of a point source produced by an optical system with increasing amounts of astigmatism. The images were produced in the paraxial image plane.

---

[6] Remember the power is the reciprocal of the focal length of the optics in meters.

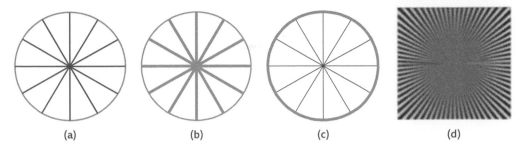

Figure 7.23 (a) Object to be imaged by optical system with astigmatism. (b) Image of wheel in tangential focal plane. (c) Image of wheel in sagittal focal plane. (d) Spoke target showing difference in resolution in the horizontal and vertical planes.

$0 \le \theta \le 2\pi$.

If we place the wheel shown in Figure 7.23a in the object plane of the optical system, then the image in the tangential field should have the appearance of Figure 7.23b. The name tangential now assumes some significance. The points along the circumference of the wheel are all associated with tangential rays and are in focus.

The spokes of the wheel are normal to the circumference of the wheel and are associated with rays in the sagittal plane, which is defined as the plane normal to the tangential plane. All of the spokes are thus imaged in the sagittal field, as shown in Figure 7.23c.

Astigmatism is a common aberration in the human eye. Almost everyone has a little bit. It arises when your cornea and/or lens is egg shaped rather than spherical. In a glass optical system the surfaces are usually spherical rather than distorted, so astigmatism is not the result of a misshapen optic but rather rays in the sagittal and tangential planes make different angles with respect to the normal of the optical surface.

A ray intercept curve is not as useful for astigmatism because the curve only displays intercepts for rays in one plane and the rays are focusing in different planes. Instead of a ray intercept plot, the z-coordinate of the tangential and sagittal image points are plotted as a function of h. The z-coordinate is found by dividing the intercept value, ε_x or ε_y, by the tangent of the ray angle, θ. The result is a pair of curves such as those shown in Figure 7.24. The shift of the off-axis image point along the optical axis is called the sagitta or sag of the lens. The curves are parabolic in shape due to the dependence upon h, Eq. (7.4).

7.2.4 Field curvature

There is another aberration term with a functional form similar to the astigmatism aberration term. It is an off-axis aberration that is associated with an image plane shift that changes with image height, h^2; i.e., the term is associated with an image plane curvature that is parabolic. This aberration is called field curvature or Petzval curvature.[7] Every point in the object is focused perfectly but on a curved image plane (Figure 7.25).

The field curvature is measured in the same way as astigmatism, by the displacement along the optical axis of the image point as a function of the image height. The curvature

---

[7] Field curvature is governed by the Petzval Theorem: Petzval Sum $= \sum_i \Phi_i / n_i$.

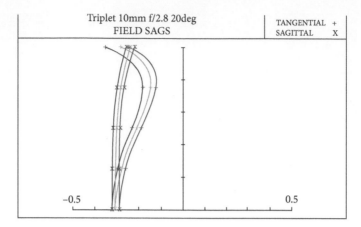

Figure 7.24 The field sags of a lens with astigmatism. The amount of astigmatism for three different wavelengths is shown.

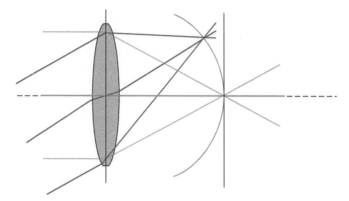

Figure 7.25 Field curvature.
By BenFrantzDale (own work) (CC-BY-SA-3.0
(http://creativecommons.org/licenses/by-sa/3.0) or
GFDL (http://www.gnu.org/copyleft/fdl.html),
via Wikimedia Commons.

is proportional to the sum of the powers of the optical elements. We can combine positive and negative lenses to reduce the field curvature, but this approach cannot be blindly applied because the lens system needs curvature to carry out its normal imaging role. Often a negative lens called a field flattener is placed at or near an image plane where it has little effect on image quality but reduces the field curvature by reducing the Petzval sum of the curvatures of the entire imaging system.

7.2.5 Distortion

The final aberration is called distortion and it is the result of the magnification changing with off-axis image distance. If we have corrected all of the other aberrations, the image for each point in the object is formed perfectly but at the wrong height above the optical axis.

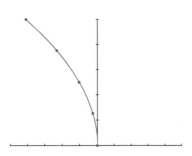

Figure 7.26 This image displays barrel distortion and is associated with negative distortion, displayed in the graph at the left.
(Image courtesy of Sean McHugh, http://www.cambridgeincolour.com/tutorials/lens-corrections.htm.)

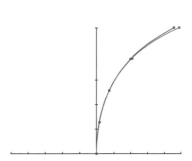

Figure 7.27 Pincushion distortion can be seen as the edges of the walls in the figure curve outward at the top and bottom. Pincushion is associated with positive distortion, displayed in graph at the left.
(Image courtesy of Sean McHugh, http://www.cambridgeincolour.com/tutorials/lens-corrections.htm.)

Barrel: If the off-axis distance increases more slowly than the object, we have *barrel* distortion, as shown in Figure 7.26.

Pincushion: If the off-axis distance increases faster than in the object, one has pincushion distortion, as shown in Figure 7.27.

Distortion is measured by measuring the fractional change in image position, and a distortion plot is shown at the left of each of the distortion figures.

7.3 **Aberration Correction**

From the simple thin lens equation

$$\frac{1}{f} = (n-1)(1/R_1 - 1/R_2)$$

you see that there are an infinite number of curvatures for a given index that will produce a desired focal length. Changing the curvatures of the lens surface (called bending in the trade) allows the designer the ability to change the aberrations produced by the lens without changing the focal length. This is the basic tool of optical design.

You cannot efficiently correct the aberrations of an optical system by choosing modifications to a design at random. To learn the methods for correcting the aberrations we have just introduced and to learn to correct higher order aberrations would require several textbooks the size of this one. There is one rule of thumb that we can give you that can be used without any detailed calculations.

Rule of Thumb: You can improve the imaging capability of a lens by orienting the side with the larger radius of curvature pointing toward the most distant object. This divides the power of the lens more equally between the object and image positions.

There is one aberration we can treat without the need to develop a detailed theory; that is, the elimination of image defects due to the presence of light containing a variety of wavelengths—called *chromatic aberration*. Previously, we have ignored any variation in response to different wavelengths. Now we will briefly review the behavior of optical materials to different wavelengths. With this knowledge we can learn how to correct the chromatic aberration in an optical system.

7.3.1   Chromatic aberration

We have assumed the properties of the dielectric constant, ε, in optical materials to be independent of wavelength. In Figure 7.28 we indicate in a schematic drawing that ε and therefore the index, n, will actually show complicated behavior as a function of wavelength. The lower curve gives the absorption properties (the imaginary part of the index) and the upper curve indicates the index of refraction over a wide range of E&M frequencies. The peak in absorption around 10^{12} Hz would be due to a quantum resonance associated with molecules and the peak around 10^{15} Hz would be associated with electronic transitions. We can work near these resonance to enhance nonlinear optical performance but usually one would stay away from a region with high loss.

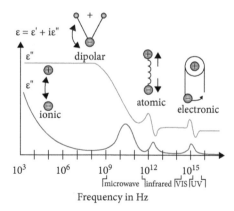

Figure 7.28 An illustration of the frequency response of various dielectric mechanisms in terms of the real and imaginary parts of the permittivity.
Image used with the consent of Professor Kenneth A. Mauritz, available from https://commons.wikimedia.org/wiki/File:Dielectric_responses.svg.

GaP is a popular III–V semiconductor used from the ultraviolet (UV) to the long wave infrared (LWIR). The optical properties over that wavelength range are shown in Figure 7.29. GaP shows resonant behavior in the UV and the infrared (IR) but between these two resonant structures the index is nearly constant. GaP would be a useful optical material over the range of wavelengths where the nearly constant index behavior occurs.

The dispersion in optical glasses, illustrated in Figure 7.30, produce wavelength-dependent optical properties similar to that of GaP. The smooth increase in index as we move from the IR

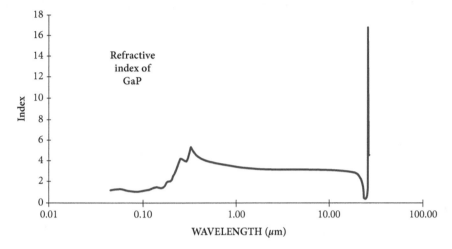

Figure 7.29 Index of refraction as a function of wavelength showing the region 1.0 μm $\leq \lambda \leq$ 10 μm, where GaP would be considered as a useful optical material.

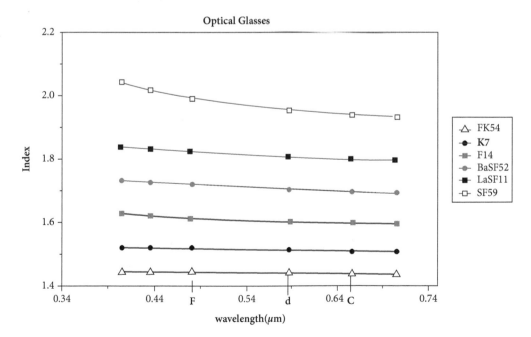

Figure 7.30 The index of refraction as a function of wavelength for a few optical glasses [5]. The materials were selected to show the range of indexes available for visible optics. Outside of the visible region, there are few indices available to the optical designer. The wavelengths labeled C, d, and F are Fraunhofer lines used to characterize the dispersion in optical glasses.

to the UV is called *normal dispersion*. *Anomalous dispersion* displays a decrease in index as we move from long to short wavelengths.

Refraction displays a wavelength dependence that leads to the dispersion of white light into individual colors by a prism, as shown in Figure 7.31.

The wavelength dependence of refraction produces a modification in the performance of a lens called primary chromatic aberration (Figure 7.32). We will present an example of aberration correction by correcting this chromatic aberration.

Chromatic aberration is a result of a variation in the focal length, or equivalently the power, of a lens due to the wavelength dependence of the index of refraction:

$$\frac{1}{f} = \Phi = (n-1)\left(\frac{1}{R_1} - \frac{1}{R_2}\right)$$

$$\delta\Phi = \delta n\left(\frac{1}{R_1} - \frac{1}{R_2}\right) = \frac{\delta n}{n_i - 1}\,\Phi = \frac{\Phi}{\nu_i}, \tag{7.6}$$

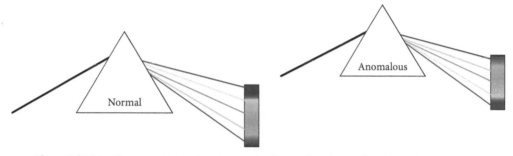

Figure 7.31 The effects on refraction by prisms made of materials with normal and anomalous dispersion.

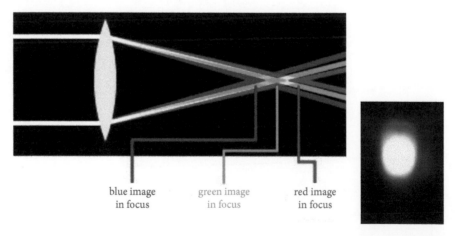

blue image in focus   green image in focus   red image in focus

Figure 7.32 On the left is longitudinal chromatic aberration with the blue focus occurring first and the red focus occurring last. On the right is an image of the best focus displaying lateral chromatic aberration, with blue light appearing on the top of the image and red light on the bottom.

where ν_i is the Abbe value for glass type "i." The definition of the Abbe value[8] is

$$\nu_D = \frac{n_D - 1}{n_F - n_C},$$ (7.7)

evaluated at the index value of the glass measured at each of the three wavelengths: D, F, and C; see Table 7.3. The change in the focal length is called the longitudinal chromatic aberration or longitudinal color. The magnification of a lens also changes with wavelength. This change is called the lateral chromatic aberration or lateral color.

The dispersion properties of glasses are measured at standard wavelengths equal to prominent absorption lines in the solar spectrum (Figure 7.33) that were labeled with letters by Joseph von Fraunhofer. These letter labels, defined in Table 7.3, are used in all discussions of dispersion in the visible region of the spectrum.

The Abbe value, defined in Eq. (7.7), is used to characterize the dispersion of a particular glass. Its definition suggests an infinitesimal change but in practice it is defined in terms of three discrete indices. The numerator, $n_d - 1$, in Eq. (7.7) is called the *refractivity* and the denominator, is called the *partial dispersion*. If you check Table 7.3 you will see that the partial dispersion spans the visible spectrum with the refractivity positioned roughly in the center.

Glasses are grouped into two types: crown and flint. A very long time ago crown glass was a kind of window glass composed of soda, lime, and silica. The glass was formed by blowing a globe and whirling it into a disk. The appearance of the disk led to the name crown (Figure 7.34).

Flint glass[9] was composed of alkalis, lead oxide, and silica and used for cut-glass tableware. It was first produced by George Ravenscroft under commission to the trade organization entitled the Worshipful Company of Spectacle Makers of London [7] (they still operate in London). Ravenscroft discovered that by adding lead to flint glass he could prevent the tendency of flint glass to darken. This discovery allowed England to overtake Venice as the leading glass producer.

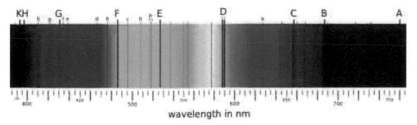

Figure 7.33 Fraunhofer lines are the dark lines in the solar spectrum. By Fraunhofer_lines.jpg: nl:Gebruiker:MaureenVSpectrum-sRGB.svg: PhroodFraunhofer_lines_DE.svg: public domain, https://commons. wikimedia.org/w/index.php?curid=7003857.

---

[8] In optics literature, the Abbe value is also known as the V-number, the constringence, or the reciprocal dispersive power.

[9] Flint glass is also known as lead glass.

Table 7.3 Standard Wavelengths

Fraunhofer designation	Wavelength (nm)	Element
r	706.5188	He
C	656.2725	H
C'	643.8469	Cd
D1	589.5923	Na
D2	588.9953	Na
d	587.5618	He
e	546.0740	Hg
F	486.1327	H
F'	479.9914	Cd
g	435.8343	Hg
h	404.6561	Hg

Today the adjectives crown and flint separate glasses into low dispersion, low index crowns that fulfill the inequalities

$$n_d > 1.60 \qquad v_d > 50$$
$$n_d < 1.60 \qquad v_d > 55,$$

and high dispersion, high index flints, which do not fulfill these inequalities.

Glass types are identified by an alphanumeric such as BK7 and by a numerical index: 517642. The letter B indicates a borosilicate and the letter K indicates a "crown"-type glass. The numerical index provides much more information about the glass type but is not easily to remember. The first three digits are the refractivity of the glass,

$$n_d - 1 = (1.51680 - 1) \times 10^3 \approx 517,$$

and the last three digits are the Abbe number:

$$v_d = (64.17) \times 10 \approx 642.$$

A lens where $\delta\Phi = 0$ is called an achromat. From Eq. (7.6), we see that a single lens cannot be an achromat unless it has zero power. We can combine several lenses to approach an achromat (achromatic lens). It is easy to show that the focal length of two lenses is given by

$$\frac{1}{f} = \frac{1}{f_1} + \frac{1}{f_2},$$

Verrerie en bois, Grande Verrerie en Plats. Intérieur d'une Halle et différentes Opérations de la Verrerie à Vitres.

Figure 7.34 Manufacturing method of crown glass. This technique was replaced by cast and rolled glass by about the 1660s [6].
By *Diderot Encyclopedia*—http://glassian.org/Making/Diderot/crown_glasshouse.html, public domain, https://commons.wikimedia.org/w/index.php?curid=18727601.

where f_1 and f_2 are the focal lengths of the individual lenses. In terms of the power of a lens, this result is written

$$\Phi = \Phi_1 + \Phi_2. \tag{7.8}$$

The variation in the power of the lens pair with wavelength is given by

$$d\Phi = \delta\Phi_1 + \delta\Phi_2 = \frac{\Phi_1}{\nu_{d1}} + \frac{\Phi_2}{\nu_{d2}}. \tag{7.9}$$

Newton believed that all glasses exhibited the same dispersion, making it impossible to produce an achromat. His belief led him to invent the reflecting telescope. A lawyer and amateur astronomer, Chester Moor Hall proved Newton wrong by discovering that crown and flint glass could be combined into a doublet with no chromatic aberration. He attempted to keep his discovery secret by using different London lens makers to produce the two

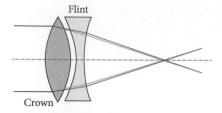

Figure 7.35 Dollond's telescope objective. This design differed from the one produced by Hall by interchanging the crown and flint. This design was more rugged than Hall's design and easier to manufacture [6]. Dollond's design therefore dominated the marketplace.

components of his doublet. Unfortunately, both lens makers were swamped with work and subcontracted the job to the same individual. That individual shared the information with John Dollond. Dollond recognized the purpose of the two lenses and applied for a patent on the idea (Figure 7.35). The court awarded Dollond the patent even though Moor Hall was the inventor. The judge ruled that the person who made an invention public for the general good deserved the patent.

By proper choice of optical glasses, we can keep the power required by Eq. (7.8) constant while reducing the variation in the power predicted by Eq. (7.9) to 0:

$$d\Phi = 0,$$
$$\frac{\Phi_1}{\nu_{d1}} = -\frac{\Phi_2}{\nu_{d2}} \tag{7.10}$$

We can solve Eq. (7.10) for the focal lengths to obtain

$$-\frac{f_1}{f_2} = \frac{\nu_{d2}}{\nu_{d1}}. \tag{7.11}$$

Because all optical glasses have an index that rises as the frequency rises (called normal dispersion), the Abbe value must be positive (a diffractive optical element would have a negative Abbe value). Equation (7.11) therefore implies that an achromat must be a combination of a positive and a negative lens. Simultaneous solutions of Eqs. (7.8) and (7.10) yield

$$\Phi_1 = \frac{\nu_{d1}}{\nu_{d1} - \nu_{d2}} \Phi, \qquad \Phi_2 = \frac{\nu_{d2}}{\nu_{d2} - \nu_{d1}} \Phi. \tag{7.12}$$

We see that the powers of the two lenses making up the achromat are inversely proportional to the difference in the Abbe values, $\Delta\nu$, of their glasses. High power, short focal length, lenses are more expensive, have a smaller maximum aperture, and have larger aberrations than low power lenses. For this reason, glasses that yield a $\Delta\nu > 20$ are usually selected in an achromat design. The properties of commercial glasses are shown in the Abbe diagram in Figure 7.36. A sample of optical glasses found in the *Schott Optical Glass Catalog* (no. 3111) are listed in Table 7.4.

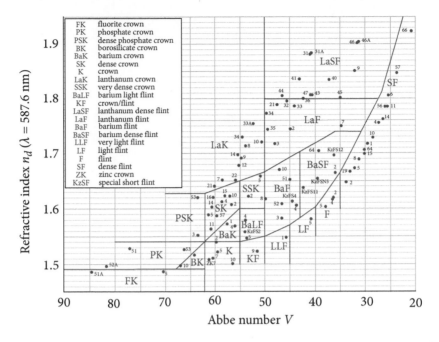

Figure 7.36 An Abbe diagram, also known as "the glass veil," plots the Abbe number against refractive index for a range of different glasses (dots).
By Abbe-diagram.png: Bob MellishAbbe-diagram.svg: Eric Bajartderivative work: Eric Bajart—Abbe-diagram.svg, CC BY-SA 3.0, *https://commons.wikimedia.org/w/index.php?curid=10472043*.

Table 7.4 Optical Glasses

Glass type	n_d	ν_d	$n_F - n_C$	n_C
BK1 510635	1.51009	63.46	0.008038	1.50763
BK7 517642	1.51680	64.17	0.008054	1.51432
SK4 613586	1.61272	58.63	0.010451	1.60954
F2 62036	1.62004	36.37	0.017050	1.61503
F4 617366	1.61659	36.63	0.016834	1.61164
SF6 805254	1.80518	25.43	0.031660	1.79609

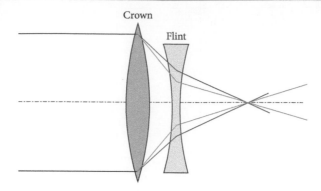

Figure 7.37 Achromat produced using the procedure outlined in the text.

We now can apply the information we have learned to construct an achromatic doublet using BK7 as the crown and F4 as the flint component. From Eq. (7.12) and the data in Table 7.4, we have

$$\Phi_1 = \frac{64.17}{64.17 - 36.63}\Phi = 2.33\Phi$$

$$\Phi_2 = \frac{36.63}{36.63 - 64.17} = -1.33\Phi.$$

The resulting achromat is shown in Figure 7.37. The power of the achromat is 43% of the power of the crown element and the flint's power is 57% of the crown. We see that the positive element overshoots the desired power and the overshoot is subtracted out by the negative element. Along with the reduction in power is a cancellation of the dispersion effects for two wavelengths.

The method of producing an achromat just described ensures that the focal length of the doublet is the same at two wavelengths, C and F light. At other wavelengths this is not the case. The chromatic aberration still present is called the *secondary spectrum*. If we wish to make the focal length the same at C, F, and d light, then we must make the variation in power equal to 0:

$$\delta\Phi_{d,C} = (n_{1d} - n_{1C})\frac{\Phi_1}{n_{1d} - 1} + (n_{2d} - n_{2C})\frac{\Phi_2}{n_{2d} - 1}$$

$$\delta\Phi_{d,C} = \left[\frac{n_{1d} - n_{1C}}{n_{1F} - n_{1C}}\right]\frac{\Phi_1}{\nu_{1d}} + \left[\frac{n_{2d} - n_{2C}}{n_{2F} - n_{2C}}\right]\frac{\Phi_2}{\nu_{2d}}. \tag{7.13}$$

The fractions in the brackets are called *relative partial dispersions*. Only if the partial dispersions of the two glasses are equal can we reduce the secondary spectrum by achromatizing the lens at three wavelengths. Abbe was the first to note that if you plot the relative partial dispersion versus ν_d for the various optical glasses, a nearly linear relation is observed between the two parameters. This means that we cannot meet the condition for reduced secondary spectrum and also fulfill the requirement that the Abbe values for the two glasses have a large difference. It is for this reason that it is impossible to reduce the chromatic aberration below that obtained by the simple calculations we have just carried out.

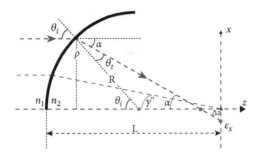

Figure 7.38 Ray trace to find spherical aberration for an acrylic plastic lens.

7.4 **Problem Set 7**

(1) Consider a light source placed at a fixed distance s from a screen, such that a lens of focal length f can be placed between the source and the screen. Show that as long as $f < f_{max}$ then there are two positions where the lens can be placed so that an image is formed on the screen, and find a value for f_{max}.

(2) A short-sighted textbook author has a far point 17 cm from his eye. What type and power of a lens does he need to correct the light so that light from infinity appears to come from 17 cm?

(3) For a particular far-sighted person, they have a near point of 102 cm. How powerful does the lens need to be if the person wants to see an object that is 27 cm away from their eye?

(4) Assume that the lens in our yard light is made of acrylic plastic with an index of $n_2 = 1.49$ (Figure 7.38). (a) Redesign the lens with this new index. (b) Using this lens, calculate the longitudinal spherical aberration.

(5) The longitudinal spherical aberration of two rays which have been traced through a system is 1.0 and 0.5; the ray slopes $(\tan \gamma)$ are 0.5 and 0.35, respectively. What are the transverse aberrations (a) in the paraxial plane, and (b) in a plane 0.2 before the paraxial plane?

(6) Find the radii of curvature for a cemented achromatic double with a focal length of 10 cm using BK7 as the crown and SF6 as the flint. Assume that the crown is equiconvex (the radius of curvature of both surfaces are the same) and set one curvature of the flint equal to the crown so they may be glued together.

(7) Redesign the lens in Problem 6 but now make the flint an equiconvex lens.

(8) Redesign the lens in Problem 6 using SK4 as the crown and F4 as the flint.

REFERENCES

1. Born, M., and E. Wolf, *Principles of Optics*, 7th edn. Cambridge, UK: Cambridge University Press, 1999.

2. Lambda Research Corporation, Littelton MA, 2008. Free limited version of optical design program for educational perposes.

3. Wyant, J. C., in *Using Webmathematica to Solve Optics Problems*. Wolfram, Wolframweb-Mathematica, 2018.

4. Welford, W. T., *Aberrations of the Symmetrical Optical System*. New York: Academic Press, 1974.

5. Schott North America (2007).

6. Guenther, B. D., *Optics and Photonics News* (10), 15–18 (1999).

7. Apothecaries' Hall, Black Friars Lane, London EC4V 6EL (2018).

8 Guided Waves

8.1 Optics in Communications

Optics grew as a result of needs in society to see images and to transmit information. One of the first needs where optics was able to aid in support of different religious orders in their preservation of sacred and secular texts for future generations. Because the printing press was not yet invented, the preservation was accomplished by monastic scribes copying and illuminating texts. The result was a selective preservation of the literary history of the West.

The problem the monks experienced was the same as experienced by people today; the vision of the scribes was not always equal to the task of making accurate copies and the problem grew as the monks aged. Around 1286 the first pair of eyeglasses was made near Pisa, Italy and by the start of the fourteenth century eyeglasses that allowed monks to make accurate copies were mass-produced and sold in Florence. By the end of the fifteenth century, low cost glasses[1] were sold throughout Europe by the thousands. The tight coupling of optics technology to the vision needs of the society could be measured by noting the volume of sales increase with the invention of the printing press (1450). The glasses were now for sale by traveling salesmen; see Figure 8.1. Another dramatic sale increase occurred with the invention of the newspaper around 1665.

Another societal need that caused a major growth in optics was communications. In the very early times you simply used a messenger to transmit information. To increase the flow of information you had to improve the roads or use a faster messenger, for example, a carrier pigeon. The invention of the printing press allowed an increase in the volume of information and thus an increase in the data rate of information transmission.

Optics first contributed to information transmission with the invention of the telescope around 1608. This allowed the English Navy to use signaling flags between ships at sea. A famous message sent by flags was by Admiral Horatio Nelson at Trafalgar in 1805: "England expects that every man will do his duty." The French copied and expanded on this transmission technique for use on land and build towers to create information channels. It was claimed that it only took two minutes to send a message. Even with this claim, carrier pigeons were used in World War I [1] and II.

Technology contributed to communication with the invention of the electric telegraph (1830s) followed by the telephone (1876), where information was transmitted over copper wires. Another important technology was the use of radio signals (1902) for information

---

[1] A pair of eyeglasses cost about one-third of a day's salary of a tradesman.

Modern Optics Simplified. B. D. Guenther. © B. D. Guenther 2020.
Published in 2020 by Oxford University Press. DOI: 10.1093/oso/9780198842859.001.0001

Figure 8.1 *A Vendor of Spectacles Showing His Wares to an Old Woman,* by Adriaen van Ostade.
Credit: *Wellcome Collection. CC BY 4.0 https://creativecommons.org/licenses/by/4.0*

transmittal through the air. Optics did not participate directly in the growth of communication technology during this stage but technology allowed the advancement of communication capability.

In 1880 William Wheeler invented the use of light pipes with reflective coatings to bring light into homes from a centralized source—just like central heating. During this same time a number of important medical applications turned to optics for solutions. For example, the use of glass rods to bring illumination into the body was introduced in 1888 and improved in 1898 by designing a surgical lamp based on glass rods.

Medicine was interested in bringing images of the internal organs out to a doctor and turned to bundles of optical fibers to accomplish this goal. Early attempts were disappointing but in 1954 Dutch scientist Abraham van Heel and British scientist Harold. H. Hopkins separately wrote papers published at the same time in the science magazine *Nature* on the use of bundles of optical fibers for image transmission. The contributions of these two papers were quite different. Hopkins figured out how to construct an ordered bundle of 10,000 fibers that could be used to transmit an image [2]. Van Heel designed a fiber bundle using fibers with a high index core and a low index coating [3]. The coating prevented cross-talk between the individual fibers. The combined contributions solved the problem of short-range image transmission for medical applications. Van Heel's idea of coating fibers also enhanced long-range communications.

The papers by Hopkins and van Heel led to the design of endoscopes for medical diagnosis but this application did not trigger much interest in communication technology. The medical applications required transmission over short distances but communications technology demanded long-distance transmission of the guided wave. Research in guided wave optics did not blossom into a full-scale research area until the patent (no. 3,711,262) written in 1970 by Robert Maurer, Donald Keck, and Peter Schultz from Corning Glass that addressed reduction of attenuation in fibers for long-distance signal transmission. The fiber design covered in this patent was capable of carrying 65,000 times more information than copper wire and had a loss of 17 dB/km at 633 nm.[2]

The Dorset (UK) police installed the first fiber-optic link in 1975. Two years later, the first live telephone traffic through fiber optics occurred in Long Beach, California. The standard cabled wire-pairs could transmit ~50 simultaneous phone calls using electrical current while an optical fiber was able to transmit 50 million.

There were problems that had to be solved before fiber optics could become the mainstay of communications. One of the first problems addressed was attenuation. The loss in the fiber mentioned above has now been reduced to 0.2 dB/km, allowing repeater station spacing to be determined by issues other than attenuation. Repeater spacing of 160 km has been demonstrated. We will discuss some of the other improvements as we learn how the optical fiber is able to guide light.

8.2 Guided Waves

In the discussion of the Fabry–Perot interferometer, a light wave was assumed to undergo multiple reflections at the boundaries of a lossless dielectric (see Figure 4.31). The wave trapped in the dielectric layer slowly died away because of the light transmitted out of the dielectric layer at each reflection from a boundary. We can eliminate the reflective loss by ensuring that the incident angle of the wave in the layer meets the requirement for total reflection (Eq. (3.43)), then the wave would not experience loss but would propagate through the dielectric layer; such a wave, confined to the layer, is called a *guided wave* and the dielectric layer is called a *waveguide*. In the early days of fiber optics, plane structures as well as fibers were studied, anticipating the use of the planar structure to create optical circuits.

In this section, we will use geometric optics to discover the properties of guided waves. The theory to be used is not a "pure" geometric theory because we use the properties of interference and of phase change upon total reflection that we derived using wave theory. The proper view of this approach is that we assume that the guided waves are plane waves and the "rays" are normal to these plane waves and indicate the direction of the wave vector.

We will discuss the requirements placed on the angle of incidence in the waveguide and briefly review techniques for coupling energy into guided wave modes of the waveguide. The discussion will be limited to a one-dimensional problem by assuming that the waveguide, Figure 8.2, has a thickness, d, in the x-direction and extends to infinity in the y- and z-directions with light propagating along the z-direction. The results for the dielectric slab of Figure 8.2 in general will not apply to optical fibers that have cylindrical symmetry. However, the results

---

[2] This is a visible laser wavelength emission from a HeNe laser.

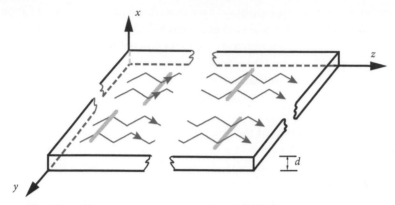

Figure 8.2 Geometry of planar waveguide. Light is propagated along the *z*-direction.

Figure 8.3 Laser fountain.
By Sebastien Forget on Wikipedia (https://upload.wikimedia.org/wikipedia/commons/c/ca/Fontaine_laser1.jpg).

derived here will apply to those rays in a fiber that pass through the cylindrical axis of the fiber. The rays that cross the fiber axis as they propagate are the *meridional rays* we met in ray tracing; all other rays are called *skew rays*. We will look at an example of a single skew ray in a fiber to see what impact a skew ray places on propagation through the guide but will avoid the complexity that accompanies development of a theory that includes skew rays.

Swiss physicist *Daniel Colladon* first demonstrated the guided wave in a light water-jet experiment in Geneva in 1841 but *John Tyndall* was given credit for the idea; evidently, his demonstration in 1854 received more publicity. Now water–light shows can be found at every resort area (Figure 8.3).

The ancient Romans knew how to heat and draw out glass into fibers of such small diameter that they became flexible. They also observed that light falling on one end of the fiber was transmitted to the other. The use of guided waves was not practical until the idea of cladding a core glass by another glass, of lower index, was proposed by Van Heel. Figure 8.3 shows the problem of a guide without cladding. With cladding or very smooth sides, there would be no illumination escaping the high index core of the fiber and the water jets in Figure 8.3 would be invisible. The water jets do not have smooth sides or cladding so we can see light scattered from the modes of the waveguide formed by the jets.

A theoretical investigation of guided waves did not appear until the publication by *Peter Debye* in 1910 [4]. The theoretical derivation we will use is a much simpler one called the zigzag method [5]. Consider a three-layer sandwich of dielectric layers with indices $n_2 > n_3 > n_1$. A ray is introduced into the region of index n_2 so that the angle of incidence of the ray satisfies the inequality

$$\sin^{-1}\left(\frac{n_3}{n_2}\right) < \theta_2 < \frac{\pi}{2}. \tag{8.1}$$

These rays, called *waveguide modes*,[3] are associated with waves trapped in the region of index n_2 that we will call the waveguide region or the guiding layer; see Figure 8.4.

8.2.1 **Guided modes**

In free space, we proved (Chapter 2) that the electric and magnetic vectors are perpendicular to the propagation vector of the electromagnetic wave. It is customary to denote such waves as *transverse electromagnetic* (TEM) waves in guided wave theory. In Chapter 4, we found that for the case of total reflection, we no longer had transverse waves; therefore, we must modify the notation for the polarization of waves undergoing total reflection. The new notation defines *TE waves* as waves with a longitudinal component of the magnetic field (these waves correspond closely to the E_N polarization for free space waves) and *TM waves* as waves with a longitudinal component of the electric field. We will limit our discussion to the TE waves because the results

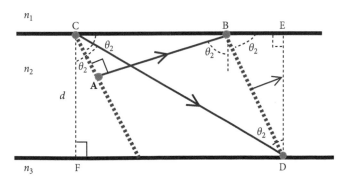

Figure 8.4 Geometry for finding the guided modes using the ray model of optics.

---

[3] Mode refers to a standing wave of an oscillating system; in one dimension, a guitar string is a good model. In a single mode each part of the system moves with the same frequency, i.e., the resonant frequency, and with a fixed phase to its neighbors. In music the modes are called harmonics or overtones.

are qualitatively the same for the two polarizations and the purpose of our discussion is to develop a qualitative understanding of guided waves.

Using Figure 8.4, we follow a plane wave propagating through the waveguide of index n_2 and thickness d. We have denoted the wavefronts associated with the rays, which are assumed planar, by dotted blue lines in Figure 8.4. One ray travels from C to D and undergoes two reflections while a second ray goes from A to B and experiences no reflection. Since the wavefronts, dotted lines AC and BD in Figure 8.4, are surfaces of constant phase traveling in parallel directions, we must have the phase difference associated with the two propagation paths equal to a multiple of 2π.

The optical path lengths associated with the distances CD and AB can be determined using the triangles drawn in Figure 8.4. Triangle CDF yields

$$\frac{d}{\overline{\text{CD}}} = \cos\theta_2.$$

Triangle ABC yields

$$\frac{\overline{\text{AB}}}{\overline{\text{CB}}} = \sin\theta_2.$$

Triangle CDE yields

$$\frac{\overline{\text{CE}}}{d} = \tan\theta_2.$$

Finally, triangle BDE yields

$$\frac{d}{\overline{\text{BE}}} = \tan\theta_2.$$

The distance along the guide axis is

$$\overline{\text{CE}} = \overline{\text{CB}} + \overline{\text{BE}}.$$

Solving for CB and substituting for CE and BE

$$\overline{\text{CB}} = d\left(\tan\theta_2 - \frac{1}{\tan\theta_2}\right).$$

The phase difference between the two paths is given by

$$n_2 k\left(\overline{\text{CD}} - \overline{\text{AB}}\right) + \delta_1 + \delta_3,$$

where δ_1 and δ_3 are the phase shifts associated with total reflection and whose values can be obtained from Eq. (3.47). We require that the total phase difference be equal to an integer multiple of 2π:

$$n_2 k\left(\overline{\text{CD}} - \overline{\text{AB}}\right) + \delta_1 + \delta_3 = m \cdot 2\pi \tag{8.2}$$

$$\frac{n_2 k d}{\cos\theta_2} - n_2 k d \sin\theta_2 \left(\tan\theta_2 - \frac{1}{\tan\theta_2}\right) + \delta_1 + \delta_3 = m \cdot 2\pi$$

$$\frac{n_2 k d}{\cos\theta_2}\left(1 - \sin^2\theta_2 + \cos^2\theta_2\right) + \delta_1 + \delta_3 = m \cdot 2\pi \tag{8.3}$$

$$\frac{n_2 k d}{\cos\theta_2} \cdot 2\cos^2\theta_2 + \delta_1 + \delta_3 = m \cdot 2\pi$$

$$2n_2 k d \cos\theta_2 + \delta_1 + \delta_3 = m \cdot 2\pi$$

This relation describes how the propagation of the guided wave varies with wavelength and is called the *dispersion relation*[4] for TE guided waves. We have neglected any effect on the plane waves arising from the finite size of the guide. It sounds a little strange since we are working in a narrow waveguide but this is an infinite plane wave result.

We see that θ_2 can only have discrete values, θ_m, determined by the integer values of m in Eq. (8.3). The rays associated with the discrete values of θ_m are the waveguide modes. Figure 8.5 displays a discrete mode propagating in a dielectric layer (xylene floating on a water surface). In this guide, $n_1 = 1.0$, $n_2 = 1.5054$, $n_3 = 1.33$.

The number of bounces the wave makes as it propagates through the layer can characterize the mode in Figure 8.5. Remember that to have total reflection, the angle θ_2 must satisfy the inequality for θ_m

$$\sin \theta_m > \frac{n_3}{n_2} \quad or \quad \cos \theta_m < \sqrt{1 - \left(\frac{n_3}{n_2}\right)^2}.$$

This means that there are a maximum number of modes, $m_{\max}$, that can propagate in the guide. This maximum value is given by

$$m_{\max} \leq \frac{2dn_2}{\lambda_0} \sqrt{1 - \left(\frac{n_3}{n_2}\right)^2} + \frac{\delta_1 + \delta_3}{2\pi}. \tag{8.4}$$

We define the quantity

$$V = kdn_2 \sqrt{1 - \left(\frac{n_3}{n_2}\right)^2} = kd\sqrt{n_2^2 - n_3^2}, \tag{8.5}$$

called the *normalized film thickness*, for the case of planar waveguides and *normalized frequency* or simply the *V-number* for optical fibers. The inequality, Eq. (8.4), can be written in terms of the *V*-number

$$m_{\max} \leq \frac{V}{\pi} + \frac{\delta_1 + \delta_3}{2\pi}.$$

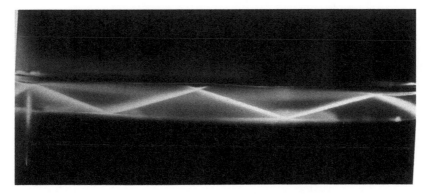

Figure 8.5 A light guide mode is shown propagating in a guiding layer of xylene bounded above by air and below by water containing a dye.

---

[4] The dispersion equation relates the frequency of the wave to its wavelength. Here it is a function of the geometrical boundary conditions of a waveguide, but it can also depend on the optical properties of the transmitting medium.

We will discover in a moment (in Eqs. (8.6) and (8.7)) that δ_1 and δ_3 are negative. Since m can only be a positive integer or zero, if $V < \delta_1 + \delta_3$, then no mode can propagate in the dielectric layer. This condition is called the *cut-off condition* and the V-number, where $2V_c = \delta_1 + \delta_3$ is defined as the cut-off value for the waveguide, V_c. We will return to this cut-off value of V in a moment.

The collection of modes can be used to create a generalized mode made from a superposition of the normal modes, as we did when we created a general temporal waveform with sines and cosines in Fourier theory in Chapter 1. Mathematically, normal mode means the modes are orthogonal to each other and one will not influence any of the other modes.[5] Since the number of modes in a guide is a function of V and $\delta_1 + \delta_3$ is limited in magnitude to be no larger than 2π, we can evaluate the contributions to the number of waveguide modes by the basic guide parameters from Eq. (8.5).

As an example of how the maximum number of modes and the normalized film thickness are calculated, we will use the xylene film floating on water in Figure 8.5. Assume the film layer is 1 mm thick. This is very thick for a conventional guide but is great when photographing the bouncing light ray. The index of xylene is $n_2 = 1.5054$ and exists between air with index $n_1 = 1.0$ and water with index, $n_3 = 1.33$. If the wavelength of the propagating light is 0.5 μm (green), the normalized film thickness is

$$V = \frac{2\pi}{5 \times 10^{-7}} \left(1 \times 10^{-3}\right) \sqrt{(1.5054)^2 - (1.33)^2} = 8859.$$

The maximum number of modes in this guide is[6]

$$m_{\max} \leq \frac{8859}{\pi} + 1 \approx 2820,$$

where we have assumed that the phase shift for a pair of reflections at the two interfaces is limited to 2π.

[6] The total number of modes is given by $m_{\max} = m + 1$.

The number of modes in the guide increase with increasing guide index, n_2, decreasing substrate index, n_3, increasing guide thickness, and decreasing wavelength.

We derived the functional form of δ_1 and δ_3 during our discussion of total reflection Eqs. (3.46) and (3.47). It is these parameters that contain the effects of polarization. For the TE waves, we can rewrite Eqs. (3.46) and (3.47) as

$$\delta_1 = -2 \tan^{-1} \frac{\sqrt{\sin^2 \theta_2 - \left(\frac{n_1}{n_2}\right)^2}}{\cos \theta_2} \tag{8.6}$$

[5] In real life, defects prevent this from being true, especially over long distances. For this reason, single mode, $m = 0$, is preferred for communication applications.

$$\delta_3 = -2 \tan^{-1} \frac{\sqrt{\sin^2 \theta_2 - \left(\frac{n_3}{n_2}\right)^2}}{\cos \theta_2} \tag{8.7}$$

Substituting these relationships into Eq. (8.3) yields a transcendental equation that must be solved to determine the values of θ_2 associated with the waveguide modes. A graphical solution to this transcendental equation will be presented after the introduction of several new variables that are often used in the wave formalism of guided wave theory to describe the observed propagation in a guide. If we assume that the ray path is collinear with the propagation vector of the wave, then this notation can be incorporated into our ray theory.

8.2.2 Propagation vector formalism

The geometric ray is parallel to the propagation vector, **k**. Thus, there is a discrete set of propagation vectors associated with the set of rays predicted by Eq. (8.3). We can decompose any one of these propagation vectors into two components, as shown in Figure 8.6. To the external observer the light associated with a particular waveguide mode appears to propagate through the guide with an effective propagation constant, β, defined in Figure 8.6. The bounds on β for waveguide modes follow directly from Eq. (8.1):

$$kn_3 < \beta < kn_2.$$

We can use the effective propagation constant to define an *effective guide index of refraction*,

$$N = \frac{\beta}{k}.$$

The limits on this effective index are

$$n_3 < N < n_2.$$

The effect of introducing the new terms β and κ in Figure 8.6 is to divide the wave, in the guide, into two components: a standing wave which exists between the two interfaces and a wave propagating along the z-axis at a velocity ω/β.

Using the newly defined variables, we may rewrite Eqs. (8.6) and (8.7) as

$$\delta_{1,3} = -2 \tan^{-1} \left[\frac{\sqrt{\beta^2 - (n_{1,3}k)^2}}{\kappa} \right]. \tag{8.8}$$

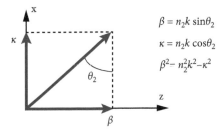

$$\beta = n_2 k \sin\theta_2$$
$$\kappa = n_2 k \cos\theta_2$$
$$\beta^2 - n_2^2 k^2 - \kappa^2$$

Figure 8.6 Definition of the propagation constant of a guided wave.

The radical in Eq. (8.8) is equal to the decay constant of an evanescent wave associated with total reflection, introduced in Chapter 3,

$$\sqrt{\beta^2 - n^2 k^2} = kn\alpha = \gamma.$$

We introduce a decay constant for each of the guide boundaries:

$$\gamma_1^2 = (kn_1\alpha_1)^2 = \beta^2 - n_1^2 k^2 = \left(n_2^2 - n_1^2\right) k^2 - \kappa^2$$
$$\gamma_3^2 = (kn_3\alpha_3)^2 = \beta^2 - n_3^2 k^2 = \left(n_2^2 - n_3^2\right) k^2 - \kappa^2. \tag{8.9}$$

Combining all of the new variables we have introduced, we may rewrite the dispersion relation, Eq. (8.3), as

$$\kappa d - \tan^{-1}\left(\frac{\gamma_1}{\kappa}\right) - \tan^{-1}\left(\frac{\gamma_3}{\kappa}\right) = m\pi. \tag{8.10}$$

8.2.2.1 Solution for asymmetric guide

To solve the dispersion relation graphically, we first take the tangent of both sides of the dispersion equation to get

$$\tan\left(\kappa d - m\pi\right) = \tan\left[\tan^{-1}\left(\frac{\gamma_1}{\kappa}\right) + \tan^{-1}\left(\frac{\gamma_3}{\kappa}\right)\right].$$

We now apply the trigonometric identity

$$\tan\left(a \pm b\right) = \frac{\tan a \pm \tan b}{1 \mp \tan a \tan b}$$

to both sides of the equation

$$\frac{\tan \kappa d \pm \tan m\pi}{1 - \tan \kappa d \tan m\pi} = \frac{\frac{\gamma_1}{\kappa} + \frac{\gamma_3}{\kappa}}{1 - \frac{\gamma_1\gamma_3}{\kappa^2}},$$

$$\tan \kappa d = \frac{\kappa\left(\gamma_1 + \gamma_3\right)}{\kappa^2 - \gamma_1\gamma_3}. \tag{8.11}$$

This is the mode equation for the TE waveguide modes and is the form of the dispersion relation usually encountered in the wave formalism of guided wave optics. We can solve this equation by plotting separately, in Figure 8.7, the functions tan(κd) and

$$f(\kappa d) = \frac{\kappa d\left(\gamma_1 d + \gamma_3 d\right)}{\left(\kappa d\right)^2 - \left(\gamma_1 d\right)\left(\gamma_3 d\right)}.$$

In Figure 8.7, the intersection of the f(κd) curve with the various tan(κd) curves are the solutions of Eq. (8.11). The value of κd at each intersection of the two functions allows the calculation of the guided ray's incident angle for that mode

$$\cos\theta_2 = \frac{(kd)\lambda}{2\pi n_2 d}.$$

Table 8.1 list the angle of incidence of a guided ray for the modes shown in Figure 8.7.

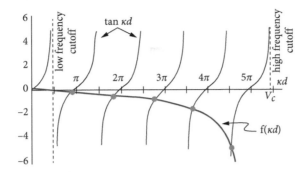

Figure 8.7 Using $n_1 = 1.0$, $n_2 = 1.62$, and $n_3 = 1.515$, the values for a sputtered glass wave guide, along with $d = 3$ μm and $\lambda = 0.63$ μm, we plot the two sides of Eq. (8.11) as separate functions. The intersections of the curves are the solutions of Eq. (8.11). Table 8.1 lists the angle of incidence for the modes identified by the points of intersection. The cut-off value for the V-number is the asymptote of the $f(\kappa d)$ curve. There are only five modes in this guide.

Table 8.1 Incident angle of guided mode

Mode	κd	θ_2
0	0.9π	86.6°
1	1.8π	83.3°
2	2.75π	79.7°
3	3.7π	76.1°
4	4.55π	72.8°

It is important to note that the intersection of $f(\kappa d)$ and $\tan(\kappa d)$ at the origin, in Figure 8.7, is not a physically realizable solution. Why not?

$$f(\kappa d) = \frac{\kappa d\,(\gamma_1 d + \gamma_3 d)}{(\kappa d)^2 - (\gamma_1 d)(\gamma_3 d)}.$$

For this to be zero we must have the numerator equal to zero, $\gamma_1 = -\gamma_3$, or $\gamma_1^2 = -\gamma_3^2$. The total reflected field from Chapter 3 is

$$E = E_0 e^{-\gamma x} e^{-ikz\sqrt{1+\alpha^2}}.$$

Using the expression for γ

$$\left(n_2^2 - n_1^2\right)k^2 - \kappa^2 = \left(n_2^2 - n_3^2\right)k^2 - \kappa^2$$
$$n_1^2 = n_3^2.$$

But for an asymmetric guide we must have $n_1^2 \neq n_3^2$. Thus, the solution at zero is not allowed for an asymmetric guide. This imposes a minimum value of κd, $\kappa d = \pi/2$, below which there are no solutions possible in an asymmetric waveguide; i.e., no guided mode is supported. This cut-off establishes the *largest* wavelength that can propagate

$$\kappa d = \pi/2$$

$$\frac{2\pi d}{\lambda_M}\sqrt{n_2^2 - n_3^2} = \pi/2$$

$$\lambda_M = 4d\sqrt{n_2^2 - n_3^2}$$

$$\lambda_M \Downarrow \qquad (n_2^2 - n_3^2)\Downarrow \qquad d\Downarrow.$$

The lack of a solution at zero can also be interpreted as stating that light cannot propagate along a direction parallel to the two boundaries of an asymmetric guide.

The curve $f(\kappa d)$ in Figure 8.7 does not extend without limit toward larger and larger values of κd but rather has a maximum value of κd, its asymptote, beyond which one of the values of γ becomes imaginary. From Eq. (8.9),

$$\gamma_1 \to \text{imaginary when } (n_2^2 - n_1^2)k^2 < \kappa^2,$$

$$\gamma_3 \to \text{imaginary when } (n_2^2 - n_3^2)k^2 < \kappa^2.$$

Our initial conditions assumed $n_3 > n_1$ so γ_3 will become imaginary first. The point of interest to us is when $\gamma_3 = 0$, i.e., just before γ_3 becomes imaginary. Physically it is at this point that propagation into the substrate is allowed and the guided modes disappear. Remember if $\gamma_3 = 0$, the evanescent wave extends infinitely far into the substrate because the penetration depth of the evanescent wave, $1/\gamma$, becomes infinite. At this point θ_2 no longer meets the condition for total reflection at the boundary between medium 2 and 3.

When $\gamma_3 = 0$

$$(n_2^2 - n_3^2)k^2 = \kappa^2.$$

We take the square root and multiply both sides by the guide thickness

$$kd\sqrt{n_2^2 - n_3^2} = \kappa d.$$

From Eq. (8.5), we have

$$V_c = \kappa d, \qquad\qquad\qquad\qquad\qquad\qquad\qquad\qquad (8.12)$$

where V_c is the cut-off value of the normalized film thickness. For a given guide thickness, d, there is a minimum wavelength that will propagate in the guide. This cut-off limit occurs for rays that have incident angles that are less than the critical angle. The V-number determines the minimum wavelength that will propagate in a guide and it decreases as we move to the left along the abscissa in Figure 8.7. The wavelength dependence can be found from

$$V_c = \kappa d$$

$$= kd\sqrt{n_2^2 - n_3^2}$$

$$= \frac{2\pi d}{\lambda}\sqrt{n_2^2 - n_3^2}.$$

The minimum wavelength occurs for a V-number equal to V_c:

$$\lambda_{\min} = \frac{2\pi d}{V_c}\sqrt{n_2^2 - n_3^2}.$$

The result is that there are two cut-off wavelengths in an asymmetric guide and we can define a bandpass of the optical guide of

$$\underbrace{\pi/2}_{\substack{f_{min} \\ \lambda_{Max}}} \leq f(\kappa d) \leq \underbrace{\kappa d = V_c}_{\substack{f_{Max} \\ \lambda_{min}}}.$$

As we have seen in our discussion, the term "cut-off" describes the inability of a waveguide to support a given propagation mode. The high-frequency cut-off value was defined as $2V_c = \delta_1 + \delta_3$ based on the mathematical requirement that the mode number be a positive integer. The high-frequency cut-off condition was given a more physical interpretation as the incident angle for which $\gamma_3 = 0$ and the evanescent field becomes a propagating field (since $n_3 > n_1$, γ_3 reaches zero before γ_1 and is thus the limiting parameter). When performing a graphic solution for the guided modes, we found that an asymmetric guide of thickness d and with indices, n_1, n_2, and n_3 has a low-frequency cut-off also. The physical origin of the existence of a low-frequency cut-off can be understood by considering the procedure used to analyze a waveguide through the use of Maxwell's equations.

To obtain wave solutions in the guide, we apply the boundary conditions introduced in Chapter 3, during the derivation of Fresnel's formula. In a symmetric guide, the boundary conditions in a plane normal to the guide interfaces would be identical at each interface. For an asymmetric guide, this would not be the case. For this reason, it is impossible for a plane wave to propagate through an asymmetric guide with its propagation vector parallel to the two interfaces. Such a wave can exist in a symmetric guide because the boundary conditions can be satisfied on the wavefront at both interfaces.

The asymmetric guide could only support guided waves for V-numbers larger than $\pi/2$. This means that an asymmetric guide cannot support guided modes for wavelengths larger than

$$\lambda \geq 4d\sqrt{n_2^2 - n_3^2}.$$

8.2.2.2 Solution for symmetric guide

A symmetric guide has the same index cladding on the top and the bottom of the guiding layer. With limitations that we will discuss, the symmetric guide could be viewed as having similar characteristics to an optical fiber. We will keep the conditions for an asymmetric guide to generate a new expression for the cut-off condition that can be used to enhance our understanding of propagation in a symmetric guide by allowing $n_1 = n_3$. Rewriting Eq. (8.12)

$$\tan V_c = \tan \kappa d = \frac{\gamma_1}{\kappa}.$$

We will express the cut-off in terms of the index of the upper and lower claddings of the guide:

$$\tan V_c = \tan_{\gamma_3 \to 0} \kappa d = \frac{\gamma_1}{\kappa} = \frac{\sqrt{(n_2^2 - n_1^2) k^2 - \kappa^2}}{\kappa}$$

The value of κ at cutoff is

$$\kappa^2_{\gamma_3 \to 0} = (n_2^2 - n_3^2) k^2$$

and

$$\tan V_c = \sqrt{\frac{n_2^2 - n_1^2}{n_2^2 - n_3^2} - 1} = \sqrt{\frac{n_3^2 - n_1^2}{n_2^2 - n_3^2}}. \tag{8.13}$$

This new expression for the cut-off condition leads to the conclusion that for a *symmetric guide*, where $n_1 = n_3$, there is no minimum value for κd, because from Eq. (8.13), V_c can equal zero. This result means that light can propagate straight down a symmetric guide parallel to the sides of the guide. (The symmetric guide would act as a low-pass filter because there is a minimum propagating wavelength but no maximum wavelength, $0 \le \kappa d \le V_c$.) To understand this behavior, we will solve the dispersion equation for a symmetric guide.

Because the indices on either side of the guide are the same, Eq. (8.9) reduces to a single equation

$$\gamma^2 = \left(n_2^2 - n_1^2\right) k^2 - \kappa^2$$
$$(\gamma d)^2 = V^2 - (\kappa d)^2, \tag{8.14a}$$

which is the equation of a circle. The dispersion relationship for the symmetric guide is

$$\kappa d - m\pi = 2\tan^{-1}\left(\frac{\gamma d}{\kappa d}\right),$$
$$\gamma d = \kappa d \tan \frac{1}{2}\left(\kappa d - m\pi\right).$$

This dispersion relationship can be expressed in two forms determined by the value of $m \cdot \pi/2$:

$$\gamma d = \begin{cases} \kappa d \tan \dfrac{\kappa d}{2} & m \mapsto even \\ -\kappa d \cot \dfrac{\kappa d}{2} & m \mapsto odd \end{cases} \tag{8.14b}$$

The TE modes are determined by the simultaneous solution of Eqs. (8.14a) and (8.14b). The graphic solution for the even modes of a symmetric guide, with properties similar to the asymmetric guide just discussed, is shown in Figure 8.8.

An intersection between Eqs. (8.14a) and (8.14b) at the origin is a physically realizable solution for a symmetric guide and results in a minimum κd of zero. It corresponds to $\theta_2 = \pi/2$, which is a ray propagating along the z-axis, parallel to the interfaces of the guide. The maximum κd, and thus the value of V_c, occurs at the value of κd where the circle intersects the abscissa.

The cut-off value of the V-number for a planar guide was given by Eq. (8.13). This relationship demonstrated that a symmetric guide had no maximum cut-off wavelength. A graphic solution for even modes of a symmetric guide illustrated this fact.

8.2.3 Wave theory of fiber guide

The zigzag theory gives us a nice visual representation of what is going on in a guide but it fails in several areas to explain propagation in a fiber. We are missing the skew modes and we have no idea about the light distribution that will exit the fiber. The light distribution is a result of interference between the wavefronts of the rays passing along the fiber. Both shortcomings can

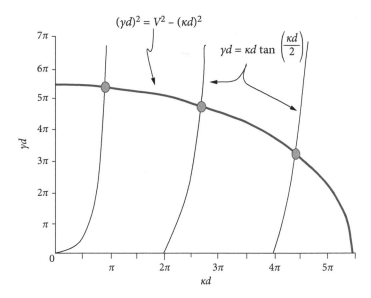

Figure 8.8 Using $n_1 = n_3 = 1.515$ and $n_2 = 1.62$, along with $d = 3$ μm and $\lambda = 0.63$ μm, we plot Eq. (8.14a). The intersections of the curves are the solutions of Eq. (8.14b) for even modes. The lowest mode is never cut off for this symmetric guide.

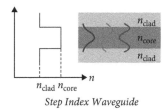

Step Index Waveguide

Figure 8.9 Classical wave analysis of an optical waveguide showing the field distribution normal to the fiber axis on the right for the first three modes.

be corrected by solving the Helmholtz equation. We are not prepared to solve this equation but we can discuss the form the solution will take and other physical properties.

We apply the paraxial approximation to Maxwell's equations and assume we can address the profile in the x-direction using the one-dimension scalar Helmholtz equation

$$\left[\frac{\partial^2}{\partial x^2} + k_0^2 n^2 - \beta^2 \right] E(x) = 0. \tag{8.15}$$

In the z-direction the wave is described by $e^{i\beta z}$ where β has discrete values.

If we solve Eq. (8.15) for a step index fiber we obtain the result shown in Figure 8.9. The field distributions of the first three modes in the fiber are shown on the right of the drawing.

It is interesting to note the mathematical similarity of this analysis with the quantum theoretical analysis of a particle in a potential well (*particle in a box*; Figure 8.10) where the independent Schrödinger equation has the same general form as the scalar Helmholtz equation

$$\left[\frac{\hbar}{2m} \frac{\partial^2}{\partial x^2} - V + E \right] \psi(x) = 0.$$

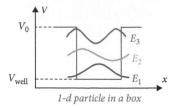

Figure 8.10 Quantum mechanical analysis of a particle in a box (square well potential).

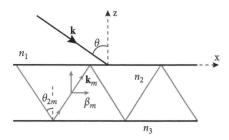

Figure 8.11 Coupling geometry for radiation illuminating the wall of an optical guide.

To solve our three-dimensional Helmholtz equation we have to work in a cylindrical coordinate system. The resulting equation is the linear second-order differential equation called the Bessel equation and its solution involves Bessel functions of integer order (sometimes called cylindrical harmonics). In optical fibers, the propagation constants of the various TE modes are obtained by graphic solutions of equations involving Bessel functions rather than the tangent functions of Eq. (8.10); therefore, for fibers, the low frequency cut-off condition is determined by the first zero of the Bessel function, $J_0(\kappa d)$, i.e., $2.405 \le \kappa d \le V_c$.

8.2.4 Coupling into guided wave modes

A waveguide mode, fiber or planar, cannot be excited by simply illuminating the guide surface. We now understand guide modes so we can examine the reasons behind the restriction on mode coupling.

From the discussion in Chapter 3 of boundary conditions for Maxwell's equations, we know that the tangential components of E and H must be continuous across a boundary. Using the geometry shown in Figure 8.11 we have

$$k_x = \frac{\omega}{c} n_1 \sin \theta$$

must equal

$$\beta_m = \frac{\omega}{c} n_2 \sin \theta_{2m}.$$

You can view this as a momentum conservation requirement. Snell's law allows us to relate the angle the illuminating wave makes on the guide surface, θ, to the angle that the guided mode makes with the boundary of the guide:

$$\sin\theta_{2m} = \frac{n_1}{n_2}\sin\theta < \frac{n_1}{n_2}.$$

The mode in the guide must meet the requirement for total reflection:

$$\theta_{2m} > \theta_c \rightarrow \sin\theta_{2m} > \frac{n_1}{n_2}.$$

There is a conflict between these two conditions, making it impossible to couple energy into the guided mode!

To couple into a planar guide, we can make use of the evanescent fields between total reflection in two materials. A prism with an index n_p greater than the guide index, n_2, is adjusted so that light incident on the prism will experience total internal reflection for angles greater than θ_p, as shown in Figure 8.12. The angle for total reflection in the prism when it is surrounded by a material of index n_1 ($n_1 = 1$ normally because the component is in air)

$$\sin\theta_p > n_1/n_p.$$

The z-component of the propagation vector in the prism is

$$k_z = (\omega/c)\,n_p\sin\theta_p > (\omega/c)\,n_1.$$

In the optical guide the total reflection angle θ_m for the mth guided mode is given by

$$\beta_m = \frac{\omega}{c}n_2\sin\theta_{2m}.$$

Total reflection requires that

$$\sin\theta_{2m} > n_1/n_2.$$

The propagation vector in the guide must exceed

$$\beta_m > \frac{\omega}{c}n_1.$$

To match the z-component of the propagation vector in the prism to that in the guide we set k_z equal to β_m. To accomplish this, we bring the prism close to guide to couple to the evanescent fields associated with the totally reflected waves,

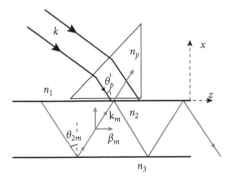

Figure 8.12 Prism coupling into an optical guide.

$$\frac{n_2}{n_p} \sin \theta_{2m} = \sin \theta_p \quad \Rightarrow \quad \theta_{2m} > \theta_p,$$

and this is obtained whenever

$$n_p > n_2.$$

The energy from the light in the prism flows into the guided mode through the evanescent fields generated by total reflection.

8.2.4.1 End coupling

A second technique for launching a waveguide mode in either a fiber or planar guide is to couple the light into the polished end of the dielectric layer—along the z-direction and perpendicular to the x,y-plane. This technique is called *end-fired coupling.*

As an example, we will consider a step-index waveguide, Figure 8.13, and use geometric optics to follow the light, introduced into a flat end surface of the guide. In Figure 8.13a, the x,y-plane of a step-index guide and its index of refraction profile, a step function, are shown. The high index region at the center is called the *core* and the lower index region surrounding the core is called the *cladding.* A representation of light coupling into the guided modes as well as the light that arrives at an angle too shallow to meet the condition of total reflection are shown in Figure 8.13b. The drawing implies that there is a maximum cone of light (acceptance cone) that can couple into guided modes of the step-index fiber. We will determine the critical cone angle by using the fiber geometry shown in Figure 8.14.

The guide will accept rays that are incident onto the end face of the guide at an angle less than a critical angle that we will denote by θ_{NA} and call the *acceptance angle.* The value of this maximum angle is obtained by using the fact that rays trapped in the guide must be incident on the core/cladding interface at an angle θ_2, Figure 8.14b, that exceeds the critical angle for

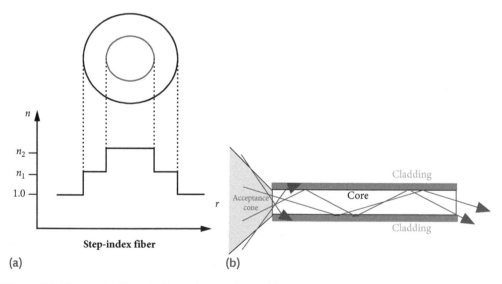

(a)   (b)

Figure 8.13 (a) Step-index fiber. The fiber is shown end on and the index of refraction profile is displayed. (b) Limitation on incident angle for end-fired coupling.

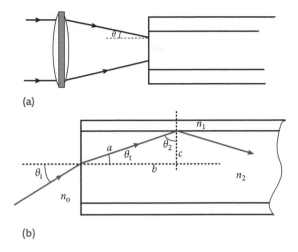

(a)

(b)

Figure 8.14 (a) Edge-coupled guide. A lens is used to focus light into the polished end of a guided wave structure. (b) Coordinates for determining the numerical aperture of a step-index optical guide.

total reflection. When this requirement is met, the light is trapped in the guide until it emerges from the other end of the fiber.

In Figure 8.14b the angle θ_2 must exceed the critical angle

$$\theta_2 \geq \theta_c = \sin^{-1}\frac{n_1}{n_2}.$$

The angle θ_2 can be written in terms of the sides of the triangle drawn in Figure 8.14b:

$$a^2 = b^2 + c^2$$

$$1 = \left(\frac{b}{a}\right)^2 + \left(\frac{c}{a}\right)^2,$$

$$\sin\theta_2 = \frac{b}{a},$$

$$\left(\frac{c}{a}\right)^2 = 1 - \left(\frac{b}{a}\right)^2 \leq 1 - \left(\frac{n_1}{n_2}\right)^2.$$

The sides of the triangle can also be related to the transmission angle θ_t at the front surface of the guide

$$\frac{c}{a} = \sqrt{1 - \left(\frac{n_1}{n_2}\right)^2} = \sin\theta_t.$$

Snell's law relates the transmission angle to the angle of incidence on the fiber face,

$$\frac{n_2}{n_0} = \frac{\sin\theta_i}{\sin\theta_t}.$$

The acceptance angle can be written in terms of the above relations:

$$\sin\theta_i \leq \sin\theta_{NA} = \frac{n_2}{n_0}\sqrt{1 - \left(\frac{n_1}{n_2}\right)^2}.$$

We define the *numerical aperture* of the guide in terms of the acceptance angle as

$$NA = n_0 \sin \theta_{NA} = n_2 \sin \theta_i$$

$$NA = n_2\sqrt{1 - \left(\frac{n_1}{n_2}\right)^2}$$

$$= \sqrt{n_2^2 - n_1^2}.$$

(8.16)

The numerical aperture is a dimensionless number that characterizes the maximum spread in angles that the fiber will accept or emit. If the fiber is in air, then $n_0 = 1$ and the NA is less than 1. The numerical aperture will determine the number of modes that a fiber can carry.

The maximum number of modes in a fiber is given by

$$m_{\max} \approx \frac{1}{2}\left(\frac{\pi d}{\lambda}NA\right)^2.$$

For a symmetric guide with these properties

$$d = 50 \ \mu m \quad n_1 = n_3 = 1.50 \quad n_2 = 1.53$$

How many modes propagate?

$$V = 2\pi(50)\sqrt{(1.53)^2 - (1.5)^2}$$
$$= 94.7$$

$$m \le V/\pi - \phi/\pi$$
$$m \approx 30.$$

Current single-mode fiber design used in communications has the following parameters:

$$\lambda_0 = 1.55 \ \mu m \quad n_2 = 1.444 \quad n_1 = 1.4365 \quad d = 2.3 \ \mu m$$

$$V_c = \frac{2\pi d}{\lambda_0}\sqrt{n_2^2 - n_1^2} = \frac{2\pi\left(2.3 \times 10^{-6}\right)}{1.55 \times 10^{-6}}\sqrt{\left[(1.444)^2 - (1.4365)^2\right]}.$$
$$= 1.35$$

Skew Rays

The numerical aperture result we derived in Eq. (8.16) will apply to rays in a planar guide or to meridional rays in a fiber but must be modified for skew rays. In addition to meridional rays, a fiber will allow the propagation of skew rays. As mentioned earlier, skew rays travel the length of the fiber without ever crossing the cylindrical axis of the fiber. Figure 8.15 shows a skew ray from two perspectives. The lower drawing is an end view of the fiber showing the skew ray propagating down the fiber without crossing the axis of the fiber.

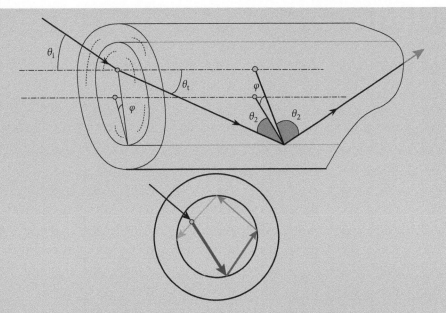

Figure 8.15 Propagation of a skew ray in an optical fiber. The ray propagates in a helical path. The upper figure identifies the geometry used to evaluate the numerical aperture for a skewed ray. The lower figure shows the propagation of a skew ray as viewed from the end of the fiber. As the ray progresses down the fiber, the arrows representing the rays decrease in thickness.

The size of the vectors representing the rays in Figure 8.15 is used to denote the depth of the rays propagating into the fiber—smaller vectors represent deeper rays. The angle between each pair of rays is 2φ, where φ is defined in the upper drawing. The skew ray is confined to a plane that makes an angle φ with the plane containing the cylindrical axis of the fiber. The component of the ray propagating in the core and parallel to the normal vector of the core–cladding interface is

$$k_t \cos \varphi \sin \theta_t.$$

The normal vector of the core–cladding interface is the radius vector from the cylindrical axis of the fiber to the point where the ray strikes the interface and, with the ray, establishes the plane of incidence. The component of the ray along the normal is given, in terms of the angle of incidence, θ_2, as

$$k_t \cos \theta_2 = k_t \sqrt{1 - \sin^2 \theta_2}.$$

Assuming that θ_2 is equal to the critical angle for total reflection and using Snell's law to replace θ_t by θ_i

$$\cos \varphi \frac{n_0}{n_2} \sin \theta_i = \sqrt{1 - \left(\frac{n_1}{n_2}\right)^2}.$$

We thus discover that the numerical aperture for a skew ray, propagating in the plane defined by φ, is given by

$$(NA)_s = \frac{NA}{\cos \varphi},$$

where the NA represents the numerical aperture for a meridional ray given by Eq. (8.16). We see that the maximum acceptance angle of a meridional ray is the minimum acceptance angle for skew rays. The skew rays act to increase the light-gathering capability of a fiber.

As an example of the application of Eq. (8.16), assume $n_0 = 1$, $n_1 = 1.5$, and $n_2 = 1.53$. The critical angle is then

$$\theta_c = \sin^{-1}\left(\frac{n_1}{n_2}\right) = \sin^{-1}\left(\frac{1.5}{1.53}\right) = 78.6°.$$

The numerical aperture is 0.303, which corresponds to a maximum incident angle of $\theta_{NA} = 17.6°$. If a skew ray is propagating in this fiber with $\varphi = 50°$, then the maximum incident angle for this skew ray is 28.1°.

The symmetry associated with the propagation of a ray in an optical fiber predicts that the angle of the output ray from a fiber will be equal to the incident angle. Because the meridional ray is confined to a plane, the ray will always emerge at a position determined by the input ray. The skew ray, however, will exit a fiber at a position that is a function of the number of reflections (bounces) it has undergone rather than upon the input ray. The skew ray will thus smooth the output distribution of light along a circumference of a circle of radius r, where r is the position of the input ray at the entrance to the fiber. An example of this smoothing is shown in Figure 8.16. In Figure 8.16a, an array of slits is shown distributed along the radius of a fiber from the center (on the right of Figure 8.16a) to the edge of the core (on the left of Figure 8.16a). The light distribution upon exiting the fiber is shown in Figure 8.16b.

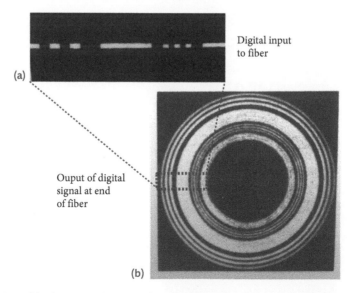

Digital input to fiber

(a)

Ouput of digital signal at end of fiber

(b)

Figure 8.16 (a) A set of slits form a one-dimensional pattern. This pattern is positioned at the fiber input with the right-most edge of the pattern centered on the fiber axis; see dotted box. (b) The output of the multimode fiber with (a) as an input displays the intensity smoothing produced by skew rays. (Courtesy of A. Tia and A. Friesem [6].)

Table 8.2 Common fiber optic connectors

Connector type	Image	Insertion loss	Characteristics
FC High-precision, ceramic ferrule connector is equipped with an anti-rotation key		0.5–1.0 dB	Preferred for single-mode fibers. Good mechanical isolation
ST Bayonet-style, keyed coupling mechanism, push and turn locking		0.3 dB	Good for field applications with multimode and single-mode fibers
SC Snap-in connector		0.2–0.45 dB	Push–pull design prevents fiber damage
LC Small form factor connector, half the size of the SC		0.1–0.15 dB	Very popular connector for single-mode fiber

8.2.5 **Dispersion of guided waves**

Another defect that can reduce the performance of a fiber communication link is dispersion. Dispersion, through a broadening of pulses as the pulses propagate, causes distortion and reduces the allowable data rates in optical fiber communication.

The effects on the propagation of light that are due to the wavelength dependence of the refractive index are called *material* (or *chromatic*) *dispersion*. Material dispersion in silica-based optical fibers is negligible at wavelengths around 1.3 μm. Most applications, requiring high data rates, attempt to take advantage of this fact by proper choice of materials in fiber construction and proper choice of light sources.

A smaller chromatic dispersion, called *waveguide dispersion*, is observed in a guide with a gradient index of refraction; it is also called *profile dispersion*. The propagation in the cladding is faster than the core because the index of the cladding is lower. This means the longer wavelengths move faster than the shorter wavelengths. The dispersion is specified as ps/km (picosecond/kilometer).

Polarization mode dispersion has several origins: stress birefringence set up in the fiber due to micro-bends or geometric asymmetry due to slightly elliptical cores.

The final source of dispersion in a guided wave is called *multimode dispersion* (some prefer to call this distortion). If a light pulse introduced into a guide excites several modes, then the energy in each mode will appear to travel at a different velocity and the modes will arrive at the end of the fiber at different times. From the zigzag model we have been using, we see that

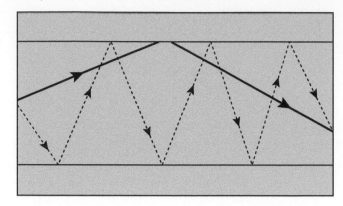

Figure 8.17 Two guide wave modes propagating in an optical guide. The different path lengths show the origin of multimode dispersion in optical guides.

the physical reason that the modes appear to travel at different velocities is that rays associated with each mode travel different optical path lengths (Figure 8.17).

A simple relation for a fiber can be extracted from our symmetric guide discussion if skew rays are ignored. The meridional rays of an optical fiber can be viewed as equivalent to a ray in a symmetric guide with a velocity of propagation given by ω/β, where

$$\beta = kn_2 \sin\theta_2.$$

The limits on θ_2 for guided waves are given by Eq. (8.1). Using the two limits of θ_2, we find the maximum difference in propagation time for a fiber of length L is

$$\Delta t = \frac{Ln_2}{c}\left(\frac{1}{\sin\theta_c} - \frac{1}{\sin\dfrac{\pi}{2}}\right) = L\frac{n_2\,(n_2 - n_1)}{n_1 c}. \tag{8.17}$$

If we code information as pulses and propagate the pulses down the fiber, we must not have any overlapping pulses. The pulse width at the start of the fiber is τ, and after propagating along the fiber is $\tau + \Delta\tau$. The minimum pulse spacing that can be used is, thus, $\Delta\tau$. Energy would be wasted if the pulse width exceeded $\Delta\tau$ so we also set the pulse width at $\Delta\tau$. From Shannon's sampling theorem[7] we need two samples for the highest frequency transmitted; thus, the maximum bandwidth of the fiber is

$$B = \omega_c \propto \frac{1}{2\Delta t}.$$

Multimode dispersion is the major source of distortion when comparing the bandwidth of the various classes of optical fibers. The allowable communication bandwidth of a fiber shown in Table 8.3 indicates the effect of multimode dispersion.

Multimode dispersion in graded index fibers is lower than step-index fibers because rays traveling at small angles relative to the fiber axis travel almost exclusively in the high-index region of the fiber and therefore at slow propagation speeds. Rays traveling at large angles travel

---

[7] A signal whose frequency content does not exceed B Hz can be completely represented by a set of points spaced by ½B seconds apart.

Table 8.3 Optical Fiber Bandwidth

Fiber type	Bandwidth
Multimode, step index	20 MHz • km
Multimode, graded index	3.5 GHz • km
Single mode, step index	100 GHz • km

in the low-index region and have high propagation velocities. The gradient in the index of refraction partially compensates for the difference in geometrical path lengths by modifying the optical path lengths of the various rays. Single-mode fibers only transmit one mode; thus, multimode dispersion does not exist. The cut-off V-number for single-mode operation in a graded index fiber is

$$V_c < 2.405 \sqrt{\frac{g+2}{g}},$$

where g is the profile parameter; see Eq. (8.19). Because of model dispersion, graded index fibers are preferred for communications applications and for the highest bandwidths, single-mode fibers are the most frequent fiber design used.

Propagation in a Graded Index Optical Fiber

Generally, the ray equation Eq. (8.15) is difficult to solve. To reduce the complexity of the problem, the assumption is made that the light rays propagate in the general direction of the z-axis and those that travel at an angle to the z-axis depart by only small angles. This is another statement of the paraxial ray approximation and allows us to write $ds \approx dz$; s is the actual path (equivalent to the approximation $\sin \theta \approx \tan \theta$). This approximation results in the paraxial ray equation or the *Eikonal equation*

$$\frac{d}{dz}\left(n\frac{d\mathbf{r}}{dz}\right) = \nabla n. \tag{8.18}$$

This equation provides a mathematical link between physical optics and geometrical optics. The mathematical technique applies to any wave equation and is encountered in quantum mechanics as the WKB (Wentzel, Kramers, Brillouin) method. Because the Schrödinger and Helmholtz equations have the same mathematical form they share solution techniques. In quantum mechanics, we use the WKB method to solve the Schrödinger equation in the classical limit where the change of potential energy within the particle's wavelength (a deBroglie wavelength) is small compared with the kinetic energy [20]. In optics we apply the same WKB method [21] to solve the Helmholtz equation in the geometrical optic limit when the index of refraction changes slowly over the distance of a wavelength.

We cannot obtain an exact solution for arbitrary refractive index profiles but a parabolic-shaped profile (see Figure 8.18) can be solved.

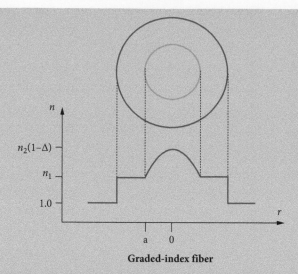

Figure 8.18 Profile of the index of refraction variation in a graded index optical fiber with a parabolic variation in the index of refraction.

$$n(r) = n_2 \left[1 - \left(\frac{r}{a} \right)^g \Delta \right], \qquad r < a, g = 2, \tag{8.19}$$

where g is the profile parameter and we have introduced the quantity Δ which is defined as

$$\Delta = \frac{n_2^2 - n_1^2}{2n_2^2}.$$

In practice the index of the cladding is almost the same as the core, resulting in an approximation for Δ of

$$\Delta = \frac{(n_2 - n_1)(n_2 + n_1)}{2n_2^2} \approx \frac{n_2 - n_1}{n_2}.$$

We will not solve the problem but simply indicate three interesting solutions obtained for a fiber.

Figure 8.19 shows the propagation of a general ray in the core of a graded-index fiber. The radii, r_1 and r_2, in Figure 8.19 denote the location of the maximum and minimum excursions of the ray (called the turning points) as it propagates along a helical path down the fiber.

To understand the trajectories of the rays as they move between the turning points, we will evaluate two limiting cases. The first assumes that the azimuth angle, ϕ, is a constant and the ray is confined to a plane. This ray is a meridional ray confined to the meridional plane defined by ϕ (Figure 8.20). The ray moves in a sinusoidal trajectory, periodically crossing the symmetry axis (the z-axis) with a period of

$$\frac{1}{\Omega} = \frac{a}{\sqrt{\Delta}}.$$

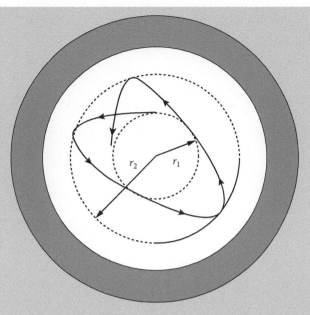

Figure 8.19 A ray projection showing the path of a general ray in the core of a graded-index fiber. The path is contained between the classical turning points. The cladding is denoted by the shaded region.

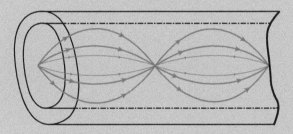

Figure 8.20 Meridional rays propagating in a graded index fiber with a parabolic index profile.

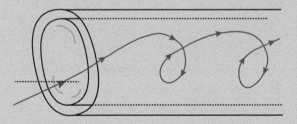

Figure 8.21 Propagation of a skew ray in a graded index fiber.

A second limiting case assumes r is a constant and the azimuth angle is a linear function of the z-coordinate. These are called helical rays because they travel on a helical path about the z-axis a fixed distance, A, from the axis. A typical helical ray is shown in Figure 8.21. A general ray has a trajectory that falls between these two limiting cases.

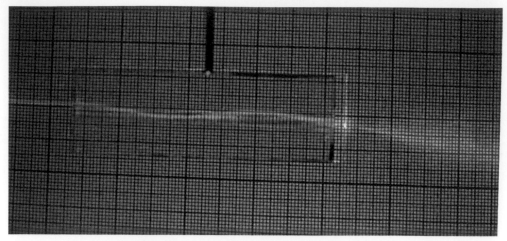

Figure 8.22 Beam of light propagating along a graded index guide.
(Courtesy of David Hamblen.)

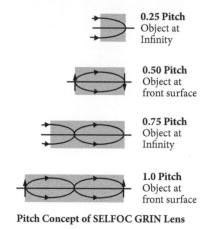

Pitch Concept of SELFOC GRIN Lens

Figure 8.23 Gradient index (GRIN) lenses. Commercially available lenses of this type are called SELFOC [7] lenses.
Image courtesy of Go!Foton, an exclusive licensee of SELFOC. SELFOC is a registered trademark of NSG Group.

A gradient index lens can be constructed using a gradient index fiber of the proper length. The gradient index core creates guided waves with trajectories shown in Figure 8.21. A small beam of light is shown propagating along a graded index guide in Figure 8.22, demonstrating the sinusoidal propagation path predicted by a solution to Eq. (8.17). By adjusting the length of this type of fiber we can construct what are called SELFOC lenses and use the fiber sections to focus, collimate, or image the meridional rays, as is shown in Figure 8.23.

8.3 Fiber Losses

An interesting question is how many bounces does the guided wave make per length L. An answer to that question will allow us to determine the losses we should expect in propagating through an optical fiber. We use the same geometry as shown in Figure 8.14b. The distance taken from entry into the fiber to the first bounce is

$$b = \frac{a}{\tan \theta_t}.$$

The distance traveled from the first bounce to the Nth bounce is

$$2Nb = \frac{2Na}{\tan \theta_t}.$$

The number of bounces per unit length is thus

$$\frac{N}{L} = \frac{1}{2a} \frac{\sin \theta_t}{\cos \theta_t},$$

The number of bounces per unit length is thus

$$\frac{N}{L} = \frac{n_0 L \sin \theta_i}{2a\sqrt{n_2^2 - n_0^2 \sin^2 \theta_i}}.$$

Using Snell's law

$$\sin \theta_t = \frac{n_0}{n_2} \sin \theta_t$$

$$\cos \theta_t = \sqrt{1 - \sin^2 \theta_t} = \sqrt{1 - (n_0/n_2)^2 \sin^2 \theta_i}.$$

To estimate how many bounces that might occur, let us assume the incident angle of light on the fiber face is θ_i=30°. The core diameter is $a = 25$ μm and the index of the core is $n_2 = 1.6$. We assume the fiber is in air so that $n_0 = 1.0$. The number of bounces is $N \approx 650$ per unit fiber length. This means the distance traveled is a lot longer than the fiber length. This fact makes losses in the fiber even more important.

The absorption as a function of wavelength in a typical fiber is shown in Figure 8.24. The absorption in manufactured glass fiber is shown for three periods in the development

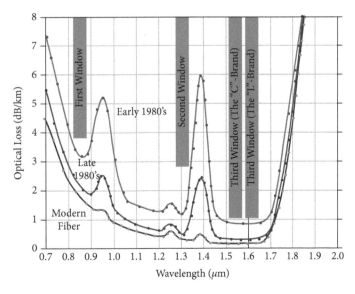

Figure 8.24 Transmission through optical fiber as a function of wavelength. Also shown are the operating windows for optical communication.

of the technology: early 1980s, late 1980s, and current. To give some perspective to the numbers we should note that in 1966 the losses in fibers were about 1000 dB/km. By 1970 contaminant removable resulted in losses around 20 dB/km, which compared well with 5–10 dB/km in coaxial cable then in use.[8] By moving the operating wavelength into the infrared at 1.3 µm, transmission shown by the first curve in Figure 8.24 was obtained. Operation at the longer wavelengths was made possible by the development of InGaAsP semiconductor lasers, resulting in losses shown by the curve labeled "Modern Fiber." The three absorption peaks apparent in the absorption curves are due to OH^- absorption at 950, 1380, and 2730 nm; the OH ions are associated with water vapor captured during manufacture of the fiber or from the atmosphere during fiber use. The general rise in attenuation at shorter wavelengths is due to Rayleigh scattering that scales as $1/\lambda^4$ and makes wavelengths below about 800 nm unusable for long haul communications. The rise above about 1.6 µm is due to molecular absorption by SiO_2 in the infrared.

The windows (low absorption regions) indicated in Figure 8.24 are associated with bands identified in Table 8.4. The bands are associated with different light sources available as transmitters in a communication system. Availability of low-cost, light-emitting diodes (LEDs) led to the use of the first, 850-nm window to provide a low-cost system with losses of about 2 dB/km. The high loss limited applications to short ranges.

InGaAsP semiconductor lasers operating around 1310 nm (second window) were developed in the late 1970s. This resulted in fiber design with losses as low as 0.5 dB/km and allowed the creation of a national optical fiber network using single-mode fibers.

In 1986 the erbium-doped fiber-based optical amplifiers allowed repeater systems operating at 1550 nm (third and fourth windows) to remove the need to down and up convert to electronic frequencies for restoration of the signals. The result was the creation of long-range (up to 100 km) communication systems.

The glass used in a fiber optic cable is ultra-pure, ultra-transparent, silicon dioxide, or fused quartz. During the glass fiber optic cable fabrication process, impurities are purposely added to the pure glass to obtain the desired indices of refraction needed to guide light. Germanium,

Table 8.4 Communication Bands in Optical Fibers

Band	Description	Wavelength range
O band	Original	1260 to 1360 nm
E band	Extended	1360 to 1460 nm
S band	Short wavelengths	1460 to 1530 nm
C band	Erbium window	1530 to 1565 nm
L band	Long wavelengths	1565 to 1625 nm
U band	Ultralong	1625 to 1675 nm

[8] As a reminder, 3 dB attenuation is equivalent to a 50% power loss and 10 dB is equivalent to a 90% power loss.

titanium, or phosphorous is added to increase the index of refraction. Boron or fluorine is added to decrease the index of refraction.

The scattering and absorption losses are called intrinsic losses. There is also extrinsic attenuation due to macrobending or microbending. Macrobending is visible bends that can be removed by adjusting the fiber position. The microbend is due to imperfections in the shape of the fiber produced during manufacturing. The design of standard fiber connectors (see Table 8.2) has reduced the losses associated with fiber interconnects, often required for field repair of a communication channel.

Other methods to reduce propagation loss in a fiber have become available with the development of new controls over material properties. We have developed the ability to manufacture periodic linear one-dimensional arrays of high- and low-index materials to produce mirrors or to stop reflection. Attempts have been made to line hollow core fibers with multilayer reflectors, making a guided wave dependent on the reflectivity of the mirrors rather than total reflection. By creating a hollow core fiber, it was hoped that the loss would be that of free space rather than glass. The first designs of these one-dimensional, periodic index, Bragg fibers [8], was made using analytic techniques. Later numerical techniques were used to design fibers with photonic crystal outer layers [9].

8.4 **Photonic Crystal Fibers**

In condensed matter physics, the ability to develop electronic devices is based on our understanding of how charge carriers move within the lattice that forms the crystal structure of the material. The atoms and molecules that form the crystal structure generate the potential within which the charge carriers move. The theory was developed by *Felix Bloch* (1905–83) in 1928. Electrons propagate as waves, and if the waves meet certain criteria, they can travel through the periodic potential created by the atoms and molecules that make up the material. Importantly, however, the structure can prohibit the propagation of certain waves in certain directions, creating gaps in the allowable electron energy through interference. It is the control of this property that is the origin of modern electronic devices.

In 1987, almost simultaneously, *Eli Yablonovitch* [10] and *Sajeev John* [11] began efforts to utilize the analogy between the propagation of electromagnetic waves in periodic dielectric structures and electrons in crystals. Yablonovitch was interested in controlling spontaneous emission through the use of bandgaps to improve the efficiency of lasers. John was interested in using light to observe Anderson localization [12].[9] Other applications that were of greater commercial success were found. One of those was the construction of fibers that did not depend on total reflection but rather on the rejection of certain waves by energy gaps in photonic crystals created in optical fibers. Initially, it was thought that applications would arise by simply mapping semiconductor theory onto the optical systems. It was quickly discovered that

[9] In addition to bandgaps effecting the propagation of electrons in crystalline solids, P. W. Anderson predicted that errors in the periodicity of a crystal would result in electronic localization but electron–electron interactions and electron–phonon scattering prevented the observation. John looked to electromagnetic radiation propagating in a periodic index medium as the ideal structure to observe localization.

polarization plays an important role in electromagnetic vector theory but not in the scalar quantum theory used for electron propagation. New calculations had to be performed to guide the construction of useful periodic index of refraction structures. Now open source software is available for these calculations [13].

Photonic crystals with periodic index of refraction in one, two, and three dimensions have been created. The first crystal was designed to operate in the microwave frequencies by Dr. Yablonovitch while at Bell Labs. A photo of one of the first constructed crystals is seen in Figure 8.25. Only one-dimensional crystals can be treated analytically and have been constructed since 1887. We evaluated simple structures in the chapter on interference where we discussed their use for mirrors and anti-reflection coatings. One-dimensional crystals have incomplete bandgaps (usually called stopbands) over limited angular range but when we turn to two- and three-dimensional crystals we can, through the use of digital numerical analysis of Maxwell's equations, discover structures with complete bandgaps existing at all angles. The major problem with these more complex structures is how to fabricate them.

Both one-dimensional [15] (called Bragg fibers; see Figure 8.26) and two-dimensional structures (Figure 8.27) have been created as the cladding of the core of optical fibers. The core is usually hollow because that is the major advantage of this design.

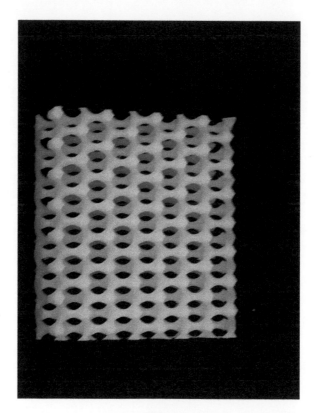

Figure 8.25 One of the first photonic crystals made by Eli Yablonovitch [14]. It was designed to operate in the microwave frequency band.
Sample of photonic crystal provided courtesy of Eli Yablonovitch; photo by author.

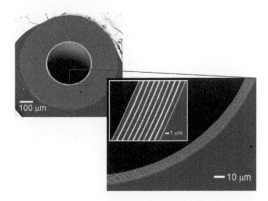

Figure 8.26 A hollow core multilayer fiber to transmit 3.5 μm radiation [16].

Figure 8.27 Hollow core photonic crystal fiber often called a Holey fiber (PCF) [17]. By Defense Advanced Research Projects Agency (DARPA)—http://www.darpa.mil/NewsEvents/Releases/2013/07/17a.aspx, Public Domain, https://commons.wikimedia.org/w/index.php?curid=27544069.

Commercial hollow core fibers have large continuous bandwidth, only a small number of core modes, zero dispersion at the design wavelength, and very little nonlinearity (if desired) or bending loss. It was hoped that the transmission loss in the hollow core fiber would be less than in current communication fibers (about 0.2 db/km). Scattering from surface defects introduced during pulling of the fiber has limited attenuation in the hollow core PCF to just over 1 db/km.

The loss is too high to use this technology as a replacement for standard single-mode optical fiber but the fact that the core is hollow does allow this fiber design to be used in a number of applications:

1. Low-loss transmission of IR signals for medical applications;

2. Transmission of high-energy laser radiation for surgery and material processing;

3. Dispersion compensation;

4. Nonlinear optical devices; and

5. Sensing applications [18].

8.5 Problem Set 8

(1) Assume that we have a step-index fiber with $n_2 = 1.6$ and $n_1 = 1.46$. Considering only meridional rays, what is the smallest value that the angle of incidence in the fiber, θ_2, can have? What is the numerical aperture when $d = 0.6\ \mu m$ and $\lambda = 1\ \mu m$?

(2) What is the maximum radius that the fiber of Problem (1) can have if single-mode operation is desired?

(3) Assume a plane wave in incident normal to the polished end of a step-index fiber, shown in Figure 8.28. Derive an equation for the minimum radius that a fiber can be bent before light is lost from the guided mode. Assume that the index of the core is 1.66 and the index of the cladding is 1.52. Using the equation just derived, what is the minimum radius?

(4) The distance between successive reflections in a fiber guide is called the skip distance, L_s. What is the skip distance for a fiber in air with a core diameter of 100 μm, core index of $n_2 = 1.5$, and cladding index of $n_1 = 1.4$? Assume the wavelength is 1550 nm.

(5) A step-index fiber has a core with an index of 1.480 and a diameter of 100 μm, The cladding has an index of 1.460. For a propagating wavelength of 850 nm, what is the numerical aperture and acceptance angle from air and how many modes can propagate?

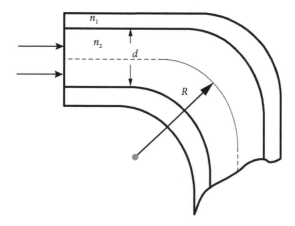

Figure 8.28 Light propagation around fiber bend.

(6) A typical single-mode optical fiber has a core diameter of 8 μm and a refractive index of 1.46. The normalized index difference is $\mid n_1 - n_2 \mid = \Delta = 0.3\%$. The cladding diameter is 125 μm. Calculate the numerical aperture and acceptance angle of the fiber. What is the single mode cut-off wavelength of the fiber?

(7) What should be the core radius of a single-mode fiber that has a core index of 1.468 and a cladding index of 1.447 and is to operate at a wavelength of 1.3 μm?

(8) Assume that we have a step-index fiber with $n_2 = 1.6$ and $n_1 = 1.46$. Considering only meridional rays, what is the smallest value that the angle of incidence in the fiber, θ_2, can have? What is the numerical aperture when $d = 0.6$ μm and $\lambda = 1$ μm?

(9) Using the parameters in Problem (8), how many modes can propagate in the fiber? What happens if we increase d to 3 μm? If we increase λ to 1.3 μm?

(10) What is the maximum radius that the fiber of Problem (8) can have if single-mode operation is desired?

(11) Calculate the phase shifts, introduced on total reflection inside a plane guide, with $n_2 = 1.53$ and $n_3 = 1.5$, and $n_1 = 1.0$. Assume the angle of incidence is $\theta_2 = 85°$.

(12) What is the critical angle i_c at the glass–cladding interface of an optical fiber whose core has a refractive index equal to 1.5 and cladding with a refractive index 1.45?

(13) In a step-index fiber in the ray approximation the ray propagating along the axis of the fiber has the shortest route, while the ray incident at the critical angle has the longest route. Determine the difference in travel time (in ns/km) for the modes defined by those two rays for a fiber with $n_{core} = 1.5$ and $n_{cladding} = 1.485$.

REFERENCES

1. Kratz, J., in *Pieces of History*. Washington, DC: US National Archives, 2018.

2. Hopkins, H. H., and N. S. Kapany, *Nature* **173**, 39–41 (1954).

3. v. Heek, A., *Nature* **173**, 39 (1954).

4. Hondros, D., and P. Debye, *Ann. Phys.* **337**, 465–76 (1910).

5. Tein, P. K., *Appl. Opt.* **10** (112395) (1971).

6. Friesem, A. A., U. Levy, and Y. Silberberg, *Proc. IEEE* **71**(2), 208–21 (1983).

7. Nippon Sheet Glass Co., Japan (2008).

8. Yeh, P., A. Yariv, and E. Marom, *J. Opt. Soc. Am.* **68** (9), 1196–201 (1978).

9. Joannopoulos, J. D., S. G. Johnson, J. N. Winn, and R. D. Meade, *Photonic Crystals: Molding the Flow of Light*, 2nd edn. Princeton NJ: Princeton University Press, 2008.

10. Yablonovitch, E., *Phys. Rev.Letts.* **58**, 2059 (1987).

11. John, S., *Phys. Rev.Letts.* **58** (23), 2486–9 (1987).

12. Anderson, P. W., *Phys. Rev.* **109**, 1492–505 (1958).

13. Johnson, S. G., in *MPB* (Massachusette Institute of Technology, 2018), pp. MPB is a free and open-source software package for computing the band structures, or dispersion relations, and electromagnetic modes of periodic dielectric structures, on both serial and parallel computers.

14. Yablonovitch, E., T. J. Gmitter, and K. M. Leung, *Phys. Rev. Letts.* **67** (17), 2295–9 (1991).

15. Vienne, G., Y. Xu, C. Jakobsen, H. J. Deyerl, J. B. Jensen, T. Srensen, T. P. Hansen, Y. Huang, M. Terrel, R. K. Lee, N. A. Mortensen, J. Broeng, H. Simonsen, A. Bjarklev, and A. Yariv, *Optics Express* **12** (15), 3500–8 (2004).

16. Temelkuran, B., S. D. Hart, G. Benoit, J. D. Joannopoulos, and Y. Fink, *Nature* **420** (6916), 650–3 (2002).

17. Russell, P., *Science* **299**(5605), 358–62 (2003).

18. Pinto, A. M. R., and M. Lopez-Amo, *J. Sensors* **2012**, 21 (2012).

9 Fraunhofer Diffraction

9.1 Introduction

Geometrical optics does a reasonable job of describing the propagation of light in free space unless that light encounters an obstacle or the light is coherent. Geometrical optics predicts that the obstacle would cast a sharp shadow. On one side of the shadow's edge there would be a bright uniform light distribution, due to the incident light, and on the other side there would be darkness. Close examination of an actual shadow edge reveals, however, dark fringes in the bright region and bright fringes in the dark region. This departure from the predictions of geometrical optics was given the name *diffraction* by Francesco Maria Grimaldi (1613–63).

There is no substantive difference between diffraction and interference. The separation between the two subjects is historical in origin and is retained for pedagogical reasons. Interference is usually associated with the intentional formation of two or more light waves that are analyzed in their region of overlap according to the procedures outlined earlier. Diffraction is usually associated with the obstruction of a single wave, by either a transparent or an opaque obstacle. The obstruction acts as the source of a multitude of scattered waves that can interfere. An example of both interference and diffraction is shown in Figure 9.1. The object is a transparent rectangular aperture and when it is illuminated with a coherent beam of light both diffraction from the edges and interference from light scattered from the edges are observed. The distinction between interference and diffraction is rather fuzzy and arbitrary as can be seen in Figure 9.1.

A rigorous diffraction theory exists that is based on Maxwell's equations and the boundary conditions associated with an obstacle [1]. The boundary conditions are used to calculate a field scattered by the obstacle. The origins of this scattered field are currents induced in the obstacle by the incident field. The scattered field is allowed to interfere with the incident field to produce a resultant diffracted field. The application of the rigorous vector theory is very difficult and for most problems, an approximate scalar theory due to Kirchhoff is used. Vector formulation is required in calculations in modern lithography for nanotechnology and microelectronics. However, when the numerical aperture is less than 0.5, the scalar theory provides remarkable agreement with experiment and is used because of its simplicity. We will limit our discussion to the scalar theory and cover a number of topics based on diffraction theory in the remaining chapters.

The approximate scalar theory is based on *Huygens' principle* (Christiaan Huygens, 1629–95), *which* states that:

Modern Optics Simplified. B. D. Guenther. © B. D. Guenther 2020.
Published in 2020 by Oxford University Press. DOI: 10.1093/oso/9780198842859.001.0001

Figure 9.1 Diffraction produced by the sharp edges of a rectangular aperture. Light scattered from the edges produce interference fringes in the center of the aperture, showing the equivalence of interference and diffraction.

Each point on a wavefront can be treated as a source of a spherical wavelet called a secondary wavelet or a Huygens wavelet. The envelope of these wavelets, at some later time, is constructed by finding the tangent to the wavelets. The envelope is assumed to be the new position of the wavefront.

Rather than simply using the wavelets to construct an envelope, the scalar theory assumes that the Huygens wavelets interfere to produce a new wavefront.

Application of the scalar theory results in an integral called the *Huygens–Fresnel* integral. The traditional interpretation given to this integral is that it is the sum of a group of Huygens' wavelets. The modern interpretation of the integral is in terms of linear system theory.

Free space is viewed, by modern theory, as a linear system and the incident wave is the input to this linear system. The Huygens wavelet is interpreted as the impulse response of free space and the Huygens–Fresnel integral is a convolution of the incident wave with the impulse response of free space.

Analytic solutions of the Huygens–Fresnel integral are usually obtained through the use of two approximations. In both approximations, all dimensions are assumed to be large with respect to a wavelength. In one approximation, the viewing position is assumed to be far from the obstruction; the resulting diffraction is called *Fraunhofer diffraction*. The second approximation, which leads to *Fresnel diffraction*, assumes that the observation point is near the obstruction.

Rather than use a more rigorous mathematical treatment of scalar diffraction theory, originally developed by *Gustav Kirchhoff* and *Arnold Johannes Wilhelm Sommerfeld* (1868–1951), we will develop a simple derivation of the Huygens-Fresnel integral based on an application of Huygens' principle and on the addition of waves to calculate an interference field. The derivation will be more descriptive and intuitive than formal.

We will apply Huygens' principle to diffraction by assuming two pinholes act as sources for Huygens' wavelets. The interference between these two sources is a generalization of the discussion of Young's interference experiment that involves the addition of the Huygens wavelet concept. The addition of the Huygens wavelets will be extended from 2 to N wavelets and then to a continuous distribution of Huygens' sources. The use of a continuous distribution leads to the Huygens–Fresnel integral for diffraction.

When using Huygens' principle, a problem arises with the spherical wavelet produced by each Huygens' source. Part of the spherical wavelet propagates backward and results in an envelope propagating backward toward the light source, but such a wave is not observed in nature. Huygens ignored this problem. Fresnel required the existence of an *obliquity factor* to cancel the backward wavelet but was unable to derive its functional form. When Kirchhoff placed the theory of diffraction on a firm mathematical foundation, the obliquity factor was generated quite naturally in the derivation. In the intuitive derivation of the Huygens–Fresnel integral, given in this chapter, we will present an argument for treating the obliquity factor as a constant at optical wavelengths. We will then derive the constant value it must be assigned.

The Huygens–Fresnel integral is not the only solution of the differential equation describing the optical system. If the differential equation we attempt to solve is the Helmholtz equation, we find another possible solution is a paraxial *Gaussian wave*, i.e., a wave whose intensity distribution, transverse to the propagation direction, is Gaussian in amplitude. We will discover that the Gaussian wave can be characterized by a *beam waist* (the diameter of the beam of light at the half-power points of the Gaussian distribution) and the *radius of curvature* of the wave's phase front. We can use the ABCD matrix to evaluate the propagation of a Gaussian wave combining geometric optics with diffraction.

The first reference to diffraction appears in the notebooks of *Leonardo De Vinci* (1452–1519) but a published description did not occur until the works of Grimaldi appeared, posthumously. Grimaldi conducted a number of experiments involving narrow light beams grazing small obstacles. Newton repeated many of the experiments of Grimaldi and claimed the diffraction was simply a new kind of refraction. He dropped the subject when he found that he could not explain the observations in terms of his particle model of light. Both Grimaldi and Newton conducted their experiments in hope of determining the nature of light but both seemed to retreat from their observations of wavelike behavior.

Young attempted to treat diffraction in terms of interference of light reflected from an edge but was convinced by Fresnel that the approach was wrong. Scientists have returned to Young's concept of diffraction and it is now an active research topic [2].

Fresnel adapted the wavelet concept that Huygens developed to explain birefringence, to explain diffraction. Fresnel enhanced Huygens' theorem by assuming interference between wavelets and thereby was able to explain diffraction from straight edges and apertures.

9.1.1 **Poisson's spot**

Fresnel was a civil engineer who pursued optics as a hobby, after a long day of road construction. Fresnel was also active politically, supporting Napoleon at an inopportune time. His detention by the police for his political views gave Fresnel time to conduct very high-quality observations of diffraction. Fresnel submitted the theory, based on his observations, to the Paris Academy in competition for a prize in 1818. *Siméon Denis Poisson* (1781–1840), one of the judges, presented an argument, as a *reductio ad absurdum,* to prove the theory was incorrect.

Poisson pointed out that a natural consequence of Fresnel's theory was the appearance of a bright spot at the center of the shadow of a circular object that was obviously impossible. Figure 9.2 shows what is now called Poisson's spot in the shadow of a ball bearing.

François Jean Dominique Arago (1786–1853), also a judge at the Paris competition and a friend of Fresnel, conducted an immediate experiment and found what can be seen in Figure 9.2. As you might expect, Fresnel was awarded the prize. (Arago and Poisson had been unaware that *Joseph-Nicolas Delisle*, a French astronomer, had observed the spot in 1715.)

G. Kirchhoff placed the theory developed by Fresnel on a mathematical basis in 1882. Kirchhoff demonstrated with his theory that the backward wavelets illustrated in Figure 9.3 were zero in one and three dimensions. The theory as developed by Kirchhoff had problems, associated with the boundary conditions. A. Sommerfeld avoided these problems by treating the case of an aperture or edge in a perfectly conducting thin screen and was the first to obtain

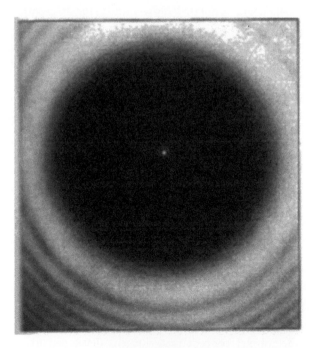

Figure 9.2 Poisson's spot in the shadow of a ball bearing. The illumination used was 488 nm from an argon ion laser. The light is very coherent, making it easy to production diffraction. The grainy structure seen in the intensity of this image is a coherent noise called speckle. It is due to interference between various beams of scattered light. There is also some digitizing noise imposed during the scanning of the original film image.

Figure 9.3 Huygens wavelets shown in gray and the resulting new propagating wavefront.

a rigorous solution for the diffraction of light waves from a conducting obstruction (in 1894). William R. Smythe extended this result to a vector formulation in 1947.

9.1.2 **Huygens' Principle**

To discover how a wave propagates through an aperture, we must solve the vector wave equation obtained from Maxwell's equations in combination with the boundary conditions associated with the aperture. This is a very difficult mathematical problem and we will be forced to discuss only approximate solutions that are based on the application of Huygens' principle to the propagation problem.

Huygens said that each point on a wavefront (i.e., the surface of constant phase) could be regarded as a source for a secondary wavelet in the form of a spherical wave with radius $c\Delta t$ (Figure 9.3). The envelope of the secondary wavelets gives the wavefront at some later time. We will ignore the problems associated with the backward propagating part of the wavelets shown in Figure 9.3.

9.2 **Fresnel Formulation**

Fresnel used Huygens' principle as a foundation for his theoretical explanation of diffraction but it was not until the theoretical derivations of Kirchhoff that the theory of diffraction was placed on a firm mathematical foundation. In this section, we will use a descriptive approach based on Huygens' principle to obtain the Huygens–Fresnel integral of diffraction.

To apply Huygens' principle to the propagation of light through an aperture of arbitrary shape, we need to develop a mathematical description of the field from an array of Huygens' sources filling the aperture. We will begin by obtaining the field from a single pinhole that is illuminated by the wave

$$\tilde{\mathbf{E}}_i(\mathbf{r}, t) = \tilde{\mathbf{E}}_i(\mathbf{r})e^{i\omega t}.$$

The theory of interference provides the rules for combining the fields from two pinholes and allows the generalization of the result to N pinholes. Finally, by letting the area of each pinhole to approach an infinitesimal value, we will construct an arbitrary aperture of infinitesimal pinholes. The result will be the Huygens–Fresnel integral.

We know that the wave obtained after propagation through an aperture must be a solution of the wave equation

$$\nabla^2 \tilde{\mathbf{E}} = \mu\varepsilon \frac{\partial^2 \tilde{\mathbf{E}}}{\partial t^2}.$$

We will be interested only in the spatial variation of the wave so we need only look for solutions of the Helmholtz equation

$$\left(\nabla^2 + k^2 \right) \tilde{\mathbf{E}} = 0.$$

The problem is further simplified by replacing this vector equation with a scalar equation,

$$\left(\nabla^2 + k^2 \right) \tilde{E}\left(x, y, z\right) = 0.$$

This replacement is proper for those cases where the field can be written as $\mathbf{n}E(x,y,z)$ (where $\mathbf{n}$ is a unit vector). In general, we cannot make the substitution $\mathbf{n}E$ for the electric field $\mathbf{E}$ because of Maxwell's equation

$$\nabla \cdot \mathbf{E} = 0.$$

Rather than working with the magnitude of the electric field, we should use the scalar amplitude of the vector potential [3]. We will neglect this complication and assume the scalar E is a single component of the vector field $\mathbf{E}$. As we pointed out earlier, we will not have any problems as long as we do not require very high-resolution spatial detail.

The pinhole is illuminated by a plane wave,

$$\tilde{\mathbf{E}}_i\left(\mathbf{r}, t\right) = \tilde{\mathbf{E}}_i(\mathbf{r})e^{i\omega t}.$$

The wave that leaves the pinhole will be a spherical wave rather than the plane wave we have used in all of our previous discussions. To make the problem easy we should convert the coordinate system to spherical coordinates (r, θ, ϕ). Because we have a point source the wave will only have radial symmetry of the form

$$\tilde{E}\left(\mathbf{r}, t\right) e^{i\omega t} = A\frac{e^{-i\delta}e^{-i\mathbf{k}\cdot\mathbf{r}}}{r}e^{i\omega t}.$$

The complex amplitude

$$\tilde{E}(\mathbf{r}) = A\frac{e^{-i\delta}e^{-i\mathbf{k}\cdot\mathbf{r}}}{r} \tag{9.1}$$

is a solution of the Helmholtz equation. The prescription for adding waves developed for interference can be applied to the problem of calculating the field at P_0, in Figure 9.4, from two pinholes: one at P_1, located a distance $\mathbf{r}_{01} = |\mathbf{r}_0 - \mathbf{r}_1|$ from P_0, and one at P_2, located a distance $\mathbf{r}_{02} = |\mathbf{r}_0 - \mathbf{r}_2|$ from P_0.

The two pinholes are a generalization of Young's interference experiment with the field at P_0, given by the superposition of the wavelets emitted from P_1 and P_2,

$$\tilde{E}\left(\mathbf{r}_0\right) = \frac{A_1}{r_{01}}e^{-i\mathbf{k}\cdot\mathbf{r}_{01}} + \frac{A_2}{r_{02}}e^{-i\mathbf{k}\cdot\mathbf{r}_{02}}. \tag{9.2}$$

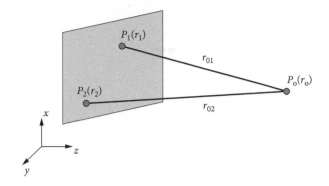

Figure 9.4 Geometry for application of Huygens' principle to two pinholes.

We have incorporated the phase δ_1 and δ_2 into the constants A_1 and A_2 to simplify the equations.

The light emitted from the pinholes is due to a wave, E_i, incident onto the screen from the left. The units of E_i are per unit area, so to obtain the amount of light passing through the pinholes, we must multiply E_i by the areas of the pinholes: $\Delta\sigma_1$ and $\Delta\sigma_2$, respectively:

$$A_1 \propto \tilde{E}_i(\mathbf{r}_1)\Delta\sigma_1 \quad A_2 \propto \tilde{E}_i(\mathbf{r}_2)\Delta\sigma_2,$$

$$\tilde{E}(\mathbf{r}_0) = C_1 \frac{\tilde{E}_i(\mathbf{r}_1)}{r_{01}}e^{-i\mathbf{k}\cdot\mathbf{r}_{01}}\Delta\sigma_1 + C_2 \frac{\tilde{E}_i(\mathbf{r}_2)}{r_{02}}e^{-i\mathbf{k}\cdot\mathbf{r}_{02}}\Delta\sigma_2, \tag{9.3}$$

where C_i is the constant of proportionality. The constant will depend on the angle that $\mathbf{r}_{0i}$ makes with the normal to $\Delta\sigma_i$. This geometrical dependence arises from the fact that the apparent area of the pinhole decreases as the observation angle approaches 90° as we discussed using Figure 3.10.

We can generalize Eq. (9.3) to N pinholes

$$\tilde{E}(\mathbf{r}_0) = \sum_{j=1}^{N} C_j \frac{\tilde{E}_i\left(\mathbf{r}_j\right)}{\mathbf{r}_{0j}}e^{-i\mathbf{k}\cdot\mathbf{r}_{0j}}\Delta\sigma_j. \tag{9.4}$$

The pinhole's diameter is assumed small compared to the distances to the viewing position but large compared to a wavelength. In the limit as $\Delta\sigma_j$ goes to 0, the pinhole becomes a Huygens source. By letting N become large, we can fill the aperture with these infinitesimal Huygens' sources and convert the summation to an integral over the aperture.

It is in this way that we obtain the complex amplitude, at the point P_0, from a wave exiting an aperture by integrating over the area of the aperture. We replace $\mathbf{r}_{0j}$, in Eq. (9.4), by $\mathbf{R}$, the position of the infinitesimal Huygens' source of area, ds, measured with respect to the observation point, P_0. We also replace $\mathbf{r}_j$ by $\mathbf{r}$, the position of the infinitesimal area, ds, with respect to the origin of the coordinate system. The discrete summation, Eq. (9.4), becomes the integral

$$\tilde{E}(\mathbf{r}_0) = \int\int_A C(\mathbf{r})\frac{\tilde{E}_i(\mathbf{r})}{R}e^{-i\mathbf{k}\cdot\mathbf{R}}\,ds. \tag{9.5}$$

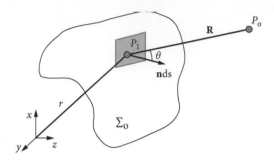

Figure 9.5 The geometry for calculating the field at P_0 using Eq. (9.4).

This is the *Huygens–Fresnel integral*. The geometry associated with Eq. (9.4) is shown in Figure 9.5, where the aperture is denoted as Σ_0, the observation point as P_0, and an arbitrary point in the aperture as P_1.

The constant $C(\mathbf{r})$ depends upon θ, the angle between $\mathbf{n}$, the unit vector normal to the aperture, and $\mathbf{R}$, shown in Figure 9.5. We want to treat $C(\mathbf{r})$ as a constant to allow us to remove it from the integrand. We will justify in the following discussion the assumption that the angular dependence of $C(\mathbf{r})$ can be ignored.

9.2.1 The obliquity factor

The parameter, $C(\mathbf{r})$, in Eq. (9.4) is called the *obliquity factor*. If we had used the theory developed by Kirchhoff and Sommerfeld, we would discover that the obliquity factor had an angular dependence given by

$$\frac{\cos(\hat{\mathbf{n}}, \mathbf{R}) - \cos(\hat{\mathbf{n}}, \mathbf{r}_{21})}{2}.$$

This obliquity factor includes the geometrical contribution to the effective area due to the incident wave arriving at the aperture, at an angle with respect to the normal to the aperture given by $\cos(\hat{\mathbf{n}}, \mathbf{r}_{21})$, where the angle is measured between $\mathbf{r}_{21}$ (the vector between the source and the observation point) and the aperture normal, $\mathbf{n}$. If the source is at infinity, then the incident wave can be treated as a plane wave, incident normal to the aperture, and we may simplify the angular dependence of the obliquity factor to

$$\frac{1 + \cos\theta}{2},$$

where θ is the angle between the normal to the aperture and the vector $\mathbf{R}$. This is the configuration shown in Figure 9.5. The obliquity factor provides an explanation of why it is possible to ignore the backward propagating wave that appears in application of Huygens' principle. For the backward wave, $\theta = \pi$, and the obliquity factor is zero.

The obliquity factor causes the amplitude of the incident and transmitted light to decrease as the viewing angle increases. This is a result of a decrease in the effective aperture area with viewing angle.

The obliquity factor increases the difficulty in calculating the Fresnel integral and it is to our benefit to be able to treat it as a constant. If we neglect its angular contribution, we need to be

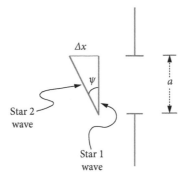

Figure 9.6 Wavefronts from two stars arriving at our telescope.

able to justify that approximation. Our justification will be based upon the resolving power of an optical system operating at visible wavelengths.

Assume we are attempting to resolve two stars that produce plane waves at the aperture of a telescope with an aperture diameter, a (Figure 9.6). The wavefronts from the two stars make an angle ψ with respect to each other (Figure 9.6):

$$\tan \psi \approx \psi = \frac{\Delta x}{a}. \tag{9.6}$$

The smallest angle ψ that can be measured is determined by the smallest length Δx that can be measured. We know we can measure a fraction of wavelength with an interferometer but, without an interferometer, we can only count the crests of the waves, leading to the assumption that $\Delta x \geq \lambda$; i.e., we can measure a length no smaller than λ and the smallest angle we can measure is

$$\psi \geq \frac{\lambda}{a}. \tag{9.7}$$

The resolution limit established by the above reasoning places a limit on the minimum separation that can be produced at the back focal plane of the telescope. We will prove later that the minimum distance on the focal plane between the images of Stars 1 and 2 is given by

$$d = f\psi. \tag{9.8}$$

From Eq. (9.7)

$$d = \frac{\lambda f}{a}.$$

The resolution limit of the telescope can also be expressed in terms of the cone angle produced when the incident plane wave is focused on the back focal plane of the lens. From the geometry of Figure 9.7 the cone angle (actually ½ cone angle) is given by

$$\tan \theta = \frac{a}{2f},$$

The separation between the two stars at the back focal plane of the lens is, thus,

$$d = \frac{\lambda}{2 \tan \theta}. \tag{9.9}$$

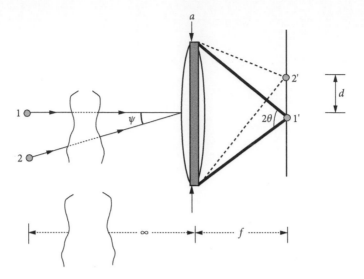

Figure 9.7 Telescope resolution.

This equation will establish the maximum value of θ to be encountered in a conventional optical system. To find a reasonable value for the minimum d we are capable of measuring, we will assume that we are diffraction limited and our resolution is determined by the Rayleigh criterion that we will discuss later. The Rayleigh criterion for minimum resolution of an optical system is

$$d = \left(\frac{1.22}{2NA}\right)\lambda.$$

This can be restated in terms of the $f/\#$ of the lens.[1]

$$f/\# = f/D \approx 1/(2NA)$$

$$d = (1.22)\,(f/\#)\,\lambda.$$

We will assume that the minimum d is 3λ—this is four times the resolution of a typical consumer grade digital camera. With this value for d in Eq. (9.9), the largest obliquity angle that should be encountered in a visible imaging system is

$$\tan\theta = \frac{\lambda}{2d} = \frac{\lambda}{6\lambda} = \frac{1}{6} = 0.167$$

$$\theta = 9.5° \approx 10°.$$

(If we had assumed that the smallest value of d was equal to λ, then $\theta = 26.6°$.) We now can make an estimate of a reasonable upper bound on the variation in the obliquity factor over the range of angles that will be encountered in a visible optical system. Table 9.1 displays the obliquity

---

[1] The number 1.22 is approximately equal to the first zero of the order-one Bessel function $J_1(x)$ divided by π. The numerical aperture, NA, was introduced for a fiber in Chapter 8, Eq. (8.15); for a lens it is proportional to twice the reciprocal of the $f/\#$ of the lens.

Table 9.1 Size of Obliquity Factor for Incident Plane Wave and Focal Length of Full Frame 35 mm Camera with Same Field of View as Shown in Figure 9.7

½ Angle θ (deg)	Obliquity factor $(1 + \cos\theta)/2$	Camera lens (focal length mm)
0	1	-
10	0.992	105
20	0.970	50
30	0.933	35
40	0.883	28

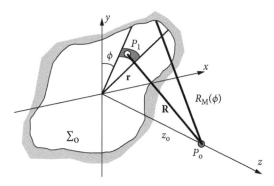

Figure 9.8 Geometry for evaluating the constant in the Fresnel integral.

factor from 0° to 40°. The obliquity factor has only changed by 0.8% over the angular variation of 0° to 10°; thus, in most situations encountered in visible optics, we are safe in treating the factor as a constant and removing it from the integrand.

The obliquity factor undergoes a 10% change when θ varies from 0° to about 40°; therefore, the obliquity factor cannot safely be treated as a constant in any optical system that involves angles that exceed 40°. This can occur in optical systems designed for wavelengths longer than visible wavelengths.

While we have shown that it is safe to ignore the variability of C, we still have not assigned a value to the obliquity factor. To find the proper value for C, we will compare the result obtained using Eq. (9.4) with the result predicted by using geometric optics.

We illuminate the aperture Σ_0 in Figure 9.8 with a plane wave of amplitude α, traveling parallel to the z-axis. Geometrical optics predicts a field at P_0, on the z-axis, a distance z_0 from the aperture, given by

$$\tilde{E}_{\text{geom}} = \alpha e^{-ikz_0}. \tag{9.10}$$

This is the amplitude of the wave in the aperture times a phase term associated with the propagation of the wave through a distance z_0. The area of the infinitesimal at P_1 (the Huygens source) is

$$ds = r \, dr \, d\phi.$$

The obliquity factor is assumed to be a constant, C, that can be removed from under the integral. The incident wave is a plane wave whose value at $z = 0$ is $E(x,y,0) = \alpha$. Using these parameters, the Fresnel integral, Eq. (9.4), can be written as

$$\tilde{E}(z_0) = C\alpha \int \int \frac{e^{-ik\cdot R}}{R} r\, dr\, d\phi. \tag{9.11}$$

The distance from the Huygens source to the observation point is

$$z_0^2 + r^2 = R^2.$$

The variable of integration can be written in terms of the distance to a point in the aperture, R,

$$r\, dr = R dR.$$

The limits of integration over the aperture extend from $R = z_0$ on the optical axis, to the maximum value of R; $R = R_M(\phi)$, the maximum value at a given value, ϕ, is not known:

$$\tilde{E}(z_0) = C\alpha \int_0^{2\pi} \int_{z_0}^{R_m(\phi)} e^{-ikR}\, dR\, d\phi. \tag{9.12}$$

The integration over R can now be carried out to yield

$$\tilde{E}(z_0) = \frac{C\alpha}{ik} e^{-ikz_0} \int_0^{2\pi} d\phi - \frac{C\alpha}{ik} \int_0^{2\pi} e^{-ikR_m(\phi)}\, d\phi. \tag{9.13}$$

The first integration in Eq. (9.13) is easy to perform. The second integration cannot be calculated because we are performing the evaluation of the Fresnel integral for a general aperture and the functional form of $R_m(\phi)$ is not known. However, the physical significance of the two integrals is apparent. The first term in Eq. (9.13) is the amplitude due to geometric optics. The second term may be interpreted as the sum of the waves diffracted by the boundary of the aperture. This diffractive term simply makes a correction to the geometric optics result contained in the first term. This is the interpretation of diffraction that Young suggested.

The aperture is irregular in shape; at least on the scale of a wavelength, thus kR_m will vary many multiples of 2π as we integrate around the aperture. For this reason, we should be able to assume the integral is near zero and ignore its value in Eq. (9.13) if we confine our attention to the light distribution on the z-axis. After neglecting the second term, we are left with only the geometrical optics component of Eq. (9.13):

$$\tilde{E}(z_0) = \frac{2\pi C}{ik} \alpha e^{-ikz_0}. \tag{9.14}$$

For Eq. (9.14) to agree with the prediction of geometric optics, Eq. (9.10), the constant C must be equal to

$$C = \frac{ik}{2\pi} = \frac{i}{\lambda}. \tag{9.15}$$

Using this result in Eq. (9.13) we obtain

$$\tilde{E}(z_0) = \underbrace{\alpha e^{-ikz_0}}_{\text{geometric}} - \underbrace{\frac{1}{2\pi} \int_0^{2\pi} \alpha e^{-ikR_m(\phi)} d\phi}_{\text{diffraction}}.$$

We see that diffraction is a correction added to the electric field transmission prediction made for the aperture by geometrical optics.

The Huygens–Fresnel integral can be written, using the value for C just derived,

$$\tilde{E}(\mathbf{r}_0) = \frac{i}{\lambda} \int \int_{\Sigma} \frac{\tilde{E}_i(\mathbf{r})}{R} e^{-i\mathbf{k}\cdot\mathbf{R}} ds. \tag{9.16}$$

This integral can be interpreted in two analogous ways. The classical interpretation views the term

$$\frac{i}{\lambda} \frac{e^{-i\mathbf{k}\cdot\mathbf{R}}}{R}$$

as a Huygens wavelet. This spherical wave of unit amplitude, when multiplied by the value of the incident wave, $E_i(\mathbf{r})$, at a point $\mathbf{r}$ in space, yields the Huygens wavelet radiated by that point. The integral Eq. (9.16) adds all of the interfering wavelets together, to produce the resultant wave.

A more modern interpretation of the Huygens–Fresnel integral is to apply linear system theory to the propagation of light through free space. In linear system theory the output of a time-invariant linear system to an input function is the convolution of the input with the impulse response of the system. We view Eq. (9.16) as a convolution integral.[2] Considering free space as the linear system, one finds that the result of propagation through free space can be calculated by convolving the incident (input) wave with the impulse response of free space.[3] Here the impulse response is

$$\frac{i}{\lambda} \frac{e^{-i\mathbf{k}\cdot\mathbf{R}}}{R}.$$

9.2.2 Approximate solutions of Huygens–Fresnel integral

The job of calculating diffraction has only begun with the derivation of the Huygens–Fresnel integral. In general, an analytic expression for the integral cannot be found because of the difficulty of performing the integration over R in Figure 9.8. Two approximations we can make to obtain analytic expressions of the Huygens–Fresnel integral will be discussed under the headings of far-field diffraction and near-field diffraction.

Rigorous solutions exist for only a few idealized obstructions. To allow discussion of the general properties of diffraction, it is necessary to use approximate solutions. Let us look for an approximate solution of the Huygens–Fresnel integral

$$\int \frac{\tilde{E}_i(\mathbf{r})}{R} e^{-i\mathbf{k}\cdot\mathbf{R}} dx.$$

for a one-dimensional aperture with an opening equal to b shown in Figure 9.9.

---

[2] We will discuss the smoothing operation, called a convolution, when we discuss imaging; see Figure 11.1.

[3] The spatial impulse response describes the propagation from a point source to the observation point. There is also a temporal impulse response that is a short duration pulse, a delta function in time.

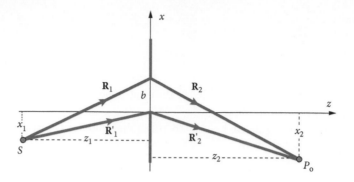

Figure 9.9 One-dimensional slit of width b used to establish Fraunhofer and Fresnel approximations.

For this diffraction component to contribute to the field at Point P_0, in Figure 9.9, the phase of the exponent must not vary over 2π when the integration is performed over the aperture shown in Figure 9.9, i.e., the optical path difference must meet the condition

$$\Delta = \mathbf{k} \cdot (\mathbf{R}_1 - \mathbf{R}_1' + \mathbf{R}_2 - \mathbf{R}_2') < 2\pi.$$

From the geometry of Figure 9.9, the two paths from the source, S, to the observation point, P_0, are

$$R_1 + R_2 = \sqrt{z_1^2 + (x_1 + b)^2} + \sqrt{z_2^2 + (x_2 + b)^2} = z_1\sqrt{1 + \left(\frac{x_1 + b}{z_1}\right)^2} + z_2\sqrt{1 + \left(\frac{x_2 + b}{z_2}\right)^2}$$

$$R_1' + R_2' = \sqrt{z_1^2 + x_1^2} + \sqrt{z_2^2 + x_2^2} = z_1\sqrt{1 + \left(\frac{x_1}{z_1}\right)^2} + z_2\sqrt{1 + \left(\frac{x_2}{z_2}\right)^2}.$$

By assuming, that the aperture, of width b, is small compared to the distances z_1 and z_2, the optical path difference between these distances, Δ, can be rewritten using the binomial expansion:[4]

$$\sqrt{1 + \delta} = 1 + \delta/2 - \delta^2/8 + \dots \quad \delta^2 < 1$$

$$\Delta = k\left(R_1 - R_1' + R_2 - R_2'\right)$$

$$\Delta \approx k\left(\frac{x_1}{z_1} + \frac{x_2}{z_2}\right)b + \frac{k}{2}\left(\frac{1}{z_1} + \frac{1}{z_2}\right)b^2 + \dots . \tag{9.17}$$

If we assume that the second term of this expansion is small, i.e.,

$$\frac{k}{2}\left(\frac{1}{z_1} + \frac{1}{z_2}\right)b^2 \ll 2\pi,$$

$$\frac{b^2}{2}\left(\frac{1}{z_1} + \frac{1}{z_2}\right) < \lambda,$$

---

[4] Binomial theorem: $(1 + x)^n = 1 + nx + [n(n - 1)/2!]\,x^2 + \dots$.

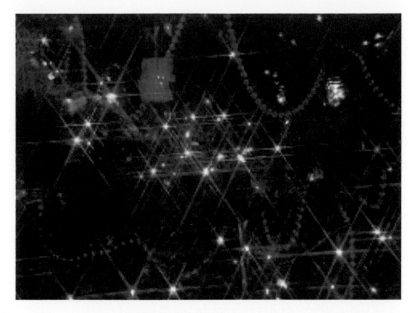

Figure 9.10 A screen is placed in front of the camera lens. Light from point sources in the field of view is diffracted by the screen and produces "stars" in the image. The stars are actually the Fraunhofer diffraction patterns of the screen, produced by each point source.
By midiman—https://www.flickr.com/photos/midiman/77259745/CC BY 2.0 https://commons.wikimedia.org/w/index.php?curid=8333178.

then we may treat the variation of the phase across the aperture as if it were linear. Physically, this means that all waves in the aperture are plane waves. The diffraction predicted by the theory based on this assumption is called far-field diffraction or *Fraunhofer diffraction*.

Fraunhofer diffraction is used by photographers and television cameramen to provide artistic highlights in an image containing small light sources. A typical image is shown in Figure 9.10. By observing a mercury or sodium streetlight, a few hundred meters away, through a sheer curtain, a Fraunhofer diffraction pattern of the fabric weave can easily be observed. Even though diffraction produced by most common light sources and objects is a small effect, it is possible to observe diffraction without special equipment. Fringes can be observed by viewing a light source through a slit formed by two fingers that are nearly touching.

9.3 Fraunhofer Diffraction

In this section, we will derive the equation describing Fraunhofer diffraction in two dimensions. We will assume the source of light is at infinity, $z_1 = \infty$ in Figure 9.11, so that the aperture, Σ, is illuminated by a plane wave traveling parallel to the z-axis. We then derive an approximate expression, analogous to Eq. (9.17), for the position vector R in Figure 9.11. We obtain an approximate expression for the position vector of the observation point, P_0, relative to the aperture, by assuming the aperture size is small, relative to the distance to the observation point.

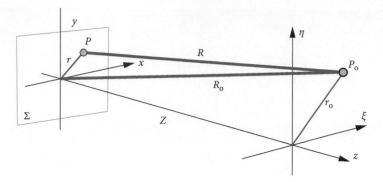

Figure 9.11 Geometry for Fraunhofer diffraction. The aperture plane is (x,y) and the observation plane is (ξ, η).

Using this approximate expression, and the paraxial approximation, we are able to formulate the Huygens–Fresnel integral as a two-dimensional Fourier transform.

In Fraunhofer diffraction, we require that the source of light and the observation point, P_0, be far from the aperture so that the incident and diffracted waves can be approximated by plane waves. A consequence of this requirement is that the entire waveform passing through the aperture contributes to the observed diffraction. For Fresnel diffraction, this is not the case, making that analysis more complicated.

The geometry to be used in this derivation is shown in Figure 9.11. The wave incident on the aperture, Σ, is a plane wave of amplitude α and the objective of the calculation is to find the departure of the transmitted wave from the geometrical optic prediction. The calculation will provide the light distribution, transmitted by the aperture, as a function of the angle the light is deflected from the incident direction. We assume that diffraction makes only a small perturbation on the predictions of geometrical optics. The deflection angles encountered in this derivation are, therefore, small and we will be able to use the paraxial approximation.

The distance from a point, P, in the aperture to the observation point P_0 of Figure 9.11 is

$$R^2 = (x - \xi)^2 + (y - \eta)^2 + Z^2. \tag{9.18}$$

From Figure 9.11 we see R_0 is the distance from the center of the aperture screen to the observation point, P_0,

$$R_0^2 = \xi^2 + \eta^2 + Z^2. \tag{9.19}$$

The difference between these two vectors is

$$\begin{aligned} R_0^2 - R^2 &= \xi^2 + \eta^2 + Z^2 - Z^2 - \left(x^2 - 2x\xi + \xi^2\right) - \left(y^2 - 2y\eta + \eta^2\right) \\ &= 2\left(x\xi + y\eta\right) - \left(x^2 + y^2\right). \end{aligned} \tag{9.20}$$

This corresponds to the difference, $\mathbf{R}_2 - \mathbf{R}_2'$, calculated in Figure 9.9. We can rewrite the difference between the two vectors as

$$R_0^2 - R^2 = (R_0 - R)(R_0 + R). \tag{9.21}$$

Using Eq. (9.20) we can write an equation for the position of point P in the aperture in terms of the aperture size, b, introduced in Figure 9.9:

$$b \geq R_0 - R = \frac{R_0^2 - R^2}{R_0 + R},$$

$$= \left[2 \left(x\xi + y\eta \right) - \left(x^2 + y^2 \right) \right] \frac{1}{R_0 + R}.$$

The reciprocal of $(R_0 + R)$ can be written as

$$\frac{1}{R_0 + R} = \frac{1}{2R_0 + R - R_0}$$

$$= \frac{1}{2R_0} \left(1 + \frac{R - R_0}{2R_0} \right)^{-1}. \tag{9.22}$$

Now using Eq. (9.22)

$$R_0 - R = \left[\frac{x\xi + y\eta}{R_0} - \frac{x^2 + y^2}{2R_0} \right] \left[1 - \frac{R_0 - R}{2R_0} \right]^{-1}.$$

If the diffraction integral is to have a finite (nonzero) value, then

$$k \mid R_0 - R \mid \ll kR_0$$
$$kb \ll kR_0.$$

This means that if we measure the aperture and distance from the aperture in units of wavelength, then the aperture size must be much smaller than the distance to the observation point. This requirement ensures that all the Huygens wavelets, produced over the half of the aperture from the center out to the edge, b, will have similar phases and will interfere to produce a nonzero amplitude at P_0. The requirement that the phase changes are small can be written as

$$\frac{1}{1 - \dfrac{R_0 - R}{2R_0}} = \frac{1}{1 - \dfrac{b}{2R_0}} \approx 1.$$

It is not the aperture size but the ratio of aperture size to observation distance that determines the far-field assumption.

By using the equation for the path difference given by Eq. (9.20), we obtain the diffraction integral

$$\tilde{E}_P = \frac{i\alpha e^{-ik \cdot R_0}}{\lambda R_0} \int \int_{\Sigma} f(x, y) \, e^{+ik\left(\frac{x\xi + y\eta}{R_0} - \frac{x^2 + y^2}{2R_0} \right)} dx dy. \tag{9.23}$$

We assume the aperture is uniformly filled by a plane wave of amplitude α. The change in the amplitude of the wave due to the change in R, as we move away from the z-axis, is neglected. This is a paraxial approximation and is equivalent to stating that the size of the aperture does not change much as we move off axis due to the great distance our observation point is from the aperture. The fact that the amplitude of the observed wave is a constant with respect to the observation point allows us to replace R in the denominator of the Huygens–Fresnel integral by R_0 and move it outside of the integral. This means that the field at the observation point from the center of the aperture is

$$\mathcal{E} = \frac{\alpha e^{-i\mathbf{k} \cdot \mathbf{R}_0}}{R_0}$$

This is a spherical wave expanding out from the center of the aperture.

We have introduced a complex transmission function, $f(x,y)$, to allow a very general aperture to be treated. If the aperture function described the variation in absorption of the aperture as a function of position, as would be produced by a photographic negative, then $f(x,y)$ would be a real function. If the aperture function described the variation in transmission of a biological sample, it might be entirely imaginary. We have not asked how we can perform the integral over $f(x,y)$. That is found to be difficult except for apertures that are separable in x and y. We will assume $f(x,y)$ is separable and essentially limit ourselves to solving one-dimensional problems. We will also initially assume $f(x)$ is a constant value across the aperture to allow us to direct our attention to the physics of the problem rather than the math.

The argument of the exponent in Eq. (9.23) is

$$ik\left(\frac{x\xi + y\eta}{R_0} - \frac{x^2 + y^2}{2R_0}\right) = i2\pi\left(\frac{x\xi + y\eta}{\lambda R_0} - \frac{x^2 + y^2}{2\lambda R_0}\right). \tag{9.24}$$

As we suggested following Eq. (9.17), we will assume the observation point, P_0, is far from the screen; we can then neglect the second term of the argument and treat the phase variation across the aperture as a linear function of position. This is equivalent to assuming that the diffracted wave is a collection of plane waves.

Mathematically, the second term in Eq. (9.24) can be neglected if

$$\frac{x^2 + y^2}{2\lambda R_0} \ll 2\pi. \tag{9.25}$$

This is called the *far-field approximation*. To get a measure of what much, much less means, let us assume that

$$\frac{r^2}{4\pi\lambda} \ll R_0 \underset{\longrightarrow}{\text{assume}} \quad \frac{r^2}{4\pi\lambda} \approx (0.01)R_0.$$

If the aperture $b = 1$ mm and the wavelength is 500 nm, then

$$\frac{r^2}{4\pi\lambda} = \frac{\left(5 \times 10^{-4}\right)}{4\pi\left(500 \times 10^{-9}\right)} = 4 \times 10^{-2}\text{m} \quad R_0 \geq 4\text{m}.$$

The far field for a 1-mm-diameter hole is 4 meters. A lens will let us replace the far-field distance by the focal length of the lens, moving the far-field observation plane to a distance compatible to the dimensions of a typical lab.

The first term in Eq. (9.24) contains the direction cosines

$$L = \frac{\xi}{R_0} \qquad M = \frac{\eta}{R_0}. \tag{9.26}$$

These cosines can remain finite even in the far field. As the unit of measure at the aperture screen, we use the illuminating wavelength, which allows the aperture coordinates to be defined as

$$X = \frac{x}{\lambda} \quad Y = \frac{y}{\lambda}. \tag{9.27}$$

We can rewrite Eq. (9.23) by using the approximation allowed by Eq. (9.25) and the definitions of Eqs. (9.26) and (9.27)

$$\tilde{E}_P(L, M) = \frac{i\alpha e^{-i\mathbf{k}\cdot\mathbf{R}_0}}{\lambda R_0} \int\int_{\Sigma} f(X, Y) e^{+i2\pi(LX+MY)} \, dX dY. \tag{9.28}$$

This is a decomposition of the diffracted wave in terms of an angular distribution of plane waves, each traveling in a unique direction[5] given by X and Y. In this form, the integral may not be immediately recognizable but, by defining two spatial frequencies, we can obtain a familiar form. The spatial frequencies in the x- and y-directions are defined as[6]

$$\omega_x = kL = -\frac{2\pi\xi}{\lambda R_0} \quad \omega_y = kM = -\frac{2\pi\eta}{\lambda R_0}. \tag{9.29}$$

We equate the spatial frequencies to coordinate positions in the observation plane by applying the paraxial approximation; this makes the mapping from the aperture to the observation plane easy to interpret. With the variables defined in Eq. (9.13), the integral becomes

$$\tilde{E}_P(\omega_x, \omega_y) = \frac{i\alpha e^{-i\mathbf{k}\cdot\mathbf{R}_0}}{\lambda R_0} \int\int_{\Sigma} f(x, y) e^{-i(\omega_x x+\omega_y y)} \, dx \, dy. \tag{9.30}$$

The Fraunhofer diffraction field, $\tilde{E}_P$, in the observation plane equals the two-dimensional Fourier transform of the aperture's transmission function, $f(x,y)$.

9.3.1 Diffraction by a rectangular aperture

We will now use Fourier transform theory to calculate the Fraunhofer diffraction pattern from a rectangular slit and will point out the reciprocal relationship between the size of the diffraction pattern and the size of the aperture.

Consider a rectangular aperture with a transmission function given by

$$f(x, y) = \begin{cases} 1 & |x| \le x_0 \\ & |y| \le y_0 \\ 0 & \text{all other } x \text{ and } y \end{cases}. \tag{9.31}$$

Because the aperture is two-dimensional, we need to apply a two-dimensional Fourier transform. By design the amplitude transmission function is separable, in x and y, so we may write the diffraction amplitude distribution from a rectangular slit illuminated by a plane wave of amplitude α as

---

[5] These plane waves are sometimes called the *natural modes* of the propagation medium.

[6] We define the spatial frequencies with a negative sign to allow equations involving spatial frequencies to have the same mathematical form as those involving temporal frequencies. The negative sign is required because ωt and $\mathbf{k}\cdot\mathbf{r}$ appear in the phase of the wave with opposite signs. Rather than use a negative spatial frequency, we could have redefined the Fourier transform using a positive exponential for spatial transforms.

Figure 9.12 Optical representation of **rect**(ξ) defined in Eq. (9.31).

$$\tilde{E}_P = \frac{i\alpha}{\lambda R_0} e^{-i\mathbf{k}\cdot\mathbf{R}_0} \int\limits_{-x_0}^{x_0} f(x)e^{-i\omega_x x}dx \int\limits_{-y_0}^{y_0} f(y)e^{-i\omega_y y}dy. \tag{9.32}$$

To keep the math simple we will reduce the problem to one dimension, replacing $f(x,y)$ with $f(x)$. An optical image of this aperture is shown in Figure 9.12. We have made the problem one-dimensional by extending the y-axis beyond the field of view.

Because $f(x)$ is a symmetric function about the origin the Fourier transform reduces to the cosine transform

$$F(\omega_x) = \int\limits_{-x_0}^{x_0} \cos\omega_x x\, dx = \frac{1}{\omega_x}[\sin\omega_x x]_{-x_0}^{x_0} = \frac{\sin\omega_x x_0}{\omega_x x_0}. \tag{9.33}$$

A mathematical interpretation of this equation is as follows: $\cos\omega x$ is a weighting function of the aperture, called the integral kernel.[7] The shape and duration of the weighting function determines the spatial average of $f(x)$ calculated by Eq. (9.33). In Figure 9.13, the cosine weighting function is plotted as a two-dimensional surface in ω_x,x-space.

The function **rect**(x) slices the weighting function, $\cos\omega x$, perpendicular to the ω_x-axis. The frequency of the weighting function is determined by the position of the slice on the ω_x-axis. The extent of the weighting function in the x-direction is determined by **rect**(x) (the x-dimension of Eq (9.31)). The value of $F(\omega_x)$ at each frequency is the area under the cosine curve. The upper curve in Figure 9.13 displays a few representative points [5] for $F(\omega_x)$. The

---

[7] The integral kernel can be thought of as the nucleus of the transform integral that maps a function between two spaces, for example from time to frequency or space to inverse space.

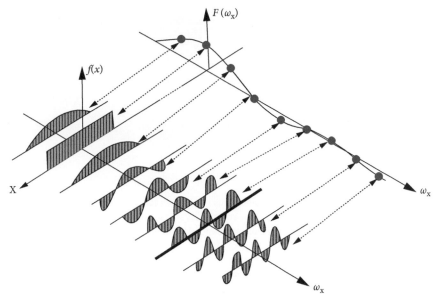

Figure 9.13 Geometrical constructions of the Fourier transform integral of a rectangular pulse. (Courtesy Jack D. Gaskill [5] approved by John Wiley and Sons Inc.)

Fourier transform of the rectangular pulse, Eq. (9.31), is the continuous frequency spectrum shown in Figure 9.14 and given by a function of the form

$$\mathrm{sinc}(x) = \frac{\sin x}{x}. \tag{9.34}$$

Since both $f(x)$ and $f(y)$ are similar functions, we can use Eq. (9.34) to write the two-dimensional function by assuming the extend in the y-direction is $|y| \leq y_0$:

$$\tilde{E}_P = i\frac{4x_0 y_0 \alpha}{\lambda R_0} e^{-i\mathbf{k}\cdot\mathbf{R}_0} \frac{\sin \omega_x x_0}{\omega_x x_0} \frac{\sin \omega_y y_0}{\omega_y y_0}. \tag{9.35}$$

The intensity distribution of the Fraunhofer diffraction produced by the rectangular aperture is

$$I_P = I_0 \frac{\sin^2 \omega_x x_0}{(\omega_x x_0)^2} \frac{\sin^2 \omega_y y_0}{(\omega_y y_0)^2}, \tag{9.36}$$

where the spatial frequencies are defined as

$$\omega_x = -\frac{2\pi \sin \theta_x}{\lambda} = -\frac{2\pi \xi}{\lambda R_0} \quad \omega_y = -\frac{2\pi \sin \theta_y}{\lambda} = -\frac{2\pi \eta}{\lambda R_0}.$$

The function Eq. (9.34) is indeterminate (0/0) at $x = 0$. Its value at $x = 0$, which is equal to 1, is found by using L'Hopital's rule as x approach 0. The area of this rectangular aperture is defined as $A = 4x_0 y_0$; we may write the maximum intensity as

$$I_0 = \left[\frac{i4x_0 y_0 \alpha e^{-i\mathbf{k}\cdot\mathbf{R}_0}}{\lambda R_0}\right]^2 = \frac{A^2 \alpha^2}{\lambda^2 R_0^2}.$$

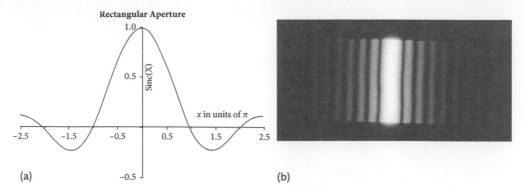

Figure 9.14 (a) Calculation of diffraction from a rectangular slit. This sinc function describes the light wave's amplitude that would exist in both the x- and y-directions. (b) Experimentally generated Fraunhofer diffraction pattern from a rectangular aperture. You can see fringes near the top and bottom due to the fact that the slit is not infinite in one dimension.

The minima of Eq. (9.36) occur when $\omega_x x_0 = n\pi$ or when $\omega_y y_0 = m\pi$. The location of the zeroes can be specified as a dimension in the observation plane or, using the paraxial approximation, in terms of an angle

$$\sin \theta_x \approx \theta_x \approx \frac{\xi}{R_0} = \frac{n\lambda}{2x_0} \qquad \sin \theta_y \approx \theta_y \approx \frac{\eta}{R_0} = \frac{m\lambda}{2y_0}.$$

The dimensions of the diffraction pattern are characterized by the location of the first zero, i.e., when $n = m = 1$, and are given by the observation plane coordinates ξ and η. The dimensions of the diffraction pattern are inversely proportional to the dimensions of the aperture. As the aperture dimension expands, the width of the diffraction pattern decreases until, in the limit of an infinitely wide aperture, the diffraction pattern becomes a delta function.

If a lens is used to display the diffraction pattern, then R_0 is replaced by the focal length of the lens, f, in all of the equations. Figure 9.14a is a theoretical plot of the diffraction pattern from a rectangular slit using Eq. (9.33). Figure 9.14b is an experimentally obtained Fraunhofer diffraction pattern using the slit shown in Figure 9.12.

9.3.2 Array theorem

There is an elegant mathematical technique for handling multiple apertures, called the array theorem. The theorem is based on the convolution integral that we will discuss in Chapter 11. The theorem uses the fact that the Fourier transform of a convolution of two functions is the product of the Fourier transforms of the individual functions:

$$\mathcal{F}\{a(t) \otimes b(t)\} = \mathcal{F}\left\{ \underbrace{\int_{-\infty}^{\infty} a(t) b(\tau - t)\, dt}_{\substack{\text{convolution} \\ \text{integral}}} \right\} = A(\omega) B(\omega). \qquad (9.37)$$

where A is the Fourier transform of a and likewise for B and b. We will prove the array theorem for one dimension, where the functions represent slit apertures. The results can be extended

Figure 9.15 The Karl G. Jansky Very Large Array centimeter wavelength radio telescope on the Plains of San Agustin in New Mexico, USA.
By user: Hajor—own work, CC BY-SA 3.0, https://commons.wikimedia.org/w/index.php?curid=138997.

to two dimensions in a straightforward way. The theory is used not only for optical systems but also in the design of non-optical systems such as focused hyperthermia systems for cancer treatment, ultrasound imaging systems, phased array radars used in modern fighter aircraft, and in the design of telescopes for radio astronomy as shown in Figure 9.15.

Assume that we have a collection of identical apertures, shown on the right of Figure 9.16. If one of the apertures is located at the origin of the aperture plane, its transmission function is $\psi(x)$. The transmission function of an aperture located at a point, x_n, can be written in terms of a generalized aperture function, $\psi(x - \alpha)$, by the use of the sifting property of the delta function[8]

---

[8] The definition of the delta function usually encountered is

$$\delta(t - t_0) = 0 \quad t \neq t_0;$$

i.e., the function is zero everywhere except at the point t_0. The integral of the delta function is

$$\int_{-\infty}^{\infty} \delta(t - t_0)\,dt = 1;$$

i.e., the area under the delta function is one. A mathematically more precise definition of a delta function is given by Eq. (9.38).

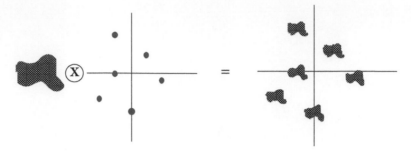

Figure 9.16 The convolution of an aperture with an array of delta functions will produce an array of identical apertures, each located at the position of one of the delta functions.

$$\Psi\left(x - x_n\right) = \int \Psi\left(x - \alpha\right) \delta\left(\alpha - x_n\right) d\alpha. \tag{9.38}$$

This convolution integral[9] will allow the application of the convolution theorem to complete the derivation of the array theorem.

The aperture transmission function representing an array of apertures will be the sum of the distributions of the individual apertures, represented graphically on the left of Figure 9.16 and mathematically by the summation

$$\Psi(x) = \sum_{n=1}^{N} \psi\left(x - x_n\right).$$

The Fraunhofer diffraction from this array is the Fourier transform of $\Psi(x)$

$$\Phi\left(\omega_x\right) = \int_{-\infty}^{\infty} \Psi(x) e^{-i\omega_x x} dx.$$

From the fact that the Fourier transform is linear: i.e., the superposition principle applies,

$$\mathcal{F}\left\{af(t) + bg(t)\right\} = aF\left(\omega\right) + bG\left(\omega\right), \tag{9.39}$$

we can write

$$\Phi\left(\omega_x\right) = \sum_{n=1}^{N} \int_{-\infty}^{\infty} \psi\left(x - x_n\right) e^{-i\omega_x x} dx.$$

---

[9] Equation (9.38) can be recognized as the convolution integral, Eq. (9.37), if we redefine the terms $t = \alpha - x_n$ and $dt = d\alpha$

$$\Psi\left(x - x_n\right) = \int \Psi\left(x - x_n - t\right) \delta(t) dt.$$

By defining $x - x_n = \tau$, we may rewrite the integral in the same form as the convolution integral which we will discuss further in Chapter 11:

We now make use of the fact that $\psi(x - x_n)$ can be expressed in terms of a convolution integral. The Fourier transform of $\psi(x - x_n)$ is, from Eq. (9.37), the product of the Fourier transforms of the individual functions that make up the convolution

$$\Phi(\omega_x) = \sum_{n=1}^{N} \mathcal{F}\{\Psi(x - \alpha)\} \, \mathcal{F}\{\delta(x - x_n)\}$$

$$= \mathcal{F}\{\psi(x - \alpha)\} \sum_{n=1}^{N} \mathcal{F}\{\delta(x - x_n)\}.$$

Again applying Eq. (9.39)

$$\Phi(\omega_x) = \underbrace{\mathcal{F}\{\psi(x - \alpha)\}}_{\text{Aperture}} \underbrace{\mathcal{F}\left\{\sum_{n=1}^{N} \delta(x - x_n)\right\}}_{\text{Array}} \tag{9.40}$$

The first transform in Eq. (9.40) is the diffraction pattern of an individual aperture and the second transform is the diffraction pattern produced by a set of point sources with the same spatial distribution as the array of identical apertures; we will call this second transform the array function. In one dimension, the array function is also called the comb function:

$$\mathbf{comb}(t) = \sum_{n=-N}^{N} \delta(t - t_n).$$

We figured out this sum in Chapter 4, Eq. (4.4), when we were preparing to calculate interference for N sources.

To summarize, the array theorem states *that the diffraction pattern of an array of similar apertures is given by the product of the diffraction pattern from a single aperture and the diffraction (or interference) pattern of an identically distributed array of point sources.*

9.3.2.1 *N* rectangular slits

We can now extract benefit from the array theorem by applying it to an array of N identical apertures, called a *diffraction grating* in optics. The Fraunhofer diffraction patterns produced by such an array has two important properties; a number of very narrow beams are produced by the array and the beam positions are a function of the wavelength of illumination and the relative phase of the waves transmitted by each of the apertures.

Because of these properties, arrays of diffracting apertures have been used in a number of applications. At radio frequencies, arrays of dipole antennas are used to both radiate and receive signals in radar and radio astronomy systems. One advantage offered by diffracting arrays at radio frequencies is that the beam produced by the array can be electrically steered by adjusting the relative phase of the individual dipoles. This design provides a large weight advantage in fighter aircraft.

An optical realization of a two-element array of radiators is Young's two-slit experiment and a realization of a two-element array of receivers is Michelson's stellar interferometer. Two optical

Figure 9.17 Hologram of Hummel figurine (Hear Ye! Hear Ye!) reconstructed using three different wavelengths. The wavelength sensitivity is due to diffraction by photographically recorded interference fringes. (Hologram courtesy of N. George.)

implementations of arrays, containing more diffracting elements, are diffraction grating and holograms (Figure 9.17). In nature, periodic arrays of diffracting elements are the origin of the colors observed on some invertebrates and birds (Figure 9.18).

Many solids are naturally arranged in three-dimensional arrays of atoms or molecules that act as diffraction gratings when illuminated by X-ray wavelengths. The resulting Fraunhofer diffraction patterns are used to analyze the ordered structure of the solids. In Figure 9.19 is the diffraction pattern generated by X-ray scattering from NaCl.

The array theorem can be used to calculate the diffraction pattern from a linear array of N slits, each of width a and separation d (Figure 9.20). The aperture function of a single slit of width a is equal to

$$\psi(x,y) = \begin{cases} 1 & |x| \le \dfrac{a}{2} \\ & |y| \le y_0 \\ 0 & \text{all other } x \text{ and } y \end{cases}. \tag{9.41}$$

The Fraunhofer diffraction pattern of the rectangular aperture has already been calculated in Eq. (9.35),

$$\mathcal{F}\{\psi(x,y)\} = K\frac{\sin\alpha}{\alpha},$$

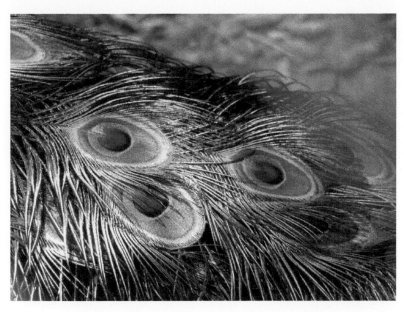

Figure 9.18 Peacock feathers displaying color due to diffraction from periodic structure made of an ordered array of melanin rods. The pigment color is brown due to melanin.
By user:AlexDuarte [public domain], from Wikimedia Commons.

where the constant K is

$$K = \frac{i2ay_0 A}{\lambda R_0} e^{-ik \cdot R_0} \frac{\sin \omega_y y_0}{\omega_y y_0},$$

A is the amplitude of the illuminating wave and the variable α is

$$\alpha = \frac{ka}{2} \sin \theta_x.$$

The array function for N sources spaced d apart is

$$A(x) = \sum_{n=1}^{N} \delta(x - x_n),$$

where $x_n = (n-1)d$. We can write their sum as a geometric series

$$\mathcal{F}\left\{ \sum_{n=-N}^{N} \delta(x - nx_0) \right\} = \sum_{n=-N}^{N} \mathcal{F}\{\delta(x - nx_0)\} = \sum_{n=-N}^{N} e^{-i\omega_x nx_0}. \tag{9.42}$$

Since this is the sum of a geometric series, we can write

$$\mathcal{F}\left\{ \sum_{n=-N}^{N} \delta(x - nx_0) \right\} = \left[\frac{e^{-iN\omega_x x_0} - 1}{e^{-i\omega_x x_0} - 1} \right] + \left[\frac{e^{iN\omega_x x_0} - 1}{e^{i\omega_x x_0} - 1} \right] - 1,$$

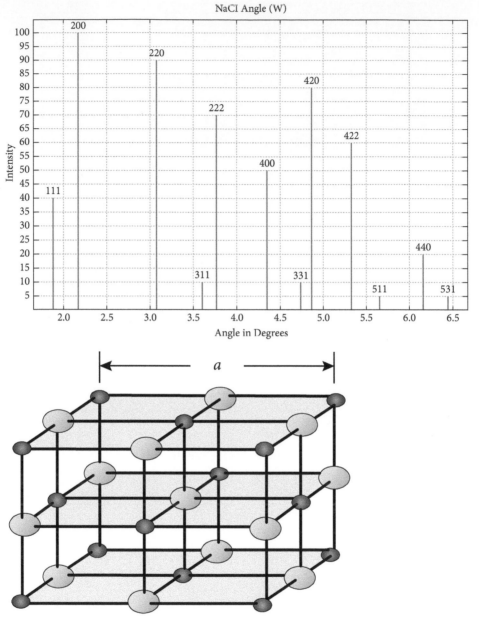

Figure 9.19 X-ray diffraction pattern of NaCl. The crystal structure is shown in the lower image with the blue spheres representing chlorine ions and the red spheres sodium ions. A = 0.563 nm. The 200 peak shown in the upper image corresponds to scattering from the {100} planes shown by the shaded areas.

Figure 9.20 Linear array of slits. Single slit rectangular aperture function is Eq. (9.41).

$$\mathcal{F}\left\{\sum_{n=-N}^{N}\delta\left(x-nx_0\right)\right\} = \frac{\cos\left(N-1\right)\omega_x x_0 - \cos N\omega_x x_0}{2sin^2\dfrac{\omega_x x_0}{2}},$$

$$\mathcal{F}\left\{\sum_{n=-N}^{N}\delta\left(x-nx_0\right)\right\} = \frac{\sin\dfrac{1}{2}\left[(2N+1)\,\omega_x x_0\right]}{\sin\left(\dfrac{\omega_x x_0}{2}\right)}. \tag{9.43}$$

A plot of Eq. (9.43) is shown in Figure 9.21 for three values of N, $N = 10$, 20, and 40. It is a periodic function made up of large primary peaks of amplitude $(2N + 1)$ surrounded by secondary peaks of decreasing amplitude as you move away from each primary peak. The width of the primary peak is

$$\omega = \frac{n\pi}{(2N+1)\,t_0}.$$

You see that the primary peaks take on the attributes of delta functions as N increases. The amplitude of the peak grows and the width decreases. From our derivation of Eq. (4.5) we can rewrite Eq. (9.43) as

$$\mathcal{F}\left\{A(x)\right\} = \frac{\sin N\beta}{\sin\beta},$$

$$\beta = \frac{kd}{2}\sin\theta_x.$$

This is simply the interference between radiation from all of the point sources making up the array.

The Fraunhofer diffraction pattern's intensity distribution in the x direction is thus given by

$$I_\theta = I_0 \underbrace{\frac{\sin^2\alpha}{\alpha^2}}_{\substack{shape\\factor}} \underbrace{\frac{\sin^2 N\beta}{\sin^2\beta}}_{\substack{granting\\factor}}. \tag{9.44}$$

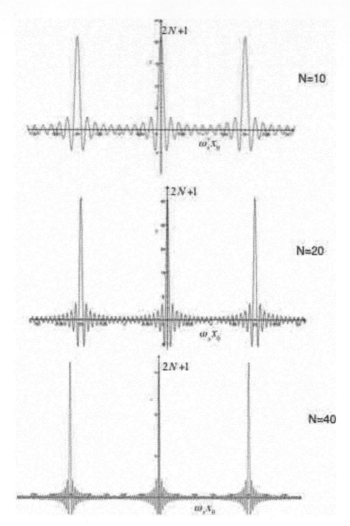

Figure 9.21 A plot of the Fourier transform of a set of 2N + 1 equally spaced delta functions where N = 10, 20, and 40. The maximum value of the Fourier transform is 2N + 1 and the first zero is inversely proportional to (2N + 1). Note that the width of the primary peaks narrows as N increases.

We have combined the variation in intensity in the y direction into the constant I_0 because we assume that the intensity variation in the x direction will be measured at a constant value of y. If we illuminate a set of apertures shown in Figure 9.20, we will generate a far-field pattern of light as is displayed in Figure 9.22. The primary peaks predicted theoretically in Figure 9.21 can be seen as bright beams of light fanning out as the light propagates from the source.

An important point to note from Eq. (9.44) is that the contribution to the transmission in the original aperture from fine structure is displayed over a large area in the diffraction pattern and vice versa for the coarse structure in the aperture.

Figure 9.22 Optical generation of the grating factor in Eq. (9.45). Another way to state the observed picture is that it is the far-field pattern of a grating.
(Courtesy of Robert Leighty, U.S. Army Engineering Topographic Laboratory.)

9.3.2.2 Young's double slit

The array theorem makes the analysis of Young's two-slit experiment a trivial exercise. The application of the array theorem will combine the effects of diffraction by a single aperture with the interference effects between multiple apertures discussed in Chapter 4 and will demonstrate that the interference between the two slits arises naturally from an application of diffraction theory. The result of this analysis will support a previous assertion that interference describes the same physical process as diffraction and the division of the two subjects is an arbitrary one.

In Chapter 4, we discussed interference from Young's two-slit experiment without considering diffraction from the slits. However, we did assume that diffraction would spread the light from each slit, causing the two waves to overlap. We can increase the sophistication of our approach and allow a lens to create the desired overlap by producing a plane wave for each source located in the focal front plane; see Figure 9.23.

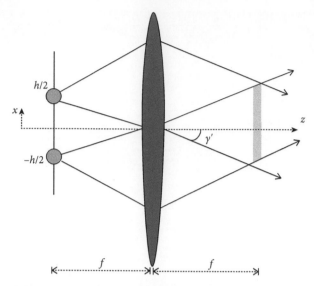

Figure 9.23 Generation of interference fringes using a two-point sources and a lens.

The ABCD matrix for a light ray propagating from the front to the back focal plane of a simple lens matrix can verify the drawing in Figure 9.23:

$$\begin{pmatrix} x' \\ \gamma' \end{pmatrix} = \begin{pmatrix} 1 & f \\ 0 & 1 \end{pmatrix} \begin{pmatrix} 1 & 0 \\ -\frac{1}{f} & 0 \end{pmatrix} \begin{pmatrix} 1 & f \\ 0 & 1 \end{pmatrix} \begin{pmatrix} x_0 \\ \gamma \end{pmatrix},$$

$$\begin{pmatrix} x' \\ \gamma' \end{pmatrix} = \begin{pmatrix} 0 & f \\ -\frac{1}{f} & 0 \end{pmatrix} \begin{pmatrix} x_0 \\ \gamma \end{pmatrix}.$$

From this we discover that

$$\gamma' = -\frac{x_0}{f}.$$

Thus, a source located in the front focal plane of the lens, at a position x_0 above the optical axis, will generate a plane wave, which makes an angle $\gamma' = -\frac{x_0}{f}$ with the optical axis in the back focal plane. In Figure 9.23 we show that the plane wave will be traveling downward at an angle γ', as predicted by the negative sign. In Figure 9.23 we have positioned two slits placed in the front focal plane of the lens at a distance h between the pair. The wave illuminating each slit is polarized along the y-axis and propagates in the x,z-plane. The plane light waves from the two slits overlap in the back focal plane and produce Young's interference. The light distribution should be given by Eq. (4.10) with the phase angle, δ, given by Eq. (4.11). To see that this is the case, consider that the two slits will act, in one dimension, as two point sources. The lens transforms light, from the two slits, into two plane waves propagating at angles γ_1 and γ_2, with respect to the z-axis. We can write the phase difference between the two plane waves using Eq. (4.60) as

$$\delta = \phi_2 - \phi_1 + k(r_2 - r_1) = \Delta\phi + \frac{2\pi x (h_1 + h_2)}{\lambda D}.$$

From Figure 9.23 we find the angle γ of the two plane waves produced is proportional to the height of the slit above and below the optical axis,

$$\gamma_1 = \frac{h_1}{2f} \quad \text{and} \quad \gamma_2 = -\frac{h_2}{2f},$$

so that if $h_1 = -h_2 = h/2$,

$$g_2 - h_2 = 2kx \sin\left(\frac{h}{2f}\right) \approx \frac{kxh}{f}, \tag{9.45}$$

which agrees with Eq. (4.59) if we replace the focal length of the lens by the distance between the slit and observation plane, D. In the back focal plane, a set of fringes will be produced with an intensity proportional to

$$\cos^2\left(\frac{kxh}{2f}\right).$$

Now we can compare this result to our array theorem prediction. The intensity of the diffraction pattern from two slits is obtained from Eq. (9.44) by setting $N = 2$,

$$I_\theta = I_0 \frac{\sin^2\alpha}{\alpha^2}\cos^2\beta. \tag{9.46}$$

The original interference analysis generated in Chapter 4 predicted that we would see a sinusoidal variation in intensity and when we evaluated the coherence of the source, we found that the intensity changed across the pattern Eq. (4.62). Now evaluating diffraction theory we find that the sinc function generated by the diffraction of a single slit modulates the interference pattern and describes the energy distribution of the overall diffraction pattern given by Eq. (9.46). Physically, α is a measure of the phase difference between points in one slit and β is a measure of the phase difference between similar points in the two separate slits. Zeroes in the diffraction intensity occur whenever $\alpha = n\pi$ or whenever $\beta = (2n + 1)\pi/2$. Figure 9.24 shows the interference maxima for three values of the ratio d/a, from the grating factor contained under the central maximum, described by the shape factor. The similarity between the results of Chapters 4 and 9 reinforce the knowledge that coherence plays a major role in the appearance of diffraction. The number of interference maxima appearing within the central diffraction maximum is given by

$$\frac{2d}{a} - 1. \tag{9.47}$$

An additional complexity we have not discussed is that both α and β are inversely proportional to the wavelength. A physical interpretation of Eq. (9.46) can be identified in Figure 9.25 where the diffraction from six slits illuminated by white light is shown. The first factor generates the diffraction associated with a single slit; it is called the *shape factor*. You can see the first sideband of the sinc function to the right of the image labeled "1st SB." At the very center of the pattern there are white fringes because the wavelength dependence of α and β are suppressed by the fact the angle dependence is near zero. As you move out in angle (to the right or left of the white central fringe) the red fringes occur at larger angles than for green

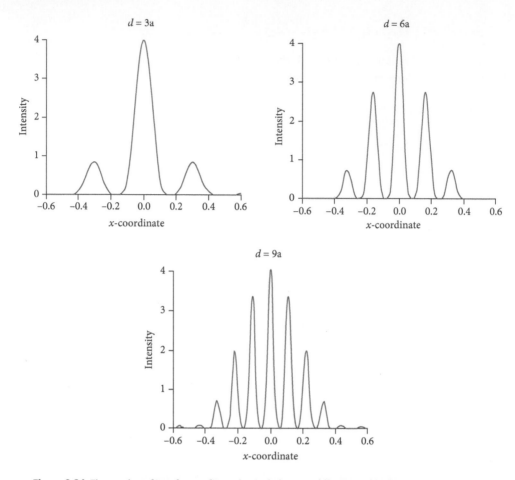

Figure 9.24 The number of interference fringes beneath the main diffraction peak of a rectangular aperture.

Figure 9.25 White light diffraction from six slits.

fringes as the inverse wavelength dependence implies. Sharp lines around the central white maximum are due to interference from the six slits. The interference contribution disappears at larger angles because the light source is not temporally coherent.

9.3.2.3 The diffraction grating

Now we will use the array theorem to derive the diffraction intensity distribution of a large number of identical apertures. We will discover that the positions of the principal maxima are a function of the illuminating wavelength. This functional relationship has led to the application of a diffraction grating to wavelength measurements.

There are two types of grating: reflection and transmission. The diffraction grating normally used for wavelength measurements is a reflection grating that contains a large number of reflecting grooves cut in a metal surface such as gold or aluminum. The CD and DVD are easily obtained versions of a reflection (or with some surgery, a transmission) grating with a period on a typical CD disk of 1.57 and 0.78 µm on a 4.7-Gb DVD.[10] The theory to be derived applies to both types of gratings but a modification must be made to the theory for reflective gratings because the shape of the grooves in the reflection grating control the fraction of light diffracted into a principal maximum. A grating whose groove shape has been controlled to enhance the energy contained in a particular principal maximum is called a *blazed grating*. The use of special groove shapes is equivalent to the modification of the phase in an antenna array at radio frequencies.

David Rittenhouse, an American astronomer, first suggested in 1785 the construction of a large number of diffracting elements into an optical device for measuring wavelength, but the idea was ignored until Fraunhofer reinvented the concept in 1819. Fraunhofer's first gratings were fine wires spaced by wrapping the wires in the threads of two parallel, threaded shafts. He later made gratings by cutting (*ruling*) grooves in gold films deposited on the surface of glass. *H. A. Rowland* made a number of well-designed ruling machines that made possible the production of large area gratings. Following a suggestion by Lord Rayleigh, *Robert Williams Wood* (1868–1955) developed the capability to control the shape of the grooves. Holographic gratings have been added to the types available commercially.

If N is allowed to assume values much larger than 2, the appearance of the interference fringes, predicted by the grating factor, changes from a simple sinusoidal variation to a set of narrow maxima, called principal maxima, surrounded by much smaller, secondary maxima, as is shown in Figure 9.21. Evaluating Eq. (9.44) as N increases we see that whenever $N\beta = m\pi$, where $m = 0, 1, 2, ...$, the numerator of the second factor in Eq. (9.44) will be zero, leading to an intensity that is zero, $I_\theta = 0$. The denominator of the second factor in Eq. (9.44) is zero when $\beta = \ell\pi$, $\ell = 0, 1, 2, ...$ The numerator and denominator are both zero whenever the ratio of m and N is equal to an integer, $m/N = 1$, resulting in an indeterminate value for the intensity, $I_\theta = 0/0$. To evaluate the observed intensity at this indeterminate point, we must apply L'Hôpital's rule,

$$\lim_{\beta \to \ell\pi} = \frac{\sin N\beta}{\sin \beta} = \lim_{\beta \to \ell\pi} \frac{N\cos N\beta}{\cos\beta} = N. \tag{9.48}$$

L'Hôpital's rule predicts that whenever

$$\beta = \frac{m}{N}\pi,$$

where (m/N) is an integer, a principal maximum in the intensity will occur with a value given by

$$I_{\theta P} = N^2 I_0 \frac{\sin^2 \alpha}{\alpha^2}. \tag{9.49}$$

---

[10] The topography is a series of pits rather than groves but for simple experiments that detail is not important.

Secondary maxima, much weaker than the principal maxima, occur when

$$\beta = \left(\frac{2m+1}{2N}\right)\pi \quad m = 1, 2, \ldots. \tag{9.50}$$

(The first value that m can have in Eq. (9.50) is $m = 1$, because when $m = 0$, m/N is an integer and we meet the condition for Eq. (9.49).) The intensity of each secondary maximum is given by

$$
\begin{aligned}
I_{\theta S} &= I_0 \frac{\sin^2 \alpha \sin^2 N\beta}{\alpha^2 \sin^2 \beta} \\
&= I_0 \frac{\sin^2 \alpha}{\alpha^2} \left[\frac{\sin^2 \left(\frac{2m+1}{2}\right)\pi}{\sin^2 \left(\frac{2m+1}{2N}\right)\pi} \right] \\
&= I_0 \frac{\sin^2 \alpha}{\alpha^2} \left[\frac{1}{\sin \left(\frac{2m+1}{2N}\right)\pi} \right]^2.
\end{aligned}
$$

The quantity $(2m+1)/2N$ is a small number due to the size of N, making it possible for the small angle approximation to be made:

$$I_{\theta S} \approx I_0 \frac{\sin^2 \alpha}{\alpha^2} \left[\frac{2N}{\pi(2m+1)} \right]^2. \tag{9.51}$$

The ratio of the intensity of a secondary maximum and a principal maximum is given by

$$\frac{I_{\theta S}}{I_{\theta P}} = \left[\frac{2}{\pi(2m+1)} \right]^2.$$

The strongest secondary maximum occurs for $m = 1$ and, for large N, has an intensity that is about 4.5% of the intensity of the neighboring principal maximum.

The position of principal maxima occurs at angles specified by the *grating formula*.[11]

$$
\begin{aligned}
\beta = \ell\pi &= \frac{m}{N}\pi \\
&= \frac{kd \sin \theta}{2} \\
&= \frac{\pi d \sin \theta}{\lambda}.
\end{aligned}
\tag{9.52}
$$

The angular positions of the principal maxima are thus

$$\sin \theta_x = \frac{m\lambda}{Nd}$$

where m is called the *interference order*; see Figure 9.26.

The model we have used to obtain this result is based on a periodic array of identical apertures. The transmission function of this array would be a periodic square wave, Eq. (1.17) and Figure 1.3. If we, for the moment, treat the grating as infinite in size, we discover that

---

[11] This same equation is found in the analysis of holography, X-ray scattering from crystals, and the acousto-optic effect. It is called the *Bragg equation*, derived by father and son William Henry and William Lawrence Bragg.

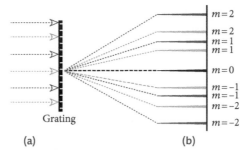

Figure 9.26 (a) An artistic rendering of interference orders of transmission grating. (b) ±2 orders showing a dark gray or a light gray line for each order. Without the color labeling of the orders it is easy to see that confusion can arise as to which peaks should be measured to discover the wavelength associated with the peak.

the principal maxima in the diffraction pattern correspond to the terms of the Fourier series describing the square wave transmission function; see Eq. (1.21).

The zero order, $m = 0$, corresponds to the a_0 term in the Fourier series and has an intensity proportional to the spatially averaged transmission of the grating. Because of its equivalence to the temporal average of a time-varying signal, the zero-order principal maximum is often called the *D.C. term* (Eq. (1.22)).

The first-order ($m = 1$), principal maximum corresponds to the fundamental spatial frequency of the grating and the higher orders correspond to the harmonics of this frequency.

The D.C. term provides no information about the wavelength of the illumination as you can see in Figure 9.25. Information about the wavelength of the illuminating light can only be obtained by measuring the angular position of the first- or higher-order principal maximum.

The fact that the grating is finite in size causes each of the principal maxima to have an angular width, Eq. (9.49), that limits the resolution with which the illuminating wavelength can be measured. To calculate the resolving power of the grating, we first determine the angular width of a principal maximum. This is accomplished by measuring the angular change, of the principal maximum's position, when β changes from $\beta = \ell\pi = m\pi/N$ to $\beta = (m + 1)\pi/N$, i.e., $\Delta\beta = \pi/N$. Using the definition

$$\beta = \frac{kd \sin\theta}{2},$$

$$\Delta\beta = \frac{\pi d \cos\theta\Delta\theta}{\lambda}.$$

The angular width is then

$$\Delta\theta = \frac{\lambda}{Nd \cos\theta}. \tag{9.53}$$

Comparing the curves shown in Figure 9.27 shows the dependence of the principal maxima on N and d. The separation of the principal maxima is displayed as a function of $\sin\theta$. The angular separation of principal maxima can be converted into a linear dimension by assuming a distance, R, from the grating to the observation plane (this may be the focal length of a mirror or a lens). In the lower right-hand curve of Figure 9.27 a distance of 2 meters was assumed. The distance R used in the design of a grating spectrometer classifies its resolving power. The

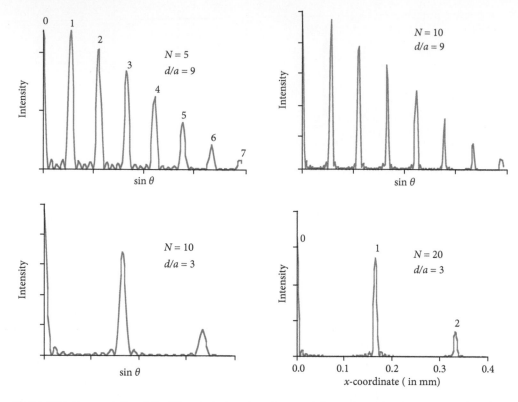

Figure 9.27 Decrease in the width of the principal maxima of a transmission grating with an increasing number of slits. The various principal maxima are called orders, numbering from 0, at the origin, out to as large as 7 in this example. Also shown is the effect of different ratios, d/a, on the number of visible orders in the figure on the lower left. Note the lower right-hand plot has its abscissa in mm, assuming an observation distance of 2 meters from the grating.

larger the value of R, the easier it is to resolve wavelength differences. For example, a 1-meter spectrometer is a higher resolution instrument than a 1/4-meter spectrometer.

The derivative of the grating formula gives

$$\Delta\lambda = \frac{d}{\ell}\cos\theta\Delta\theta$$

$$\Delta\theta = \frac{\ell\Delta\lambda}{d\cos\theta}.$$

Equating this result to Eq. (9.53) yields

$$\frac{\ell\Delta\lambda}{d\cos\theta} = \frac{\lambda}{Nd\cos\theta}.$$

The resolving power of a grating is therefore

$$\frac{\lambda}{\Delta\lambda} = N\ell. \tag{9.54}$$

The improvement of resolving power with N can be seen in Figure 9.27. A grating 2 inches wide and containing 15,000 grooves per inch would have a resolving power in second order

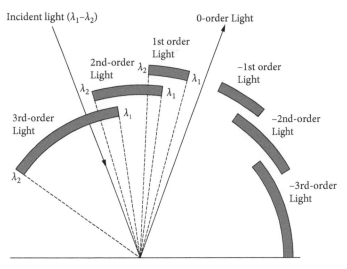

Figure 9.28 Free spectral range of a grating

($\ell = 2$) of 6×10^4. At a wavelength of 600 nm, this grating could resolve two waves, differing in wavelength by 0.01 nm.

The diffraction grating is limited by overlapping orders (see Figures 9.26 and 9.28), as was the Fabry–Perot interferometer discussed in Chapter 4. If two wavelengths, λ and $\lambda + \Delta\lambda$, have successive orders that are coincident, then $(m + 1)\lambda = m(\lambda + \Delta\lambda)$. The minimum wavelength difference for which this occurs is defined as the *free spectral range* of the diffraction grating

$$(\Delta\lambda)_{SR} = \frac{\lambda}{m}.$$

This result is identical to the result obtained for the Fabry–Perot interferometer, Eq. (4.56).

We have been discussing amplitude transmission gratings. Amplitude transmission gratings are of less utility because they waste light. The light loss is from a number of sources:

(1) Light is diffracted simultaneously into both positive and negative orders (the positive and negative frequencies of the Fourier transform). The negative diffraction orders contain redundant information and waste light.

(2) In an amplitude transmission grating, light is thrown away by the opaque portions of the slit array.

(3) The width of an apertures leads to a shape factor,

$$\text{sinc}^2 \alpha = \frac{\sin^2 \alpha}{\alpha^2},$$

for a rectangular aperture, which modulates the grating factor and causes the amplitude of the orders to rapidly decrease. This can be observed in Figure 9.27, $d/a = 3$, where the second order is very weak. Because of the loss in intensity at higher orders, only the first few orders ($\ell = 1, 2,$ or 3) are ever useful. Utilizing higher-order principal maxima increases the resolution capability of the grating but because of the shape factor there is a decrease in diffracted light intensity, reducing the signal to noise at higher orders.

(4) The location of the maximum in the diffracted light, i.e., the angular position for which the shape factor is a maximum, coincides with the location of the principal maximum due to the zero-order interference. This zero-order maximum is independent of wavelength and not of much use.

One solution to the problems created by transmission gratings would be the use of a grating that modified only the phase of the transmitted wave. Such gratings would operate using the same physical processes as a microwave-phased array antenna, where adding a constant phase shift to each antenna element controls the location of the shape factor's maximum. The construction of an optical transmission phase grating with a uniform phase variation, across the aperture of the grating, is very difficult. For this reason, a second approach, based on the use of reflection gratings, is the practical solution to the problems listed above.

By tilting the reflecting surface of each groove of a reflection grating (Figure 9.29), the position of the shape factor's maximum can be controlled. Problems (1), (3), and (4) are eliminated because the shape factor maximum is moved from the optical axis out to some angle with respect to the axis. If we compared this to a radiofrequency signal transmission, we could say we put the signal on a carrier (here a spatial carrier). The use of reflection gratings also removes Problem (2) because all of the incident light is reflected by the grating.

Robert Wood, in 1910, developed the technique of producing grooves of a desired shape in a reflective grating. Gratings with grooves shaped to enhance their performance at a particular wavelength are said to be *blazed* for that wavelength. The physical properties on which blazed gratings are based can be understood by using Figure 9.29. The shape of the grooves is controlled so that the groove faces can be treated as an array of mirror surfaces. The normal to each of the groove faces makes an angle θ_B with the normal to the ungrooved grating surface. We can measure the angle of incidence and the angle of diffraction with respect to the grating normal or with respect to the groove normal, as shown in Figure 9.29.

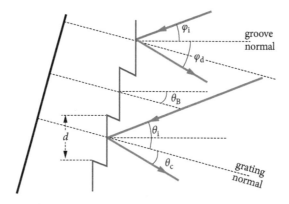

Figure 9.29 Geometry for a blazed reflection grating.

From Figure 9.29, we can write a relationship between the angles:[12]

$$\theta_i = \varphi_i - \theta_B \qquad -\theta_d = -\varphi_d + \theta_B.$$

The blaze angle provides an extra degree of freedom that will allow independent adjustment of the angular location of the principal maxima of the grating factor and the zero-order, single-aperture, diffraction maximum. To see how this is accomplished, we must determine, first, the effect of off-axis illumination of a diffraction grating.

In the discussion of spatial coherence, we mentioned briefly the effect of an off-axis source on the interference pattern produced by two slits. We saw, in Eq. (4.65), that the interference pattern is shifted by off-axis illumination. Off-axis illumination is easy to incorporate into the equation for the diffraction intensity from an array. To include the effect of an off-axis source, the phase of the illuminating wave is modified by changing the incident illumination from a plane wave of amplitude, E, traveling parallel to the optical axis, to a plane wave with the same amplitude, traveling at an angle θ_i to the optical axis:

$$\tilde{E}e^{-ikx\sin\theta_i}.$$

(Because we are interested only in the effects in a plane normal to the direction of propagation, we ignore the phase associated with propagation along the z-direction, $kz\cos\theta_i$.) The off-axis illumination results in a modification of the parameter for single-aperture diffraction from

$$\alpha = \frac{ka}{2}\sin\theta_d$$

to

$$\alpha = \frac{ka}{2}(\sin\theta_i + \sin\theta_d), \tag{9.55}$$

and for the multiple-aperture interference from

$$\beta = \frac{kd}{2}\sin\theta_d$$

to

$$\beta = \frac{kd}{2}(\sin\theta_i + \sin\theta_d), \tag{9.56}$$

where we have relabeled θ_x, in the equations for α and β, as θ_d, the angle the diffracted light makes with the grating normal.

The zero-order, single-aperture, diffraction peak occurs when $\alpha = 0$. If we measure the angles with respect to the groove face, this occurs when

$$\alpha = \frac{ka}{2}(\sin\varphi_i + \sin\varphi_d) = 0.$$

The angles are therefore related by

$$\sin\varphi_i = -\sin\varphi_d,$$
$$\varphi_i = -\varphi_d.$$

---

[12] The sign convention is in accord with that used in geometric optics; positive angles are those measured in a counterclockwise rotation from the normal to the surface. Therefore, θ_B is a negative angle.

We see that the single-aperture diffraction maximum (the shape factor's maximum) occurs at the same angle that reflection from the groove faces occurs. We can write this result in terms of the angles measured with respect to the grating normal

$$\theta_i = -(\theta_d + 2\theta_B).$$ (9.57)

The blaze condition requires the single-aperture diffraction maximum to occur at the ℓth principal maximum, for wavelength λ_B. At that position

$$\ell\pi = \frac{2\pi d}{\lambda_B}(\sin\theta_i + \sin\theta_d),$$

$$\ell\lambda_B = 2d\sin\frac{1}{2}(\theta_i + \theta_d)\cos\frac{1}{2}(\theta_i - \theta_d).$$ (9.58)

For the special geometrical configuration called the *Littrow condition* (Figure 9.30), where $\theta_i = \theta_d$, we find that Eq. (9.58) leads to the equation

$$\ell\lambda_B = 2d\sin\theta_B.$$

A moment's thought will reveal that the physical significance of the blaze condition in the Littrow configuration is that the groove face must be normal to the incident wave so that $\theta_d = \theta_B$. By adjusting the blaze angle, the single-aperture diffraction peak can be positioned on any order of the interference pattern, remembering that you cannot go past 90°. Typical blaze angles are between 15° and 30° but gratings are made with larger blaze angles.

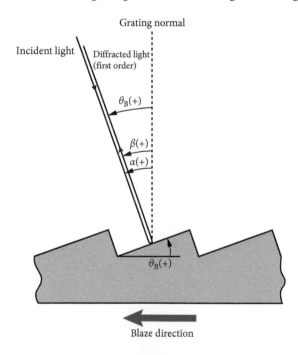

Figure 9.30 Littrow condition for a grating where the incoming and the diffracted beams are parallel to the groove normal. This is a condition that maximizes the efficiency of a blaze grating.

Diffraction from a Circular Aperture

We have been treating diffraction as a one-dimensional problem. We have assumed that we had a simple rectangular aperture function and could therefore easily separate the effects of the x- and y-dimensions. In most optical systems, however, we expect to see a circular aperture. To handle diffraction from a circular aperture we use the cylindrical geometry shown in Figure 9.31 to convert the Huygens–Fresnel integral from rectangular to cylindrical coordinates. To transform to the new coordinate system, we make use of the following equations. At the aperture plane

$$x = s \cdot \cos \varphi \quad y = s \cdot \sin \varphi$$
$$f(x, y) = f(s, \varphi) \quad dxdy = sdsd\varphi \tag{9.59}$$

At the observation plane

$$\xi = \rho \cos \theta \quad \eta = \rho \sin \theta. \tag{9.60}$$

In the new, cylindrical, coordinate system at the observation plane, the spatial frequencies are written as

$$\omega_x = -\frac{k\xi}{R_0} = -\frac{k\rho}{R_0} \cos \theta$$
$$\omega_y = -\frac{k\eta}{R_0} = -\frac{k\rho}{R_0} \sin \theta. \tag{9.61}$$

From Figure 9.31 we see that the observation point, P, can be defined in terms of the angle ψ, where

$$\sin \psi = \frac{\rho}{R_0}.$$

This allows an angular representation for the size of the diffraction pattern if it is desired.

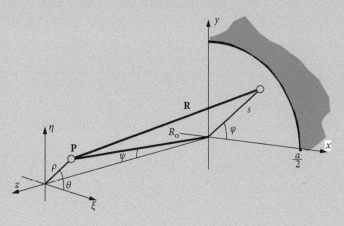

Figure 9.31 Geometry for diffraction from a circular aperture.

Using Eqs. (9.59) and (9.61), we may write

$$
\omega_x x = \omega_y y = -\frac{ks\rho}{R_0}\left(\cos\theta\cos\varphi + \sin\theta\sin\varphi\right),
$$
$$
= -\frac{ks\rho}{R_0}\cos(\theta-\varphi). \tag{9.62}
$$

The Huygens–Fresnel integral can now be written in terms of cylindrical coordinates as

$$
\tilde{E}_P = \frac{i\alpha}{\lambda R_0} e^{-i\mathbf{k}\cdot\mathbf{R}_0} \int_0^{\frac{a}{2}}\int_0^{2\pi} f(s,\varphi)\, e^{-ik\frac{s\rho}{R_0}\cos(\theta-\varphi)}\, s\,ds\,d\varphi. \tag{9.63}
$$

We can demonstrate the use of Eq. (9.63) by using it to calculate the diffraction amplitude from a clear aperture of diameter a, defined by the equation

$$
f(s,\varphi) = \begin{cases} 1 & s\le\frac{a}{2},\text{all }\varphi \\ 0 & s>\frac{a}{2} \end{cases}.
$$

The circular symmetry of this problem is such that not only is the function separable in the two dimensions but, because it has circular symmetry, it is independent of φ; thus,

$$
f(x,y) \Rightarrow f(s,\varphi) = f(s)g(\varphi) = f(s)
$$
$$
\mathcal{F}\{f(s,\varphi)\} = F(\rho,\theta) = F(\rho)
$$

$$
\mathbf{F}(\rho,\varphi) = \int_0^{2\phi} d\theta \int_0^\infty f(r)e^{i\rho r(\cos\theta\cos\varphi+\sin\theta\sin\varphi)} r\,dr,
$$
$$
= \int_0^\infty f(r)r\,dr \int_0^{2\phi} e^{-i\rho r\cos(\theta-\varphi)}\, d\theta. \tag{9.64}
$$

The second integral belongs to a class of functions called the Bessel function defined by the integral

$$
\mathbf{J}_n(r\rho) = \int_0^{2\pi} e^{i[r\rho\sin(\theta-n\theta)]} d\theta.
$$

The integral in Eq. (9.64) corresponds to the $n=0$, zero-order, Bessel function. Using this definition, we can write Eq. (9.64) as

$$
F(\rho) = \int_0^\infty f(r)\mathbf{J}_0(r\rho)r\,dr. \tag{9.65}
$$

This transform is called the Fourier–Bessel transform or the *Hankel zero-order transform* and is a two-dimensional Fourier transform of a circularly symmetric function. With this definition we can apply Eq. (9.65) to a simple circular symmetric function, sometimes called the *top-hat* function:

$$f(x,y) = \begin{Bmatrix} 1 & \sqrt{x^2 + y^2} \le 1 \\ 0 & \text{all other } x,y \end{Bmatrix} = f(r,\theta) = f(r) = \begin{cases} 1 & r \le 1 \\ 0 & \text{all other } r \end{cases}.$$

The transform of the top-hat function is

$$F(\rho) = \int_0^1 J_0(r\rho)\, r\, dr.$$

We use the identity

$$xJ_1(x) = \int_0^x \alpha J_0(\alpha)\, d\alpha$$

to obtain

$$F(\rho) = \frac{J_1(\rho)}{\rho}. \tag{9.66}$$

The diffraction pattern amplitude for a circular aperture is therefore

$$\tilde{E}_P = \frac{i\alpha}{\lambda} e^{-ik\cdot R_0} \left[\frac{\pi a}{k\rho} J_1 \left(\frac{ka\rho}{2R_0} \right) \right]. \tag{9.67}$$

A plot of the function in the bracket is given in Figure 9.32a. If we define

$$u = \frac{ka\rho}{2R_0},$$

then the spatial distribution of intensity in the diffraction pattern can be written in a form known as the *Airy formula,* named after the British astronomer George Biddell Airy (1801–92),

$$I = I_0 \left[\frac{2J_1(u)}{u} \right]^2, \tag{9.68}$$

where we have defined

$$I_0 = \left(\frac{\alpha A}{\lambda R_0} \right)^2.$$

A is the area of the aperture,

$$A = \pi \left(\frac{a}{2} \right)^2.$$

As we did for the rectangular aperture, we can replace R_0 by the focal length of a lens if that is how we plan to display the far field. The intensity pattern described by Eq. (9.68) and shown in Figure 9.32b is called the Airy pattern. The intensity at $u = 0$ in Eq. (9.68) is the same as was

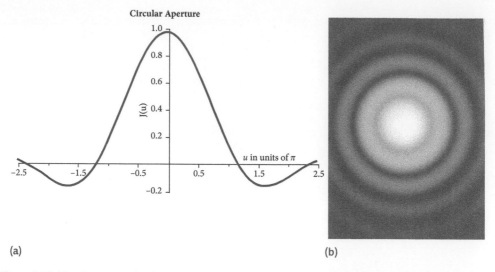

Figure 9.32 (a) Diffraction amplitude from a circular aperture. The observed light distribution is constructed by rotating this Bessel function around the optical axis. (b) Experimentally obtained Fraunhofer diffraction pattern from a circular aperture.

obtained for a rectangular aperture of the same area, Eq. (9.36), because, in the limit,

$$\lim_{u \to 0} \left[\frac{2J_1(u)}{u} \right] = 1.$$

The location of the first zero is used to characterize the size of the diffraction pattern. The circular area defined by the first zero of Eq. (9.68) is called the *Airy disk*. The Airy disk, formed by a lens of diameter a uniformly illuminated by a plane wave, is obtained by solving

$$u = \frac{ka\rho}{2f} = \frac{\pi a\rho}{\lambda f} = 1.22\pi$$

for ρ. The diameter of the Airy disk, 2ρ, which we will define as w, is given by

$$2\rho = w = \frac{2.44\lambda f}{a}.$$

For the Airy pattern, 84% of the total area is contained within the Airy disk, and 91% of the light is contained within the circle bounded by the second minimum at 2.233π. The intensities in the secondary maxima of the diffraction pattern of a rectangular aperture are much larger than that in the Airy pattern of a circular aperture. The peak intensities, relative to the central maximum, of the first three secondary maxima of a rectangular aperture are 4.7, 1.6, and 0.8%, respectively. For a circular aperture, the same quantities are 1.7, 0.04, and 0.02%, respectively.

To get a feeling for the sizes associated with the Airy disk we can calculate the size produced on the retina when the pupil is uniformly illuminated. The eye's aperture in daylight is 2.2 mm and the distance to the retina is 20 mm. For green light with a wavelength of $\lambda = 550$ nm, the

Airy disk is given by

$$r/d = 1.22r/d$$

$$2r = \frac{2(1.22)\left(550 \times 10^{-9}\right)\left(20 \times 10^{-3}\right)}{2.2 \times 10^{-3}}$$

$$= 0.01 \text{ mm} = 10 \text{ μm}$$

This calculation makes one think that diffraction is a small effect but that is because the light normally does not travel very far. What if we shine a laser pointer with a beam spread of 3 mrad on the Moon? Multiply that angle by the distance to the Moon, which is about 385,000 km and you get a spot that is almost 1,200 km wide, more than one-third the diameter of the Moon.

9.4 **Gaussian Beams**

We have developed a theory to explain diffraction that decomposes a beam of light into a distribution of infinite plane waves using Fourier theory. A plane wave extends to infinity and we assumed that each plane wave had a homogeneous transverse electric field. This means the Poynting vector is also homogeneous. Since the wavefront is infinite, it would have infinite power. We have ignored this complexity by assuming the wave had finite extent and though that fact means we do not really have plane waves, our results are still pretty good. There is another approach to the problem that works well for the resonant modes in a Fabry–Perot cavity, propagation of laser beams, and guided waves in optical fibers. In this new approach we replace an integral solution of the wave equation, the Huygens–Fresnel integral, with a differential solution obtained by applying the paraxial approximation to the wave equation, i.e., assume that the wave is traveling in the z-direction and the variation in the transverse plane is slow with respect to z. Mathematically, we will ignore the second-order z-derivatives. The differential equation that must be satisfied, to determine the spatial behavior of the paraxial wave in free space, is the Helmholtz equation. We ensure that the energy in the wave is finite by using a mathematical trick called analytical continuation, which prevents the Gaussian from extending to infinity. We will not actually do the math but we will highlight where we perform the trick.

One of the possible solutions of the paraxial Helmholtz equation is a propagating wave with a Gaussian transverse amplitude. The mathematical forms for both temporal and spatial functions are

$$f(t) = A\sqrt{\frac{\pi}{\alpha}}e^{-\frac{t^2}{4\alpha}}\cos\omega_0 t \quad \rightarrow \quad \text{Temporal pulse with carrier } \omega_0 \text{ centered at zero time and width } \alpha$$

$$f(x) = ae^{-(x-b)^2/2\sigma^2} \quad \rightarrow \quad \text{spatial pulse with peak height } a \text{ spatial pulse centered at } b \text{ and width } \sigma$$

We are only interested in the second equation, the spatial distribution in the x,y-plane perpendicular to the direction of propagation. A reason that waves with Gaussian wave shapes are of interest is that the Fourier transform of a Gaussian is another Gaussian. This means that the general form of the distribution does not change as the wave propagates; only its dimensions change. The result is that the propagation of a Gaussian wave can be described using simple equations from geometrical optics and two parameters of the Gaussian wave. We will not need the Huygens–Fresnel integral.

We assume that the solution is a scalar wave of the form

$$\tilde{E}(r) = \Psi(x, y, z)\, e^{-ikz},$$ (9.69)

propagating nearly parallel to the z-direction (paraxial) as a solution of the Helmholtz equation; i.e., the wave does not propagate in the x- or y-direction. If we substitute Eq. (9.69) into the Helmholtz equation we obtain

$$\left(\frac{\partial^2\Psi}{\partial x^2} + \frac{\partial^2\Psi}{\partial y^2} + \frac{\partial^2\Psi}{\partial z^2}\right) e^{-ikz} + k^2\Psi e^{-ikz} - 2ik\frac{\partial\Psi}{\partial z}e^{-ikz} - k^2\Psi e^{-ikz} = 0.$$ (9.70)

The paraxial assumption can be met if Ψ changes very slowly with z (linearly with z will do it). This allows $\partial^2\Psi/\partial z^2$ to be ignored. The resulting scalar wave equation is called the *paraxial wave equation* and works out to angles of 30°, as can be seen in Figure 6.3 and Table 6.1:

$$\frac{\partial^2\Psi}{\partial x^2} + \frac{\partial^2\Psi}{\partial y^2} - 2ik\frac{\partial\Psi}{\partial z} = 0.$$ (9.71)

The paraxial wave equation and its solutions lead to a description that can be shown to be equivalent to the Fresnel description of diffraction, Eq. (9.3). The mathematical manipulations required to obtain the characteristic parameters of a Gaussian wave are long and the derivation is not needed to utilize the Gaussian wave formalism in the analysis of optical systems [7]:

$$\Psi = \underbrace{\left[\frac{w_0}{w(z)}\right] e^{-\frac{x^2+y^2}{w(z)^2}}}_{\text{amplitude}} \underbrace{e^{-ik\frac{x^2+y^2}{2R(z)}}}_{\text{paraxial wave}} \underbrace{e^{-i\phi(z)}}_{\text{phase}}.$$ (9.72)

This is a paraxial spherical wave propagating away from the origin, with a wavefront having a radius of curvature $R(z)$, a distance z from the origin, as is shown in Figure 9.33. The origin is located at what is called the beam waist.

All we need to characterize the propagation of Eq. (9.72) are two parameters: the radius of curvature and the beam waist. We will discover how these two terms arise as we explore the terms contained in Eq. (9.72).

The assumption of a paraxial wave implies that the spherical wave can be approximated by a *paraxial, spherical wave* of the form

$$\frac{1}{R}e^{-ikR} = \frac{1}{R}e^{-ik\sqrt{z^2+x^2+y^2}}$$
$$\approx \frac{1}{R}e^{-ikz}e^{-ik\frac{x^2+y^2}{2z}}$$

where were have assumed that $z^2 \gg x^2 + y^2$. The paraxial assumption allows the substitution $z \approx R$ at large values of z:

$$\frac{1}{R}e^{-ikR} \approx \frac{1}{z}e^{-ikz}e^{-ik\frac{x^2+y^2}{2R}}.$$

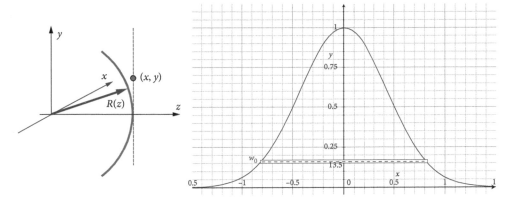

Figure 9.33 Propagation of a spherical wave. If point (x,y) is not too far from the z-axis, we may approximate this wave by a nearly plane wave called the paraxial, spherical wave. The intensity distribution in the x-dimension of the x,y-plane is shown on the right.

The sign convention is such that the beam shown in Figure 9.33 has a positive radius of curvature.[13] The solution of the wave equation, Eq. (9.72), has a Gaussian amplitude distribution shown on the right of Figure 9.33. Why is this significant? As we said earlier, the Fourier transform of a Gaussian is a Gaussian with reciprocal dimensions so that the Gaussian beam will expand but retain forever the same form.

Mathematically the distribution is given in two dimensions by

$$\left[\frac{w_0}{w(z)}\right] e^{-\frac{x^2+y^2}{w(z)^2}}.$$

The parameter w of the Gaussian distribution is the half-width of the Gaussian function measured at the point where the amplitude is $1/e$ of its maximum value; see the dotted red line in Figure 9.33. The top and bottom curves in Figure 9.34 trace the path that $w(z)$ takes as the wave propagates. We introduce a new variable

$$q = \tilde{q}_0 + z, \tag{9.73}$$

where $\tilde{q}_0$ is a purely imaginary constant whose role is to prevent a singularity in the solution that will occur at $z = 0$. This is where the analytic continuation magic takes place. The imaginary constant prevents the wave from extending to infinity in the transverse direction. It is proportional to the minimum half-width of the Gaussian function called *the minimum beam waist*, w_0, given by

$$\tilde{q}_0 = \frac{i\pi w_0^2}{\lambda} \tag{9.74}$$

---

[13] This convention is consistent with the sign convention for the radius of curvature of optical surfaces, defined in Chapter 6. The radius of curvature of an optical surface is measured from the surface to the center of curvature. Here, the coordinate origin is at the beam waist so the radius is measured, in the opposite sense, from the center of curvature to the phase front.

From Eq. (9.73),

$$\tilde{q} = \tilde{q}_0 + z = z + \frac{i\pi w_0^2}{\lambda}$$

The minimum beam waist occurs when $z = 0$; thus, the minimum beam diameter establishes the origin of the coordinate system.[14] The real valued constant

$$q_0 \equiv \frac{k w_0^2}{2} = \frac{\pi w_0^2}{\lambda} = z_R \tag{9.75}$$

is half of the *confocal parameter* (its significance will be identified in a few moments).

There is an extra phase term in Eq. (9.72),

$$\phi(z) = \tan^{-1}\left(\frac{\lambda z}{\pi w_0^2}\right) = \tan^{-1}\left(\frac{z}{z_R}\right), \tag{9.76}$$

where $\phi(z)$ is the phase difference between an ideal plane wave, shown by the dotted line through point (x,y) in Figure 9.33, and the "nearly plane" spherical wave of this theory. This phase shift is called the Gouy phase shift. When you pass through the focus (as the beam radius R changes from $-\infty$ to ∞), this phase shift equals π.

The waves described by Eq. (9.72) can be characterized by using only two simple parameters: the beam waist, w, and the radius of curvature of the phasefront of the wave, R. The *beam waist* was defined in Eq. (9.74); see the right-hand curve in Figure 9.33. The *radius of curvature* describes the radius of curvature of the phasefront of the wave, as measured from the position of minimum beam waist.

From Eq. (9.72) we can extract equations that describe the evolution of the two basic parameters as the beam propagates along the z direction. The radius of curvature as a function of propagation distance is given by

$$R(z) = z\left[1 + \left(\frac{\pi w_0^2}{\lambda z}\right)^2\right]. \tag{9.77a}$$

The radius of curvature at the beam waist, $z = 0$, is ∞. The beam's *spot size* evolution *is* given by

$$w(z)^2 = w_0^2\left[1 + \left(\frac{\lambda z}{\pi w_0^2}\right)^2\right]. \tag{9.77b}$$

The curve created by connecting the $1/e$ points of the transverse amplitude of the Gaussian beam, along its propagation path, is described by Eq. (9.77b) and shown as the dotted line on the right in Figure 9.34. The curve is a hyperbola along the wave's propagation path. At large z, the asymptotic representation of Eq. (9.77) is a straight line, the geometric ray,

$$w(z) = \left(\frac{\lambda}{\pi w_0}\right) z,$$

[14] The sign convention is Cartesian convention used throughout the text.

originating at the origin (the beam waist where $w = w_0$) and propagating in the positive z-direction. The ray is inclined, with respect to the z-axis, at the *diffraction angle*,

$$\theta = \frac{\lambda}{\pi w_0}. \tag{9.78}$$

The physical interpretation given to Eq. (9.78) is that diffraction causes a wave of diameter $2w_0$ to spread by an amount

$$\frac{2z_R\lambda}{\pi w_0},$$

after propagating a distance z_R—the *Rayleigh range*. The significance of the confocal parameter can be identified by substituting the confocal parameter of Eq. (9.75) into Eq. (9.77); the resulting equation,

$$w(z)^2 = w_0^2\left[1 + \left(\frac{z}{q_0}\right)^2\right],$$

demonstrates that when $z = z_R = q_0$ the wave's width increases, from

$$w_0 \Rightarrow \sqrt{2}w_0.$$

The Rayleigh range (Figure 9.34) measures the distance over which the light beam can be treated as a collimated beam. At the distance z_R from the beam waist, the radius of curvature is $2z_R$ and the wave looks like it originates from a point source once we have passed this position. The distance $b = 2z_R$ is called the confocal parameter and is a measure of the depth of focus of the Gaussian wave, i.e., the distance over which the beam can be treated as a collimated plane wave. The confocal parameter thus characterizes a Gaussian beam's convergent or divergent

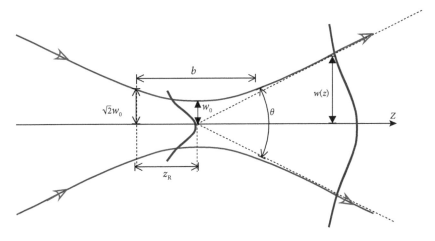

Figure 9.34 The propagation of a Gaussian beam. The upper and lower lines are the hyperbola, given by Eq. (9.77), and the dashed straight lines are asymptotes of the hyperbola, i.e., the geometric rays, inclined at the diffraction angle θ, calculated from Eq. (9.78). Also shown is the amplitude distribution at the beam waist and in the far field where geometric rays can be used. The Rayleigh range is z_R. Early papers used b as the confocal parameter or depth of focus. Modern papers use z_R.

properties. To emphasize the connection with the divergence of the beam, we rewrite the Rayleigh range as

$$q_0 = \frac{w_0}{\theta} = \frac{\pi w_0^2}{\lambda}$$

The larger the value of the beam waist, w_0, the smaller the beam will spread due to diffraction. This is why a megaphone works for a cheerleader. The megaphone changes the effective aperture, producing sound from the small diameter of a human mouth, about 50 mm, to a much larger diameter, something over 30 cm. The propagation properties of laser beams are closely modeled by Eq. (9.77). For this reason, lasers are often characterized by their *divergence angle*, Eq. (9.78).

The diffraction angle can be used to calculate the beam diameter at a distance z from the beam waist,

$$w(z)^2 = w_0^2 + \theta^2 z^2 \tag{9.79}$$

9.4.1 Gaussian beam propagation

If we divide $R(z)$ by $w(z)$, we can use the result to obtain expressions for w_0 and z in terms of R and w

$$w_0^2 = \frac{w^2}{\left[1 + \left(\dfrac{\pi w^2}{\lambda R}\right)^2\right]} \tag{9.80}$$

$$z = \frac{R}{\left[1 + \left(\dfrac{\lambda R}{\pi w^2}\right)^2\right]}. \tag{9.81}$$

From Eq. (9.80) we see that, at the minimum beam waist, the phase front of the Gaussian wave is a plane, i.e., $R = \infty$:

$$w_0^2 = \frac{w^2}{\left[1 + \left(\dfrac{\pi w^2}{\lambda R}\right)^2\right]}$$

These equations allow us to use the ABCD law and geometrical optics to describe the propagation of a diffracting Gaussian wave. The mathematics required are the simple functions derived using geometrical optics. This is made possible because the Fourier transform of a Gaussian wave is another Gaussian wave differing only in scale. For Gaussian waves we no longer need to use the Huygens–Fresnel integral, Eq. (9.16), to describe propagation.

9.4.2 The ABCD Law

We will establish a relationship between the Gaussian beam parameters we have just introduced and geometrical optics that will allow the calculation of Gaussian beam parameters after the wave has passed through an optical system. We first discover how a thin lens changes the radius

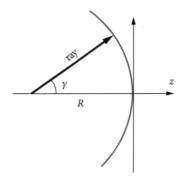

Figure 9.35 Geometry for Gaussian wave in geometrical optics.

of curvature of a Gaussian wavefront. We will learn that the radius of curvature and the complex beam parameter are governed by the same propagation equations. This finding will allow the construction of an ABCD law for a Gaussian wave.

In Chapter 6, we found the ABCD matrix, which relates the input and output parameters of an optical system, using the paraxial approximation

$$\begin{pmatrix} x_2 \\ \gamma_2 \end{pmatrix} = \begin{pmatrix} A & B \\ C & D \end{pmatrix} \begin{pmatrix} x_1 \\ \gamma_1 \end{pmatrix},$$

$$x_2 = Ax_1 + B\gamma_1 \quad \gamma_2 = Cx_1 + D\gamma_1.$$

The variable x_1 is the coordinate position, above the optical axis, of the ray entering the optical system, x_2 is the coordinate position of the ray leaving the system, and the γ's are the ray slopes. The ray slope for a Gaussian wave of radius R shown in Figure 9.35 is

$$\gamma = \frac{dx}{dz} = \tan \gamma \approx \frac{x}{R}, \tag{9.82}$$

so that

$$R = \frac{x}{\gamma}. \tag{9.83}$$

The radius of curvature of the Gaussian wave leaving the optical system described by the ABCD matrix is given by

$$
\begin{aligned}
R_2 &= \frac{x_2}{\gamma_2} \\
&= \frac{\gamma_1 \left(A \frac{x_1}{\gamma_1} + B \right)}{\gamma_1 \left(C \frac{x_1}{\gamma_1} + D \right)} \\
&= \frac{AR_1 + B}{CR_1 + D},
\end{aligned}
\tag{9.84}
$$

where R_1 is the radius of curvature of the Gaussian wave entering the optical system.

To determine the radius of curvature of the phase front of a Gaussian wave after it has passed through a simple lens, we substitute into Eq. (9.83) the ABCD matrix for a thin lens (Figure 6.10)

$$R_2 = \frac{AR_1 + B}{CR_1 + D} = \frac{R_1}{-\dfrac{R_1}{f} + 1}, \text{where} \quad \begin{matrix} A = 1 & B = 0 \\ C = -1/f & D = 1 \end{matrix}$$

$$\frac{1}{R_2} = \frac{1 - \dfrac{R_1}{f}}{R_1} = \frac{1}{R_1} - \frac{1}{f}. \tag{9.85}$$

When the Gaussian wave is propagating through free space, we use the ABCD matrix for translation (Figure 6.10) to obtain the radius of curvature of the phase front, R_2, after the wave has propagated a distance d:

$$R_2 = R_1 + d. \tag{9.86}$$

These results suggest that the Gaussian wave's complex size parameter, q, will provide a description of the propagation of a Gaussian beam through an optical system.[15] For a Gaussian wave propagating through free space, from the beam waist to a position z, the complex size parameter is, from Eq. (9.73),

$$\tilde{q}_1 = \tilde{q}_0 + z.$$

If we propagate from z to $(z + d)$ the q-parameter becomes

$$\tilde{q}_2 = \tilde{q}_1 + d. \tag{9.87}$$

The complex size parameter obeys the same rule as the radius of curvature for a wave propagating in free space.

To analyze the effects of a simple lens on a Gaussian wave, recall that the complex size parameter q can be written as

$$\frac{1}{\tilde{q}} = \frac{1}{z + \dfrac{i\pi w_0^2}{\lambda}}$$

$$= \frac{z - \dfrac{i\pi w_0^2}{\lambda}}{z^2 + \left(\dfrac{\pi w_0^2}{\lambda}\right)^2}.$$

Using Eq. (9.77), we can rewrite this as

$$\frac{1}{\tilde{q}} = \frac{1}{R} - \frac{i\lambda}{\pi w^2}. \tag{9.88}$$

For a thin lens, the spot size w is the same at the front and back surfaces of the lens (remember that the front and back vertices of a thin lens define planes of unit magnification); thus, $w_2 = w_1$. The beam radius of curvature should change according to Eq. (9.86), allowing us to write Eq. (9.88) as

---

[15] The ABCD formalism extends easily to higher-order modes. In fact, the higher-order modes have the same w, R, and q as the fundamental mode we are discussing; only the phase ϕ is different.

$$\frac{1}{\tilde{q}_2} = \frac{1}{R_2} - \frac{i\lambda}{\pi w_2^2}$$

$$= \left(\frac{1}{R_1} - \frac{1}{f}\right) - \frac{i\lambda}{\pi w_1^2}.$$

Rearranging the terms yields

$$\frac{1}{\tilde{q}_2} = \frac{1}{\tilde{q}_1} - \frac{1}{f}. \tag{9.89}$$

Comparing Eqs. (9.85) and (9.89) again leads us to the conclusion that the complex beam parameter q plays a role corresponding to that played by the radius of curvature R of a spherical wave. This should not surprise us because Eq. (9.88) defines the real part of $1/q$ as being equal to $1/R$. We can rename q as the *complex curvature* of a Gaussian wave.

Because of the formal equivalence between q and R, Eq. (9.84) can be used to write

$$\tilde{q}_2 = \frac{A\tilde{q}_1 + B}{C\tilde{q}_1 + D}. \tag{9.90}$$

Equation (9.90) allows a Gaussian beam to be traced through any optical system. Several examples of the application of Eq. (9.90) will show the usefulness of this result.

9.4.3 Thin Lens

As the first example of the use of Eq. (9.90), a Gaussian beam will be followed through a thin lens. Assume that a plane wave uniformly illuminates a lens of diameter, D. Because it is a plane wave, $R_1 = \infty$ and because the aperture of the lens is uniformly illuminated, $w = D/2$. The q-parameter at the left surface of the lens is given by

$$\frac{1}{\tilde{q}_1} = 0 - \frac{4i\lambda}{\pi D^2}.$$

The ABCD matrix for a thin lens can be used to calculate the q-parameter, after passing through the lens

$$\tilde{q}_2 = \frac{\tilde{q}_1}{1 - \frac{\tilde{q}_1}{f}},$$

$$\frac{1}{\tilde{q}_2} = -\frac{1}{f} - \frac{i\lambda}{\pi\left(\frac{D}{2}\right)^2}.$$

After passing through the thin lens, the beam waist remains the same size, $w = D/2$ (remember this is the plane of unit magnification), but the radius of curvature of the phase front becomes

$$R_2 = -f.$$

That is, the beam is now converging with the center of curvature of the wavefront on the right of the wavefront.

To find the location of the minimum beam waist, we use Eq. (9.81):

$$z = \frac{-f}{1 + \left(\dfrac{4\lambda f}{\pi D^2}\right)^2} \approx -f.$$

The minus sign signifies that the current position (the back of the lens) is to the left of the minimum beam waist; i.e., the minimum beam waist is found to the right of the lens, a distance f from the lens. The size of the minimum beam waist is given by Eq. (9.80),

$$w_0^2 = \frac{D^2/4}{1 + \left(\dfrac{\pi D^2}{4\lambda f}\right)^2} \approx \left(\frac{4\lambda f}{\pi D^2}\right)^2.$$

The conclusion of this analysis is that parallel light filling the aperture of a thin lens is brought to a focus at the back focal plane of the lens. Diffraction by the aperture of the lens prevents the beam from being focused to a spot smaller than the minimum beam waist. The focal spot size is inversely proportional to the lens aperture and linearly proportional to the focal length of the lens.

9.4.4 Applications of Gaussian waves

Now we will explore two problems involving the design of laser systems utilizing Gaussian waves to carry out the design.

9.4.4.1 Stable Fabry–Perot resonator

A laser is an optical oscillator and has all of the attributes of an electronic oscillator, i.e., a source of optical gain that overcomes all losses in the device and a source of optical feedback to allow the oscillations to grow in amplitude. Our objective is to design the optical feedback. The optical feedback can be obtained by constructing a Fabry–Perot resonator, as shown in Figure 9.36. The ABCD matrix will be used to determine the importance of the three parameters, R_1, R_2, and d, on the design of the resonator. We will discover that the parameters must meet a "stability condition" for a resonant mode to exist in the Fabry–Perot resonator. The stability condition demands that the desired ray associated with a particular resonator mode be trapped inside the resonator so that no light leaks out and the mode can grow by repeated passes through a gain medium. The modes of this resonator are Gaussian waves, so we can use Eq. (9.90) to establish that the mode is stable and the q values after one pass through the cavity are equal to the original q value:

$$\tilde{q} = \frac{A\tilde{q} + B}{C\tilde{q} + D}.$$

A, B, C, and D are elements of the ABCD matrix for the Fabry–Perot resonator of Figure 9.36. Our first step is to discover the values of the elements of the ABCD matrix for the Fabry–Perot.

The resonator consists of two mirrors, which we have given spherical curvature, separated by a distance, d, as is shown in Figure 9.36. The resonator performs a role similar to a resonant cavity in a microwave system by producing large internal fields with small power input. It does

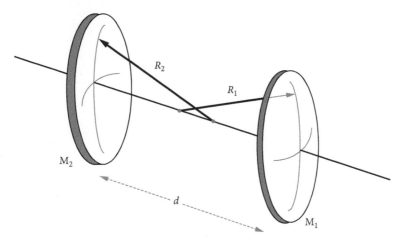

Figure 9.36 A Fabry–Perot resonator made of mirror M_1 with radius of curvature R_1 separated by a distance d from mirror M_2 with radius of curvature R_2.

this by trapping energy in the region between the mirrors. The trapped energy is distributed over the eigenmodes of the cavity (also called the resonant modes of the cavity).

The ABCD matrix can be defined by using the matrices defined in Figure 6.10. The reflection matrix for a mirror of radius R, in a medium of index n, is

$$\mathbf{M}_i = \begin{pmatrix} 1 & 0 \\ \dfrac{-2}{R_i} & -1 \end{pmatrix}.$$

The element M_{22} is negative because the angle of reflection is measured in the opposite sense to the angle of incidence. The translation matrix is given by

$$\mathbf{T}_1 = \begin{pmatrix} 1 & -d \\ 0 & 1 \end{pmatrix},$$

where we assume that the index between the two mirrors is $n = n_1$.

To follow a light ray through the Fabry–Perot resonator in Figure 9.36, we start the light ray just to the left of the surface of mirror M_1 traveling left toward mirror M_2. Let the origin of the coordinate system be at the vertex of M_2. Mirror M_1 acts as an object that is to be imaged by M_2 and the ray is followed through its reflection at mirror M_2 using the matrix for M_2, which has a positive spherical curvature, R_2. There is another translation over a distance d that brings the ray to mirror M_1 where it is reflected using the matrix M_1.

The ABCD matrix of the entire system is obtained by multiplying the matrices together in the matrix equation

$$\begin{pmatrix} x_1 \\ \gamma_1 \end{pmatrix} = \begin{pmatrix} 1 & 0 \\ \dfrac{2}{R_1} & -1 \end{pmatrix} \begin{pmatrix} 1 & d \\ 0 & 1 \end{pmatrix} \begin{pmatrix} 1 & 0 \\ -\dfrac{2}{R_1} & -1 \end{pmatrix} \begin{pmatrix} 1 & -d \\ 0 & 1 \end{pmatrix} \begin{pmatrix} x_0 \\ \gamma_0 \end{pmatrix}$$

$$\begin{pmatrix} x_1 \\ \gamma_1 \end{pmatrix} = \begin{pmatrix} 1 - \dfrac{2d}{R_2} & d\left[\dfrac{2d}{R_2} - 2\right] \\ \dfrac{2}{R_1} + \dfrac{2}{R_2}\left[1 - \dfrac{2d}{R_1}\right] & 1 - \dfrac{4d}{R_1} - \dfrac{2d}{R_2}\left[1 - 2\dfrac{d}{R_1}\right] \end{pmatrix} \begin{pmatrix} x_0 \\ \gamma_0 \end{pmatrix}, \tag{9.91}$$

where

$$A = 1 - \frac{2d}{R_2} \qquad\qquad B = d\left[\frac{2d}{R_2} - 2\right]$$

$$C = \frac{2}{R_1} + \frac{2}{R_2}\left[1 - \frac{2d}{R_1}\right] \qquad D = 1 - \frac{4d}{R_1} - \frac{2d}{R_2}\left[1 - 2\frac{d}{R_1}\right].$$

We want to solve

$$\tilde{q} = \frac{A\tilde{q} + B}{C\tilde{q} + D},$$

for $1/q$ to obtain

$$\frac{1}{\tilde{q}} = \frac{(D-A) \pm \sqrt{(D-A)^2 + 4BC}}{2B}.$$

The ABCD determinant for the resonator must equal 1 because the index of refraction is a constant in the resonator; i.e., $AD - BC = 1$. This fact allows the equation for $1/q$ to be simplified:

$$\frac{1}{\tilde{q}} = \frac{(D-A)}{2B} \pm \frac{i\sqrt{1 - (D+A)^2/4}}{B}.$$

This equation is in the standard form of the q-parameter, Eq. (9.88):

$$\frac{1}{\tilde{q}} = \frac{1}{R} - \frac{i\lambda}{\pi w^2}$$

The real part of the equation for the q-parameter can be extracted to discover the radius of curvature of the Gaussian wave in the Fabry–Perot resonator

$$R = \frac{2B}{D-A}.$$

In order to ensure that the q-parameter be complex, we require that

$$\left|\frac{D+A}{2}\right| < 1.$$

We are led to the conclusion that if a stable, reproducing, Gaussian wave exists in a Fabry–Perot resonator, then the radii of curvature of the mirrors making up the resonator match the wave's wavefront curvature. Figure 9.37 shows a typical Gaussian wavefront curvature. This wave would be a mode of a Fabry–Perot resonator if the mirrors of the resonator were positioned so that their curvature matched the wavefront curvature, shown by the gray lines in Figure 9.37. Where do we place the mirrors? The curvature of the beam at a position z from the beam waist is given by

$$R = z + q_0^2/z$$

$$z = R/2 \pm 1/2\sqrt{R^2 - 4q_0^2}.$$

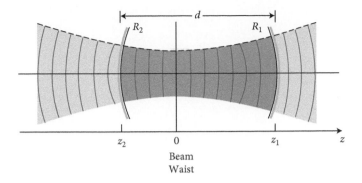

Figure 9.37 The beam waist occurs at the point where the radius of curvature of the phase is infinite.

We need the value of q_0, which we can get by using the fact that the waist is somewhere along the mirror spacing, d:

$$d = z_1 - z_2 = R_1/2 \pm 1/2\sqrt{R_1^2 - 4q_0^2} - R_2/2 \mp 1/2\sqrt{R_2^2 - 4q_0^2}.$$

Solving for q_0 gives

$$q_0^2 = \frac{d(d - R_1)(d + R_2)[(R_1 - R_2) - d]}{[(R_1 - R_2) - 2d]^2}.$$

To make the problem simpler we will assume a symmetric cavity where $-R_1 = R_2 = R$. Remember that the sign convention of Gaussian waves places the origin at the beam waist rather than the vertex of the optical component. This should explain the reason for the sign on R_1:

$$q_0^2 = d/4\,(2R - d).$$

The mirror's diameter is selected so that it intercepted, say, 99% of the beam at the mirror position on the optical axis; this requirements ensures that light does not leak out. You accomplish this by calculating the spot size on the mirror using Eq. (9.77b):

$$w^2 = w_0^2 \left[1 + \left(\frac{\lambda z}{\pi w_0^2}\right)^2\right] = \frac{\lambda}{2\pi}\sqrt{d(2R - d)}\left[1 + \left(\frac{d}{\sqrt{d(2R - d)}}\right)^2\right]$$

$$= \frac{Rd\lambda}{\pi\sqrt{d(2R - d)}}$$

To intercept 99% of the energy in the beam you want the diameter of the mirror to be $\pi\sqrt{2}w_0$.

Laser Cavity

On the final example, we will analyze a commercial HeNe laser designed to operate at $\lambda = 632.8$ nm. The optical layout of the laser cavity is shown in Figure 9.38. We will derive the ABCD matrix for this design.

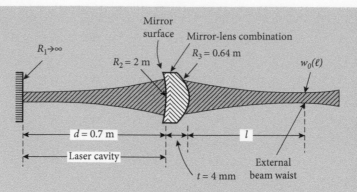

Figure 9.38 Commercial HeNe laser cavity design.

Inside the laser cavity, the phase front curvature of the Gaussian wave must match the curvature of the mirrors in the cavity, as we pointed out in Figure 9.37. This means that at the plane mirror, on the left of Figure 9.38, the radius of curvature is infinite. From Eq. (9.76) we see that the beam waist always occurs at the point where the radius of curvature is infinite (see Figure 9.37). We will make all of our measurements from the beam waist, which we now know to be at the plane mirror. The second mirror is a lens whose concave surface has a reflective dielectric coating. At mirror 2

$$z = 0.7 \, \text{m} \quad R = 2 \, \text{m}.$$

We use Eq. (9.77) to find the size of the beam waist, w_0. From w_0 we can calculate the complex beam parameter, q_1.

Light leaves the laser cavity through the lens; whose concave surface serves as one of the Fabry–Perot mirrors. To locate the beam waist and find its size outside the cavity, we must calculate the ABCD matrix. From left to right in Figure 9.38 we have the following matrices:

(1) *The propagation in the laser cavity from mirror 1 to mirror 2*: We assume the index of refraction in the cavity is $n_1 = 1.0$:

$$\begin{pmatrix} 1 & 0.7 \\ 0 & 1 \end{pmatrix}.$$

(2) *Refraction at the surface of mirror 2*: We assume the index of refraction of the lens, which also serves as mirror surface 2, is $n_2 = 1.5$:

$$\begin{pmatrix} 1 & 0 \\ \dfrac{n_2 - n_1}{n_2 R_2} & \dfrac{n_1}{n_2} \end{pmatrix}.$$

(3) *Propagation of light through the glass between the surfaces of mirror 2*: The mirror thickness is $t = 4$ mm:

$$\begin{pmatrix} 1 & \dfrac{t}{n_2} \\ 0 & 1 \end{pmatrix}$$

(4) *Refraction at the back surface of mirror 2*: We assume the index of refraction outside the laser is $n_1 = 1$:

$$\begin{bmatrix} 1 & 0 \\ \dfrac{n_1 - n_2}{n_1 R_3} & \dfrac{n_2}{n_1} \end{bmatrix}.$$

(5) *Propagation to the beam waist outside of the laser cavity*: This matrix contains the quantity of interest, d:

$$\begin{bmatrix} 1 & d \\ 0 & 1 \end{bmatrix}.$$

The ABCD matrix for the system is obtained by multiplying all of the above matrices. The resultant matrix is then used in Eq. (9.89) to find q_2. The beam waist found outside the laser cavity is a minimum beam waist so that, at the minimum waist, $R(z) = \infty$. This means that q_2 must be completely imaginary:

$$\tilde{q}_2 = -\frac{i\lambda}{\pi w_0^2}.$$

9.5 Problem Set 9

(1) In a double-slit Fraunhofer diffraction experiment, missing orders occur at those values of $\sin \theta$ which satisfy, at the same time, the condition for interference maxima and the condition for diffraction minima. Show that this leads to the condition $(d/a) = $ integer, where a is the slit width and d is the distance between slits. Show the approximate relation $d \sin \theta = m\lambda$ as the condition for interference maxima.

(2) The Fraunhofer pattern of a double slit, under $l = 650$ nm illumination, appears at the back focal plane of a 80-cm focal length lens. Using measurements taken from Figure 9.39 calculate the width and spacing of the slits.

(3) How many lines must be ruled on a transmission grating so that it will just resolve the sodium doublet (589.592 and 588.995 nm) in the first-order spectrum?

(4) A 10-watt argon-ion laser emits a wavelength of 488 nm. Its beam waist is 2 mm (w_0). How far must it propagate before diffraction enlarges the beam to 4 mm?

(5) What does an increase in wavelength do to a Fraunhofer diffraction pattern?

(6) What happens when we increase the focal length of the lens used to produce Fraunhofer diffraction?

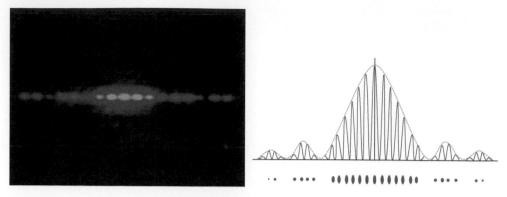

Figure 9.39 The Fraunhofer diffraction pattern from a double slit. Use the information gathered from measurements of this pattern to determine the width and spacing of the slits. The image has been enlarged by a factor of 10 so scale your measurement accordingly.

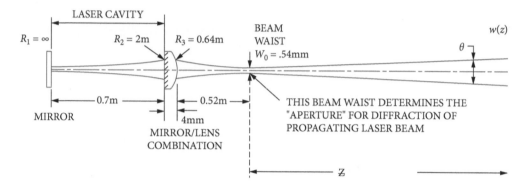

Figure 9.40 Design of commercial HeNe laser.

(7) Complete the details of the calculation for the laser cavity design shown in Figure 9.40. You will find a second beam waist, the image of the first, outside the cavity. Can you think of any reasons the optical designer placed this waist outside the cavity?

(8) We wish to couple energy from an argon laser ($\lambda = 488$ nm), with a beam waist of 350 μm, into a single-mode fiber. The mode's beam waist is 2 μm. What are the separations between the laser waist and the fiber face if as a lens we use a 10× microscope objective? Note: the focal length of a microscope objective is given by

$$f = \frac{160 \text{ mm}}{\text{power}}.$$

(9) We wish to relocate the position of a beam waist but not change its size. The lens that performs this operation is called a *relay lens*. Derive a relationship for the distance between the two waist locations, $z_1 + z_2$, and the focal length of the lens. Remember $w_{01} = w_{02}$.

(10) What is the Airy disk diameter in terms of the $f/\#$ of a lens?

(11) A beam from a HeNe laser ($\lambda = 632.8$ nm) is traveling along the z-axis and has a waist $w_0 = 0.1$ mm at $z = 0$.

    1. What is the beam spot at $z = 10$ cm?

    2. What is the radius of curvature of the wavefront R at $z = 10$ cm?

    3. At what distance z_1 is $R = 2$ m?

(12) A Gaussian beam from a HeNe laser is incident on a lens with a focal length of 10 cm. At the lens, before being refracted, the radius of curvature of the wave is infinite and its waist is 2 mm. The total power of the beam is 5 mW. What is the maximum intensity of the beam (in W/m^2) at the focal point of the wave?

REFERENCES

1. Born, M., and E. Wolf, *Principles of Optics*, 7th edn. Cambridge, UK: Cambridge University Press, 1999.

2. Rubinowicz, A., "The Miyamoto-Wolf Diffraction Wave," in E. Wolf (Ed.), *Progress in Optics*, pp. 192–240. Amsterdam: North_Holland, 1965.

3. Haus, H. A., *Waves and Fields in Optoelectronics*. Solid State Physical Electronics. Englewood Cliffs, NJ: Prentice Hall, 1983..

4. John (CC BY-SA 2.5 (http://creativecommons.org/licenses/by-sa/2.5)), v.W.C., *Diesel spill on a road*, Dieselrainbow.jpg, Editor. Wikimedia Commons, 2007.

5. Gaskill, J. D., *Linear Systems, Fourier Transforms and Optics*. New York: Wiley, 1978.

6. Kogelnik, H., and T. Li, "Laser Beams and Resonators." *Applied Optics* **5**(10): 1550–67 (1966).

7. Yariv, A., *Optical Electronics*, 3rd edn. New York: Holt, Rinehart, and Winston, 1985.

10 Fresnel Diffraction

10.1 Introduction

In the approximate solution to the Huygens–Fresnel integral, discussed in Chapter 9, it was assumed that the phase of the wavefront in the aperture was a linear function of the aperture's coordinates. This assumption led to Fraunhofer diffraction theory. We now return to the Huygens–Fresnel integral to discuss another approximate solution of the integral. The approximate solution, called *Fresnel diffraction*, assumes the phase of the wavefront in the aperture has a quadratic dependence upon aperture coordinates.[1] The curvature of the wavefront increases the mathematical difficulty of the diffraction problem over that of Fraunhofer diffraction. We will discover that, contrary to Fraunhofer diffraction, only a portion of the wave in the aperture contributes to Fresnel diffraction. That portion of the aperture that contributes to the diffraction amplitude lies near the line connecting the source and observation point, which means that, as with Fraunhofer diffraction, we are limiting our considerations to small departures from geometrical optics.

We will reformulate the Huygens–Fresnel integral,

$$\int \frac{\tilde{E}_i}{R} e^{-i\mathbf{k}\cdot\mathbf{R}} \, dx,$$

using the approximation derived in the previous chapter, Eq. (9.17)

$$\Delta \approx k\left(\frac{x_1}{z_1} + \frac{x_2}{z_2}\right)b + \frac{k}{2}\left(\frac{1}{z_1} + \frac{1}{z_2}\right)b^2 + \dots .$$

In this chapter, we will retain both terms of the expansion. For those apertures whose transmission functions can be separated into two independent one-dimensional functions of the aperture coordinates, i.e., $h(x, y)=f(x)g(y)$, the integral, with the quadratic approximation, can be separated into two integrals, called Fresnel integrals.

The Fresnel integrals have been evaluated numerically and a table of the integrals will be used to calculate the Fresnel diffraction from a straight edge. A plot of the Fresnel integrals, called the Cornu spiral, can be used to obtain a graphical solution of Fresnel diffraction. The graphical technique will be demonstrated by obtaining diffraction from a straight edge and comparing the results to the solution obtained using the tabulated data.

---

[1] This treatment is often called the near-field approximation. This leads to confusion because of the use of the near field when discussing multipole radiation fields that decay faster than the square law fields of dipole radiation. These fields are labeled non-radiative or reactive fields mentioned in Chapter 2, following Eq. (2.9). We will try eliminate confusion by using the name Fresnel diffraction.

Modern Optics Simplified. B. D. Guenther. © B. D. Guenther 2020.
Published in 2020 by Oxford University Press. DOI: 10.1093/oso/9780198842859.001.0001

Fresnel developed a technique for the solution of the Huygens–Fresnel integral based upon a geometrical construction called Fresnel zones. The Fresnel zone approach provides a unique insight into the diffraction process and can provide a partial description of Fresnel diffraction, from circularly symmetric objects. The concept of Fresnel zones will be used to discuss diffraction from a circular aperture and a disk and to obtain additional insight on Fermat's principle.

10.2 **Fresnel Approximation**

In Fraunhofer diffraction, the phase of the wave in the aperture is assumed to vary linearly across the aperture. This would occur if, for example, a plane wave was incident upon the aperture at an angle with respect to the optical axis. In Fresnel diffraction, we replace the assumption of a linear phase variation with a quadratic phase variation. This is equivalent to assuming that a spherical wave of amplitude α from a point source at position (x_s, y_s, Z') illuminates the aperture; see Figure 10.1.

The Huygens–Fresnel integral for the geometry shown in Figure 10.1 is the following integral where we continue to treat the obliquity factor as a constant that has been removed from the integral:

$$\tilde{E}_P = \frac{i\alpha}{\lambda} \iint_{\Sigma} f(x,y) \; \frac{e^{-i\mathbf{k}\cdot(\mathbf{R}+\mathbf{R}')}}{RR'} \; dx \, dy. \tag{10.1}$$

As we have stated before, the integral is nonzero only when the phase of the integrand is stationary. For Fresnel diffraction, we can ensure that the phase is nearly constant if R does

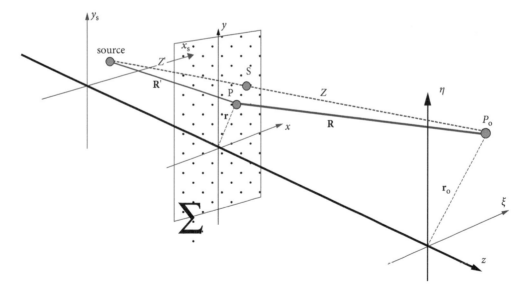

Figure 10.1 Geometry for Fresnel diffraction.

not differ appreciably from Z, or R' from Z'. This is equivalent to stating that only the wave in the aperture, around a point **S**, called the *stationary point*[2] (Figure 10.1), will contribute to E_p. Physically, only light propagating over paths nearly equal to the path predicted by geometrical optics (obtained from Fermat's principle) will contribute to E_p. The method of stationary phase is a rather sophisticated approach to solving oscillatory integrals using asymptotic analysis to determine the limiting behavior of the integral. You may run into the more general method called the method of steepest descent. Trying to develop an understanding of calculating the asymptotic behavior of a function would take us far from our discussion of optics so we will attempt to create a simple execution of the approximation.

Using the same procedure we used for Fraunhofer diffraction to approximate the distance from the source to the observation point, we now retain the quadratic terms of the binomial expansion:

$$R \approx Z + \frac{(x-\xi)^2 + (y-\eta)^2}{2Z} \tag{10.2a}$$

$$R' \approx Z' + \frac{(x_s-x)^2 + (y_s-y)^2}{2Z'} \tag{10.2b}$$

$$R + R' \approx Z + Z' + \left[\frac{(x-\xi)^2}{2Z} + \frac{(x_s-x)^2}{2Z'}\right] + \left[\frac{(y-\eta)^2}{2Z} + \frac{(y_s-y)^2}{2Z'}\right].$$

The integral obtained using these terms is quite complicated because it assumes that the source wave and the diffracted wave are both spherical.

A less complicated expression of the Huygens–Fresnel integral would be obtained if the wave, incident on the aperture, were a plane wave. The Huygens–Fresnel integral

$$\tilde{E}_P = \frac{i\alpha}{\lambda} \iint_\Sigma f(x,y) \frac{e^{-i\mathbf{k}\cdot\mathbf{R}}}{R} \, dx \, dy$$

would be rewritten, using the approximate expression for R given by Eq. (10.1):

$$\tilde{E}_P = \frac{i\alpha}{\lambda} \frac{e^{-ikZ}}{Z} \iint_\Sigma f(x,y) \, e^{-\frac{ik}{2Z}\left[(x-\xi)^2+(y-\eta)^2\right]} \, dx \, dy. \tag{10.3}$$

The physical interpretation of this equation states that when a plane wave illuminates the obstruction, the field at point P is a spherical wave, originating at the aperture a distance Z away from P,

$$\frac{e^{-ikZ}}{Z}$$

[2] The stationary point, **S**, is the point where the line, connecting the source and observation positions, intersects the aperture plane.

The amplitude and phase of this spherical wave are modified by an integral, with a quadratic phase dependent on the obstruction's spatial coordinates.

By defining three new parameters,

$$\rho = \frac{ZZ'}{Z+Z'} \quad or \quad \frac{1}{\rho} = \frac{1}{Z} + \frac{1}{Z'},$$ (10.4a)

$$x_0 = \frac{Z'\xi + Zx_s}{Z+Z'},$$ (10.4b)

$$y_0 = \frac{Z'\eta + Zy_s}{Z+Z'},$$ (10.4c)

the more general expression for Fresnel diffraction of an incident spherical wave can be placed in the same format as the expression obtained for an incident plane wave. We can use a little geometry in one dimension to discover the origin of these new parameters.

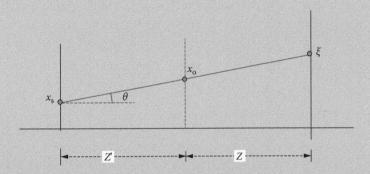

Figure 10.2 Geometry for evaluation of parameters used in equation for Fresnel diffraction.

The parameters x_0 and y_0 are the coordinates, in the aperture plane, of the stationary point, **S**, which lies on a line joining the source and the observation point. To illustrate the derivation of the stationary point's coordinates, the x coordinate will be obtained, using the geometry in Figure 10.2:

$$\tan\theta = \frac{\xi - x_s}{Z+Z'} \quad and \quad \tan\theta = \frac{x_0 - x_s}{Z'}.$$

If we equate the two relations and solve for x_0, we obtain Eq. (10.4b).

The parameters in Eq. (10.4) can be used to express, after some manipulation, the spatial dependence of the phase, in the integrand of Eq. (10.1):

$$R + R' = Z + Z' + \frac{(\xi - x_s)^2 + (\eta - y_s)^2}{2(Z+Z')} + \left[\frac{(x - x_0)^2 + (y - y_0)^2}{2\rho}\right].$$

A further simplification to the Fresnel diffraction integral can be made by obtaining an expression for the distance, D, between the source and the observation point. We will demonstrate the derivation, in one dimension, by using Figure 10.2. The total distance from x_s, the source position, to ξ, the observation position, in Figure 10.2 is

$$\sqrt{(Z+Z')^2 + (\xi - x_s)^2} \approx Z + Z' + \frac{(\xi - x_s)^2}{2(Z+Z')}.$$

If this one-dimensional calculation is extended to two dimensions, the distance between the source and the observation point can be defined as

$$D = Z + Z' + \frac{(\xi - x_s)^2 + (\eta - y_s)^2}{2(Z+Z')}. \tag{10.5}$$

We use this definition of D to also write

$$\frac{1}{ZZ'} = \frac{Z+Z'}{ZZ'} \frac{1}{Z+Z'} \approx \frac{1}{\rho D}.$$

Using the parameters we have just defined in Eqs. (10.4) and (10.5), we may rewrite Eq. (10.1) as

$$\tilde{E}_P = \frac{i\alpha}{\lambda \rho D} e^{-ikD} \iint f(x,y) e^{-\frac{ik}{2\rho}\left[(x-x_0)^2 + (y-y_0)^2\right]} dx\, dy. \tag{10.6}$$

By introducing the variables, x_0, y_0, ρ, and D, the physical significance of the general expression of the Huygens–Fresnel integral can be understood. At point P, a spherical wave

$$\frac{e^{-ikD}}{D}$$

originating at the source a distance D away is observed. This wave would be observed if no obstruction were present. Because of the obstruction, the amplitude and phase of the spherical wave are modified by the integral in Eq. (10.6). The modification of the spherical wave made by the integral is called Fresnel diffraction.

Equation (10.6) can be shown to be equivalent to Eq. (10.1), by taking the limit as the source is moved to ∞,

$$\lim_{Z' \to \infty} \rho = Z \qquad \lim_{Z' \to \infty} x_0 = \xi \qquad \lim_{Z' \to \infty} y_0 = \eta.$$

The spherical wave, containing the variable Z', becomes a plane wave as Z' approaches infinity,

$$\lim_{Z' \to \infty} \frac{\alpha e^{-ikD}}{\rho D} = \lim_{Z' \to \infty} \left[\frac{e^{-ikZ}}{Z} \frac{\alpha e^{-ikZ'}}{Z'} \right] = \frac{\alpha e^{-ikZ}}{Z}.$$

Mathematically, Eq. (10.6) is an application of the method of *stationary phase*, a technique developed in 1887 by Lord Kelvin to calculate the form of a boat's wake. The integration is nonzero only in the region of the critical point we have labeled **S**. Physically, the light distribution at the observation point is due to wavelets from the region around **S**. The phase variations of light coming from other regions in the aperture are so rapid that the value of the integral over those spatial coordinates is zero.

The calculation of the integral for Fresnel diffraction is more complicated than Fraunhofer diffraction because, when the observation point is moved, we must recalculate the integral over a new region in the aperture.

10.2.1 **Rectangular apertures**

If the aperture function, $f(x,y)$, is separable in the spatial coordinates of the aperture, then we can rewrite Eq. (10.6) as

$$\tilde{E}_P = \frac{i\alpha}{\lambda \rho D} e^{-ikD} \int_{-\infty}^{\infty} f(x)e^{-ig(x)}\, dx \int_{-\infty}^{\infty} f(y)e^{-ig(y)}\, dy,$$

$$\tilde{E}_P = A\big[C(x) - iS(x)\big]\big[C(y) - iS(y)\big].$$

The spherical wave from the source is represented by

$$A = \frac{i\alpha}{2D} e^{-ikD}.$$

10.2.1.1 Fresnel integral

If we treat the aperture function as a simple constant that can be removed from the integral, C and S are integrals of the form

$$C(x) = \int_{x_1}^{x_2} \cos\big[g(x)\big]\, dx, \tag{10.7a}$$

$$S(x) = \int_{x_1}^{x_2} \sin\big[g(x)\big]\, dx. \tag{10.7b}$$

The integrals, $C(x)$ and $S(x)$, have been evaluated numerically and are shown in Table 10.1 as well as in collections of mathematical tables [1, 2]. To use the tabulated values for the integrals, Eq. (10.7) must be written in a general form that matches the form used in most tables:

$$C(w) = \int_0^w \cos\left(\frac{\pi u^2}{2}\right) du, \tag{10.8}$$

$$S(w) = \int_0^w \sin\left(\frac{\pi u^2}{2}\right) du. \tag{10.9}$$

Before everyone had a computer at their fingertips, it was necessary to use tables or graphic techniques to calculate functions such as Eqs. (10.8) and (10.9); now it is possible to write a routine to calculate these functions [3].

The variable u is a dummy variable for an aperture coordinate, measured relative to the stationary point, **S**, in units of $\sqrt{\dfrac{\lambda \rho}{2}}$:

Table 10.1 Fresnel Integrals

w	$C(w)$	$S(w)$	w	$C(w)$	$S(w)$	w	$C(w)$	$S(w)$
0	0	0	3.6	0.588	0.4923	6.2	0.4676	0.5398
0.1	0.1	0.0005	3.7	0.542	0.575	6.25	0.4493	0.4954
0.2	0.1999	0.0042	3.8	0.4481	0.5656	6.3	0.476	0.4555
0.3	0.2994	0.0141	3.9	0.4223	0.4752	6.35	0.524	0.456
0.4	0.3975	0.0334	4	0.4984	0.4204	6.4	0.5496	0.4965
0.5	0.4923	0.0647	4.1	0.5738	0.4758	6.45	0.5292	0.5398
0.6	0.5811	0.1105	4.2	0.5418	0.5633	6.5	0.4816	0.5454
0.7	0.6597	0.1721	4.3	0.4494	0.554	6.55	0.452	0.5078
0.8	0.723	0.2493	4.4	0.4383	0.4622	6.6	0.469	0.4631
0.9	0.7648	0.3398	4.5	0.5261	0.4342	6.65	0.5161	0.4549
1	0.7799	0.4383	4.6	0.5673	0.5162	6.7	0.5467	0.4915
1.1	0.7638	0.5365	4.7	0.4914	0.5672	6.75	0.5302	0.5362
1.2	0.7154	0.6234	4.8	0.4338	0.4968	6.8	0.4831	0.5436
1.3	0.6386	0.6863	4.9	0.5002	0.435	6.85	0.4539	0.506
1.4	0.5431	0.7135	5	0.5637	0.4992	6.9	0.4732	0.4624
1.5	0.4453	0.6975	5.1	0.4998	0.5624	6.95	0.5207	0.4591
1.6	0.3655	0.6389	5.2	0.4389	0.4969	7	0.5455	0.4997
1.7	0.3238	0.5492	5.25	0.461	0.4536	7.05	0.4733	0.536
1.8	0.3336	0.4508	5.3	0.5078	0.4405	7.1	0.4887	0.4572
1.9	0.3944	0.3734	5.35	0.549	0.4662	7.15	0.5393	0.5199
2	0.4882	0.3434	5.4	0.5573	0.514	7.25	0.4601	0.5161
2.1	0.5815	0.3743	5.45	0.5269	0.5519	7.35	0.516	0.4607
2.2	0.6363	0.4557	5.5	0.4784	0.5537	7.45	0.5156	0.5389
2.3	0.6266	0.5531	5.55	0.4456	0.5181	7.55	0.4628	0.482
2.4	0.555	0.6197	5.6	0.4517	0.47	7.65	0.5395	0.4896
2.5	0.4574	0.6192	5.65	0.4926	0.4441	7.75	0.476	0.5323
2.6	0.389	0.55	5.7	0.5385	0.4595	7.85	0.4998	0.4602
2.7	0.3925	0.4529	5.75	0.5551	0.5049	7.95	0.5228	0.532

w	C(w)	S(w)	w	C(w)	S(w)	w	C(w)	S(w)
2.8	0.4675	0.3915	5.8	0.5298	0.5461	8.05	0.4638	0.4859
2.9	0.5624	0.4101	5.85	0.4819	0.5513	8.15	0.5378	0.4932
3	0.6058	0.4963	5.9	0.4486	0.5163	8.25	0.4709	0.5243
3.1	0.5616	0.5818	5.95	0.4566	0.4688	8.35	0.5142	0.4653
3.2	0.4664	0.5933	6	0.4995	0.447	∞	0.5	0.5
3.3	0.4058	0.5192	6.05	0.5424	0.4689			
3.4	0.4385	0.4296	6.1	0.5495	0.5165			
3.5	0.5326	0.4152	6.15	0.5146	0.5496			

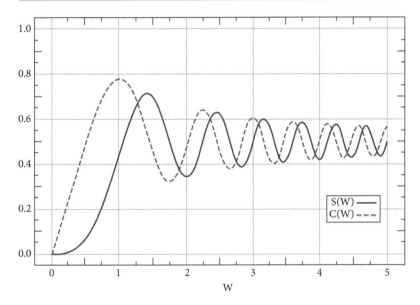

Figure 10.3 Normalized Fresnel integrals Eqs. (10.10) and (10.11).

$$u = \sqrt{\frac{2}{\lambda \rho}} (x - x_0) \qquad \text{or} \qquad u = \sqrt{\frac{2}{\lambda \rho}} (y - y_0). \tag{10.10}$$

The parameter w in Eqs. (10.8) and (10.9) and Figure 10.3 specifies the location of the aperture edge relative to the stationary point **S**. The parameter w is calculated using Eq. (10.10).

An example will clarify the method of calculating the parameter, w, for use in evaluating the Fresnel integrals. Assume the diffracting aperture is a rectangular aperture, defined as

$$f(x, y) = \begin{cases} 1 & \begin{array}{l} x_1 \leq x \leq x_2 \\ y_1 \leq y \leq y_2 \end{array} \\ 0 & \text{all other } x \text{ and } y \end{cases}.$$

The right edge of the aperture is

$$w_{x_2} = \sqrt{\frac{2}{\lambda \rho}} (x_2 - x_0)$$

and the upper edge of the aperture is

$$w_{y_2} = \sqrt{\frac{2}{\lambda \rho}} (y_2 - y_0).$$

When the observation point is moved, the coordinates, x_0 and y_0, of $\mathbf{S}$ change and the origin of the aperture's coordinate system moves. New values for w must therefore be calculated for each observation point.

The values from the Fresnel integral table are used to calculate the light wave's amplitude, E_P, and intensity, I_P,

$$I_P = A^2 \left\{ \left[C\left(w_{x_2}\right) - C\left(w_{x_1}\right) \right]^2 + \left[S\left(w_{x_2}\right) - S\left(w_{x_1}\right) \right]^2 \right\}$$
$$\times \left\{ \left[C\left(w_{y_2}\right) - C\left(w_{y_1}\right) \right]^2 + \left[S\left(w_{y_2}\right) - S\left(w_{y_1}\right) \right]^2 \right\}. \tag{10.11}$$

10.2.1.2 Cornu spiral

The calculation of the two integrals is simplified because the integrals are odd functions,

$$C(-w) = -C(w) \qquad\qquad S(-w) = -S(w).$$

This is evident in the plot of $C(w)$ versus $S(w)$ in Figure 10.4.

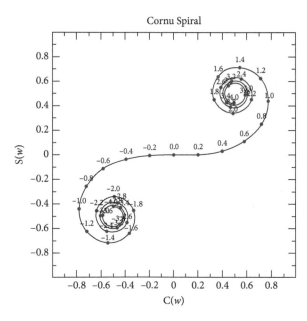

Figure 10.4 Cornu spiral obtained by plotting the values for $C(w)$ versus $S(w)$ from Table 10.1, Fresnel integrals. The numbers indicated along the arc of the spiral are values of w.

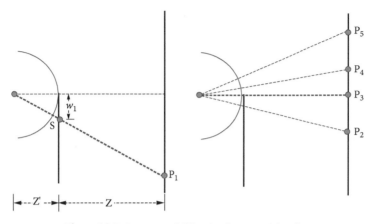

Figure 10.5 Geometry of diffraction from a straight edge.

The plot of $S(w)$ versus $C(w)$, shown in Figure 10.4,[3] is called the Cornu spiral[4] in honor of *Marie Alfred Cornu* (1841–1902), who was the first to use this plot for graphical evaluation of the Fresnel integrals.

To use the Cornu spiral, the limits w_1 and w_2 of the aperture are located along the arc of the spiral. The length of the straight-line segment drawn from w_1 to w_2 gives the magnitude of the integral. For example, if there were no aperture present, then, for the x-dimension, $w_1 = -\infty$ and $w_2 = \infty$. The length of the line segment from the point $(-1/2, -1/2)$ to $(1/2, 1/2)$ would be the value of E_P, i.e., $\sqrt{2}$. An identical value is obtained for the y-dimension, so that

$$\tilde{E}_P = 2A = \frac{\alpha e^{-ikD}}{D}.$$

This is the spherical wave that would be seen when no aperture is present.

An example will clarify the use of the Cornu spiral and the table of Fresnel Integrals. Assume that the obstruction is an infinitely long straight edge reducing the problem to one dimension. The straight edge is assumed to block the negative half-plane with its edge located at $x_1 = 0$. The other limit to be used in Eq. (10.7) is $x_2 = \infty$; i.e., the straight edge is treated as an infinitely wide slit with one edge located at infinity. Figure 10.5 shows the geometry of the problem. A set of points, P_1–P_5, on the observation screen are selected. The origin of the coordinate system in the aperture plane is relocated to a new position (given by the position of **S**) when a new observation point is selected. The value of w_1, the position of the edge with respect to **S**, must be recalculated using Eq. (10.10) for each observation point. The distance from the origin to the straight edge, w_1, is positive for P_1 and P_2, zero for P_3, and negative for P_4 and P_5.

Figure 10.6 shows the geometrical method used to calculate the intensity values at each of the observation points in Figure 10.5 using the Cornu spiral. The numbers labeling the

---

[3] If we multiply the vertical axis of Figure 10.4 by $-i$, the plot will represent the complex amplitude of Eq. (10.6).

[4] Also known as the clothoid or Euler's spiral in math circles.

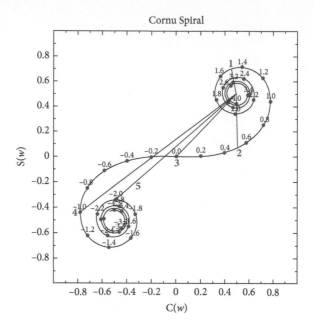

Figure 10.6 Use of Cornu spiral to solve problem of diffraction around a straight edge.

straight-line segments in Figure 10.6 are associated with the labels of the observation points, P_i, in Figure 10.5.

To obtain an accurate calculation of Fresnel diffraction from a straight edge, a table of Fresnel integrals (Table 10.1) can be used.[5] Before the calculation can proceed, Eq. (10.11) must be modified to apply to the geometry of this problem

$$I_P = I_0 \left\{ \left[\frac{1}{2} - C(w_1) \right]^2 + \left[\frac{1}{2} - S(w_1) \right]^2 \right\},$$

where $I_0 = 2A^2$. Table 10.2 shows the values extracted from the table of Fresnel integrals and used in the modified version of Eq. (10.11) to find the relative intensity at various observation points.

The result obtained by using either method for calculating the light distribution in the observation plane, due to the straight edge in Figure 10.5, is plotted in Figure 10.7 on top of an experimentally obtained Fresnel diffraction pattern from a straight edge. The relative intensities at the observation points depicted in Figure 10.5 are labeled on the diffraction curve of Figure 10.7.

The same procedure can be used to calculate the light distribution of a slit of finite width. The only modification that must be made to the straight edge calculation, in order to treat a slit, is the recalculation of w_2 for each observation point. The calculation of w_2 is accomplished by using the same procedure used to calculate w_1, in the straight edge problem. For Fresnel diffraction from a slit, the arc length on the Cornu spiral $s = (w_2 - w_1)$ is a constant, proportional to the slit width, $(x_2 - x_1)$,

---

[5] An online calculator of the Fresnel integrals can be found at https://keisan.casio.com/menu/system/000000000930.

Table 10.2 Fresnel Integrals for Straight Edge

w_1	$C(w_1)$	$S(w_1)$	I_P/I_0	Point in Figure10.5
∞	0.5	0.5	0	
2.0	0.4882	0.3434	0.01	
1.5	0.4453	0.6975	0.021	1
1.0	0.7799	0.4383	0.04	
0.5	0.4923	0.0647	0.09	2
0	0	0	0.25	3
−0.5	−0.4923	−0.0647	0.65	
−1.0	−0.7799	−0.4383	1.26	4
−1.2	−0.7154	−0.6234	1.37	
−1.5	−0.4453	−0.6975	1.16	
−2.0	−0.4882	−0.3434	0.84	5
−2.5	−0.4574	−0.6192	1.08	
−∞	−0.5	−0.5	1.0	

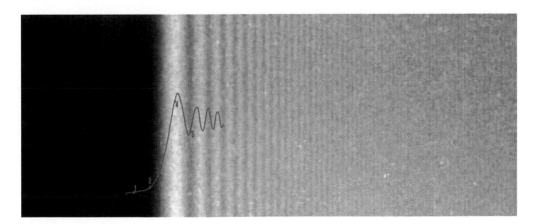

Figure 10.7 Light distribution around a straight edge with the edge located at $w = 0$. Fresnel diffraction pattern from an edge produced by illuminating edge with a plane wave from a HeNe laser.
Provided by C. C. Jones [4].

$$s = w_2 - w_1$$

$$= \sqrt{\frac{2}{\lambda \rho}} \left[(x_2 - x_0) - (x_1 - x_0) \right]$$

$$= \sqrt{\frac{2}{\lambda \rho}} (x_2 - x_1).$$

The position of the center of this arc segment moves because the location of the center of the aperture, relative to **S,** changes with the observation position. The center of the arc's position is given by

$$\frac{w_2 + w_1}{2} = \frac{1}{2} \sqrt{\frac{2}{\lambda \rho}} \left[(x_2 - x_0) + (x_1 - x_0) \right]$$

$$= \sqrt{\frac{2}{\lambda \rho}} \left[\left(\frac{x_2 + x_1}{2} \right) - x_0 \right].$$

The square of the length of the chord spanning the arc gives the intensity of light at the observation point. Fresnel diffraction experimentally generated by a rectangular aperture is shown in Figure 10.8. The aperture is not infinite in length; fringes on the right and left sides of the images are due to the edges of the aperture in the y-dimension.

Note that the shape of the Fresnel diffraction from the rectangular aperture in Figure 10.8 changes with distance from the aperture. This differs dramatically from Fraunhofer diffraction where the size of the pattern changes but the shape does not.[6] The beginning of the Fraunhofer diffraction pattern (the start of the far field) for the rectangular aperture is first seen in Figure 10.8g.

The calculations for apertures that are more complex than simple rectangles, with uniform amplitude transmission, are very demanding and are best left to the specialist. The calculations usually require numerical evaluation of Eq. (10.6). The reader can get an idea of the next degree of difficulty in diffraction problems by referring to Born and Wolf [5], where examples are given of the use of Fresnel diffraction to calculate the optical fields, near the focal plane of optical systems, including systems with aberrations.

10.3 **Fresnel Zones**

Fresnel used the geometrical construction, now called the Fresnel zone, to evaluate the Huygens–Fresnel integral. It can be used for either circular or rectangular apertures. We will apply the technique to circular apertures because the area of a circle depends on only one variable, the radius. The Fresnel zone is a mathematical construct serving the role of a Huygens's source in the description of wave propagation. Assume that at time, t, a spherical wavefront from a source at P_1 has a radius of R'. To determine the field at the observation point, P_0, due to this wavefront, a set of concentric spheres of radii $Z, Z + \lambda/2, Z + 2\lambda/2, ..., Z + j\lambda/2 \ldots$ are

---

[6] True when no lens is used to generate the far field. When we use a lens to create the Fraunhofer pattern in the focal plane, the pattern rapidly changes as we move out of the focal plane in either direction, as described by Fresnel diffraction.

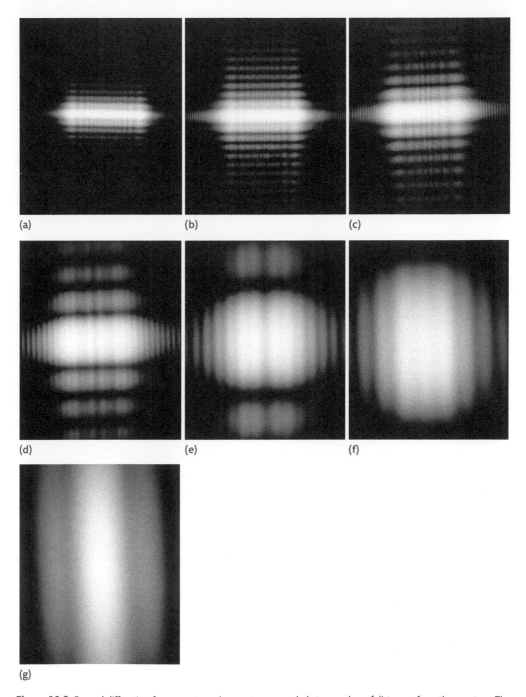

Figure 10.8 Fresnel diffraction from a rectangular aperture, recorded at a number of distances from the aperture. The distances were selected by calculating the number of zones that would exist in the short dimension if the aperture were circular. The zones match the number of zones used in Figure 10.19: (a) 5, (b) 4, (c) 3, (d) 2, (e) 1, (f) 1/2, and (g) 1/3 zones present in the aperture. In (a) the light distribution is roughly the shape of the aperture while in (g) the shape is determined by the far-field theory of Chapter 9; i.e., the light distribution size is inversely proportional to the dimension of the slit.

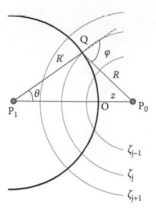

Figure 10.9 Construction of Fresnel zones for a spherical wave.

constructed, where Z is the distance from the wavefront to the observation point on the line connecting P_1 and P_0 (see Figure 10.9). These spheres divide the wavefront into a number of zones, $\zeta_1, \zeta_2, \ldots, \zeta_j, \ldots$, called *Fresnel zones*, or *half-period zones*.

We treat each zone as a circular aperture illuminated from the left by a spherical wave of the form

$$\tilde{E}_j\left(\mathbf{R'}\right) = \frac{Ae^{-i\mathbf{k}\cdot\mathbf{R'}}}{R'} = \frac{Ae^{-ikR'}}{R'}.$$

R' is the radius of the spherical wave. We only keep the radial component of the spherical wave because of symmetry in θ and ϕ. The field at P_0 due to the jth zone is obtained by using Eq. (9.5)

$$\tilde{E}_j\left(P_0\right) = \frac{A}{R'}\, e^{-i\mathbf{k}\cdot\mathbf{R'}} \iint\limits_{\zeta_j} C(\varphi)\frac{e^{-i\mathbf{k}\cdot\mathbf{R}}}{R}\, ds. \tag{10.12}$$

For the integration over the jth zone, the surface element is

$$ds = R'^2 \sin\theta\, d\theta\, d\phi. \tag{10.13}$$

The limits of integration extend over the range

$$Z + (j-1)\frac{\lambda}{2} \le R \le Z + j\frac{\lambda}{2}.$$

The variable of integration is R; thus, a relationship between R' and R must be found. This is accomplished by using the geometrical construction shown in Figure 10.9.

A perpendicular of length y is drawn from the point Q on the spherical wave to the line connecting P_1 and P_0; see Figure 10.10. In the notation of this chapter, the distance from the source to the observation point is $Z' + Z$ and the distance from the source to the plane of the zone is the radius of the incident spherical wave, $Z' = R'$. The distance from P_1 to P_0 can be written

$$x_1 + x_2 = R' + Z.$$

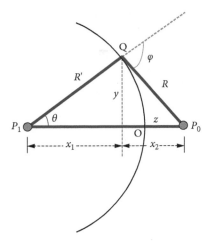

Figure 10.10 Geometry for finding R.

The distance from the observation point to the zone is

$$R^2 = x_2^2 + y^2 = [(R' + Z) - x_1]^2 + (R' \sin\theta)^2$$
$$= R'^2 + (R' + Z)^2 - 2R' (R' + Z) \cos\theta.$$

The derivative of this expression yields

$$R \, dR = R' (R' + Z) \sin\theta \, d\theta. \tag{10.14}$$

Substituting Eq. (10.14) into Eq. (10.13) gives

$$ds = \frac{R'}{R' + Z} R \, dR \, d\phi. \tag{10.15}$$

The integration over ϕ is accomplished by rotating the surface element about the $P_1 P_0$ axis. After integrating over ϕ, between the limits of 0 and 2π, we obtain

$$\tilde{E}_j (P_0) = \frac{2\pi A}{R' + Z} e^{-ik\cdot R'} \int_{Z+(j-1)\frac{\lambda}{2}}^{Z+\frac{j\lambda}{2}} e^{-ik\cdot R} C(\varphi) \, dR. \tag{10.16}$$

We assume that R', $Z \gg \lambda$ so that the obliquity factor is a constant over a single zone, i.e., $C(\varphi) = C_j$. To demonstrate that this assumption is reasonable, a list of obliquity factors for the first 12 zones are shown in Table 10.3. The parameters used to calculate these factors were $Z = 1$ m and $\lambda = 500$ nm.

As can be seen in Table 10.3, the obliquity factor only changes two parts in 10^{-7} across one zone. Applying the assumption that the obliquity factor is a constant over a zone allows the integral in Eq. (10.16) to be calculated

Table 10.3 Obliquity Factor

Zone	Angle	Obliquity
0	0	1.0
1	0.057295768	0.99999975
2	0.081028435	0.9999995
3	0.099239139	0.99999925
4	0.114591464	0.999999
5	0.128117124	0.99999875
6	0.140345249	0.9999985
7	0.151590163	0.99999825
8	0.162056667	0.9999980
9	0.171887016	0.99999775
10	0.181184786	0.9999975
11	0.190028167	0.99999725
12	0.198477906	0.999997

$$\tilde{E}_j(P_0) = \frac{2\pi i C_j A}{k(R'+Z)}\, e^{-ik(R'+Z)} e^{-ikj\frac{\lambda}{2}} \left[1 - e^{-ik\frac{\lambda}{2}}\right]. \tag{10.17}$$

Using the identity, $k\lambda = 2\pi$ and the definition for the distance between the source and observation point, $D = R' + Z$, Eq. (10.17) can be simplified as

$$\tilde{E}_j(P_0) = 2i\lambda(-1)^j\, \frac{C_j A}{D}\, e^{-ikD}. \tag{10.18}$$

Not only is the obliquity factor C_j a constant over a single zone but it is a very slowly varying parameter of j, as is shown in Figure 10.11, where the obliquity factor can be seen to change by less than two parts in 10^4 when j changes by 500. We are therefore justified in treating the absolute value of Eq. (10.18) as a constant, as j varies over a large range of values.

The physical reasons for the behavior predicted by Eq. (10.18) are quite easy to understand. The distance from P_0 to a zone changes by only $\lambda/2$ as we move from zone to zone and the area of a zone is almost a constant, independent of the zone number; thus, the amplitudes of the Huygens's wavelets from each zone should be approximately equal. The alternation in sign, from zone to zone, is due to the phase change of the light wave from adjacent zones because the propagation path for adjacent zones differs by $\lambda/2$.

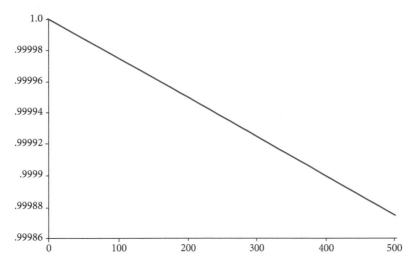

Figure 10.11 Obliquity factor as a function of the zone number. For this example, $Z = 1$ m and the wavelength was 500 nm.

To find the total field strength at P_0 due to N zones, the collection of Huygens's wavelets is added

$$E(P_0) = \sum_{j=1}^{N} \tilde{E}_j (P_0)$$

$$= \frac{2i\lambda A}{D} e^{-ikD} \sum_{j=1}^{N} (-1)^j C_j. \tag{10.19}$$

To evaluate the sum, the elements of the sum are regrouped and rewritten as

$$-\sum_{j=1}^{N} (-1)^j C_j = \frac{C_1}{2} + \left(\frac{C_1}{2} - C_2 + \frac{C_3}{2} \right) + \left(\frac{C_3}{2} - C_4 + \frac{C_5}{2} \right) + \dots.$$

Because the C's are very slowly varying functions of j, even, as shown in Figure 10.11, out to 500 zones, we are justified in setting the quantities in parentheses equal to 0. With this approximation, the summation can be set equal to one of two values, depending upon whether there are an even or odd number of terms in the summation:

$$-\sum_{j=1}^{N} (-1)^j C_j = \begin{cases} \dfrac{1}{2}(C_1 + C_N) & N \text{ odd} \\ \dfrac{1}{2}(C_1 - C_N) & N \text{ even} \end{cases}.$$

For very large N, the obliquity factor approaches zero, $C_N \to 0$, as is demonstrated in Figure 10.12. Thus, the theory has led to the conclusion that the total field produced by an unobstructed wave is equal to one half the contribution from the first Fresnel zone, i.e., $E = E_1/2$. Stating this result in a slightly different way, we see that a surprising result has been obtained—*the contribution from the first Fresnel zone is twice the amplitude of the unobstructed wave!*

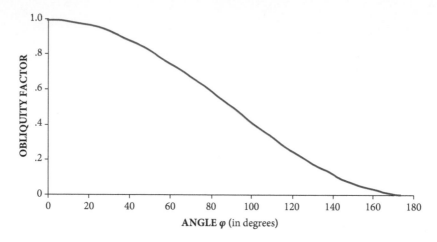

Figure 10.12 The obliquity factor as a function of the angle defined in Figure 10.9.

The zone construction can be used to analyze the effect of the obstruction of all or part of a zone. For example, by constructing a circular aperture with a diameter equal to the diameter of the first Fresnel zone, we have just demonstrated that it is possible to produce intensity at the point P_0 equal to four times the intensity that would be observed if no aperture were present. To analyze the effects of a smaller aperture, we subdivide a half-period zone into a number of subzones such that there is a constant phase difference between each sub zone. To add the waves from each of the subzones, the vector approach to adding waves is used. The individual vectors form a curve known as the *vibrational curve.*Figure 10.13a shows a vibrational curve produced by the addition of waves from nine subzones of the first Fresnel zone. The vibrational curve in Figure 10.13b is an arc with the appearance of a half-circle. The radius of curvature of the arc is defined as

$$\frac{\Delta s}{\Delta \alpha} = \rho,$$

where s and α are defined as the arc length between two subzones and the phase difference between two subzones, respectively. (See Figure 10.15b for a geometrical representation of these parameters.) If the radius of curvature of the arc were calculated, we would discover that it is a constant, except for the contribution of the obliquity factor,

$$\rho = \frac{\lambda R'}{R' + Z} \, C(\varphi).$$

Because the obliquity factor for a single zone is assumed a constant, the radius of curvature of the vibration curve is a constant over a single zone. If we let the number of subzones approach infinity (Figure 10.13b), the vibrational curve becomes a semicircle whose chord is equal to the wavelet produced by zone one, i.e., E_1. If we subdivide additional zones, and add the subzone contributions, we create other half-circles whose radii decrease at the same rate as the obliquity factor. The vibrational curve for the total wave is a spiral, constructed of semicircles that converge to a point halfway between the first half-circle. The length of the vector from the start to the end of the spiral is $E = E_1/2$, as we derived above.

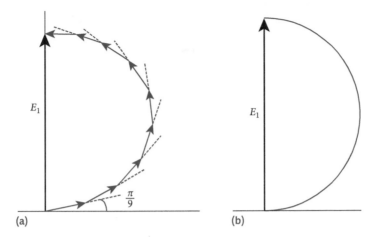

Figure 10.13 (a) Vector addition of waves from nine subzones from the first Fresnel zone. (b) Vector addition of waves from an infinite number of subzones from the first Fresnel zone.

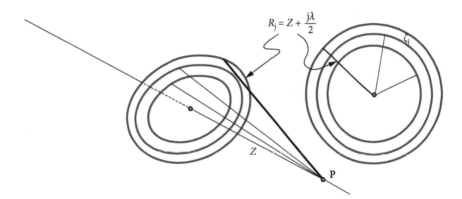

Figure 10.14 Construction of Fresnel zones for a plane wave.

To appreciate the utility of the vibrational curve, we will use the spiral just described to evaluate the Fresnel diffraction due to a circular aperture and a circular disk.

10.3.1 Incident plane wave

The Fresnel zones have been defined as spherical rings constructed upon a spherical wave. By replacing the incident spherical wave with a plane wave, the problem can be greatly simplified. With an incident plane wave, the zone is defined on a plane, removing the need for solid geometry. The use of an incident plane wave does not reduce, however, the physical insight provided by the zone construction.

Plane wave Fresnel zones consist of a set of concentric rings, constructed by drawing circles of radius R_j; see Figure 10.14. The nth plane wave Fresnel zone shown in Figure 10.14 has a radius of

$$r_n^2 = R_n^2 - Z^2$$

$$r_n^2 = \left(Z + \frac{n\lambda}{2}\right)^2 - Z^2 = nZ\lambda + \left(\frac{n\lambda}{2}\right)^2. \tag{10.20}$$

For small values of n, the nth zone has a radius of

$$r_n \approx \sqrt{nZ\lambda}. \tag{10.21}$$

In the discussion of Fresnel zones, when it is necessary to calculate the area of a zone, it will be assumed that the incident wave is a plane wave. This assumption will reduce the mathematical complexity of the examples.

10.3.2 Circular aperture

The vector addition technique, described in Figure 10.13, can be used to evaluate Fresnel diffraction, at a point P_0, from a circular aperture and yields the intensity distribution along the axis of symmetry of the circular aperture. The zone concept will also allow a qualitative description of the light distribution normal to this axis.

To develop a quantitative estimate of the intensity, at an observation point, on the axis of symmetry of the circular aperture, we construct a spiral (Figure 10.15) to represent the Fresnel zones of a spherical wave incident on the aperture. The point B on the spiral shown in Figure 10.15a corresponds to the portion of the spherical wave unobstructed by the screen. The length of the chord, AB, represents the amplitude of the light wave at the observation point P_0. As the diameter of the aperture increases, B moves along the spiral, in a counterclockwise fashion, away from A. AB first reaches a maximum when B reaches the point labeled A_1 in Figure 10.15; the aperture has then uncovered the first Fresnel zone. *At this point, the amplitude is twice what it would be with no obstruction! Four times the intensity!!*

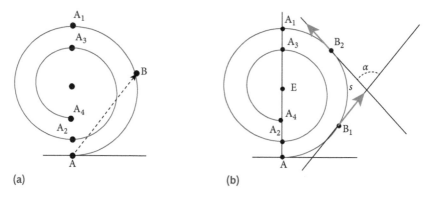

(a)   (b)

Figure 10.15 (a) Vibration curve for determining the Fresnel diffraction from a circular aperture. The change of the diameter of the half-circles making up the spiral has been greatly exaggerated for easy visualization. The actual changes are those displayed in Figure 10.11. (b) B_1 and B_2 are two points on the incident wave surface, a distance R_1 and R_2 from the observation point. The arc length, s, is the arithmetic sum of the amplitude of the wavelets from B_1 and B_2. The phase difference between the wavelets from B_1 and B_2 is α.

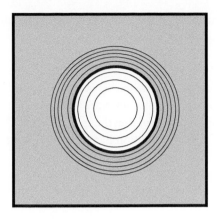

Figure 10.16 Aperture with four Fresnel zones exposed.

If the aperture's diameter continues to increase, B reaches the point labeled A_2 in Figure 10.15a and the amplitude is very nearly zero; two zones are now exposed in the aperture. Further maxima occur when an odd number of zones are in the aperture and further minima when an even number of zones is exposed. Figure 10.16 shows an aperture containing four exposed Fresnel zones. The amplitude at the observation point would correspond to the chord drawn from A to A_4 in Figure 10.15a.

The aperture diameter can be fixed and the observation point P_0 can move along, or perpendicular to, the axis of symmetry of the circular aperture. As P_0 moves away from the aperture, along the symmetry axis, i.e., as Z increases, the radius of the Fresnel zones increases without limit. (See Eq. (10.24) to calculate the zone radius for an incident plane wave.) If Z is large enough, the aperture radius, a, will be smaller than the radius of the first zone,

$$Z_{\text{max}} = \frac{a^2}{\lambda}. \tag{10.22}$$

For values of Z that exceed Eq. (10.22), Fraunhofer diffraction will be observed. At Z_{max}, the light intensity is a maximum, given by the chord length from A to A_1 in Figure 10.15. If $\lambda = 500$ nm and $a = 0.5$ mm, then this maximum occurs when $Z = 0.5$ m.

If we start at Z_{max} and move toward the aperture along the axis, Z will decrease in value, and a point will be reached when the intensity on axis becomes a minimum. The value of Z where the first minimum in intensity is observed is equal to

$$Z_{\text{min}} = \frac{a^2}{2\lambda}. \tag{}$$

In Figure 10.16 the chord would extend from A to A_2.

As the observation point, P_0, is moved along the axis toward the aperture, and Z assumes values less than Z_{min}, the point B in Figure 10.15 spirals inward toward the center of the spiral and the intensity cycles through maximum and minimum values. The cycling of the intensity, as the observation point moves toward the aperture, will not continue indefinitely. At some point, the field on the axis must approach the field observed without the aperture. It is easy to demonstrate that the theory predicts this behavior by calculating the distance between maximum and minimum values of intensity.

10.3.2.1 Intensity near the aperture

We can derive an expression for the distance between the maximum and minimum intensity along the symmetry axis of the circular aperture. We will demonstrate that this distance decreases, as the observation point approaches the aperture, until the distance between maxima is equal to a wavelength. Once that position has been reached, the on-axis intensity can be treated as a constant.

Using the geometry of Figure 10.14

$$R_n^2 = Z^2 + a^2,$$

where a is the aperture radius. We define the distance

$$q_n = R_n - Z$$

as the increase in distance, from the wavefront to the observation point, if we move out from the center of the wavefront to the nth zone. The change in q, as we move between two adjacent zones, is

$$\Delta q = \frac{\lambda}{2}.$$

Since

$$R_n^2 = \left(q_n + Z\right)^2,$$

we may write

$$\left(q_n + Z\right)^2 = Z^2 + a^2.$$

Differentiating this equation yields

$$\begin{aligned}
\Delta Z &= -\frac{q_n + Z}{q_n}\Delta q \\
&= -\frac{R_n}{R_n - Z}\Delta q \\
&= -\frac{R_n}{R_n - Z}\frac{\lambda}{2}.
\end{aligned} \tag{10.23}$$

When the observation point is very near the aperture, and Z is not very large compared to the aperture radius a, then $Z \approx R_n \approx a$. We have already stated that

$$R_n^2 - Z^2 = (R_n - Z)(R_n + Z) = a^2,$$

so that

$$R_n - Z \approx \frac{R_n}{2}.$$

Substituting this result into Eq. (10.23) demonstrates that ΔZ is on the order of a wavelength. Thus, when the observation point gets close to the aperture, the cycles, between intensity maxima, occur over a distance equal to a wavelength. The intensity changes are then unobservable

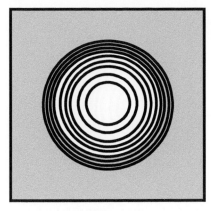

Figure 10.17 The Fresnel zones in the aperture of Figure 10.16 as the observation point moves to the left.

and the on-axis intensity is a constant. Figure 10.19 displays experimental data supporting this result on axis.

10.3.2.2 Off-axis intensity

We can also determine in a qualitative way the light distribution perpendicular to the axis of a circular aperture using Fresnel zones. Assume that a plane wave is incident upon a circular aperture of radius a and that the observation point is located a distance

$$Z = \frac{a^2}{4\lambda}$$

away from the aperture. There are four Fresnel zones in the aperture at this observation point and the light intensity on axis is very nearly zero; see Figure 10.16. As the observation point is moved off-axis to the left, the fourth zone becomes partially obstructed and the fifth zone begins to appear; see Figure 10.17.

Since the negative contribution to the amplitude is decreasing, and the positive contribution is increasing, the light intensity increases as the observation point moves off-axis. In moving off axis, successive zones appear and disappear in the aperture and a set of maxima and minima are observed in the plane perpendicular to the symmetry axis. Because of the cylindrical symmetry of the aperture, lines of constant intensity will be circles about the axis.

If the observation point is moved far off axis (Figure 10.18), the central zone becomes obscured, the contributions from the many exposed partial zones cancel one another, and the intensity falls to 0. If we pick any of the images in Figure 10.19, we see bright rings supporting this result.

10.3.2.3 Opaque screen

If the screen containing a circular aperture of radius a is replaced by an opaque disk of radius a, the intensity distribution on the symmetry axis behind the disk is found to be equal to the value that would be observed with no disk present. This prediction was first derived by Poisson, to demonstrate that wave theory was incorrect; however, experimental observation

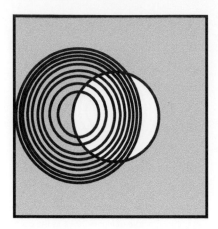

Figure 10.18 The Fresnel zones exposed just as we move into the geometrical shadow region.

supported the prediction and verified the theory. We will develop a qualitative derivation of this prediction here.

We construct a spiral, shown in Figure 10.20, similar to that for the circular aperture in Figure 10.16. The point B on the spiral represents the edge of the disk. The portion of the spiral from A to B does not contribute, because that portion of the wave is covered by the disk and the zones associated with that portion of the wave cannot be seen from the observation point. The amplitude at P_0 is the length of the chord from B to Z shown in Figure 10.20. If the observation point moves toward the disk, then B moves along the spiral toward Z. There is always intensity on axis for this configuration, though it slowly decreases until it reaches 0 when the observation point is on the disk; this corresponds to point B reaching point Z on the spiral. There are no maxima or minima observed as the disk diameter, a, increases or as the observation distance changes. If the observation point is moved perpendicular to the symmetry axis, a set of concentric bright rings are observed. The origin of these bright rings, shown in Figure 10.21 for circular and rectangular symmetry, can be explained using Fresnel zones, in a manner similar to that used to explain the bright rings observed in a Fresnel diffraction pattern from a circular aperture.

10.3.3 Zone plate

In the construction of Fresnel zones, each zone was assumed to produce a Huygens's wavelet out of phase with the wavelets produced by its nearest neighbors. If every other zone were blocked, then there would be no negative contributions to Eq. (10.22). The intensity on axis would be equal to the square of the sum of the amplitudes produced by the in-phase zones—exceeding the intensity of the incident wave. An optical component made by the obstruction of either the odd or the even zones could therefore be used to concentrate the energy in a light wave.

We will use the geometry shown in Figure 10.22 to determine how to construct an aperture that would block every other Fresnel zone. We will neglect the curvature of the wavefront, shown by the dotted line in Figure 10.22. By ignoring the wavefront curvature, we introduce aberrations into any optical component designed with this theory.

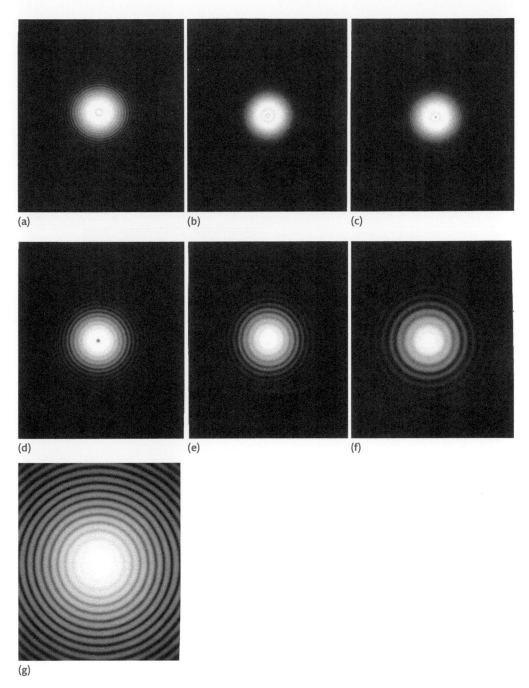

Figure 10.19 Fresnel diffraction from a circular aperture measured at positions from the aperture corresponding to (a) 5, (b) 4, (c) 3, (d) 2, (e) 1, (f) 1/2, and (g) 1/3 zones present in the aperture.

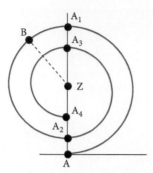

Figure 10.20 Vibration curve for opaque disk.

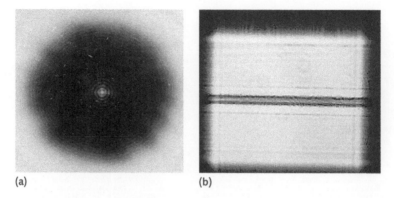

(a)   (b)

Figure 10.21 Poisson's spot observed in the shadow of a ball bearing (a) and a thin wire (b). Both are illuminated by a plane wave from a HeNe laser.

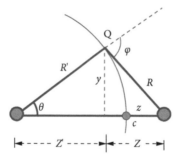

Figure 10.22 Geometry for the design of a zone plate.

The radius of an arbitrary zone is given by y and can be found by using the Pythagorean theorem

$$y^2 = R'^2 - Z'^2 \qquad\qquad y^2 = R^2 - Z^2.$$

The difference in the path length from the center of the wavefront to the point Q, a height y above the optical axis, can be obtained from these two equations

$$R' - Z' = \frac{y^2}{R' + Z'} \qquad\qquad R - Z = \frac{y^2}{R + Z}.$$

We will assume that $R' \approx Z'$ and $R \approx Z$, allowing these equations to be rewritten as

$$R' - Z' \approx \frac{y^2}{2Z'} \qquad\qquad R - Z \approx \frac{y^2}{2Z}.$$

When $y = r_n$, the radius of the nth Fresnel zone, then the difference in propagation paths for a light ray propagating from the source to the opening's edge and then to point P_0 and a light ray propagating from the source P_1 to P_0 along the axis is given by

$$(R' + R) - (Z' + Z) = \pm\frac{n\lambda}{2}.$$

By constructing an obstruction in the shape of a ring with a minimum radius given by $y = r_1$, we prevent the out-of-phase light wave from reaching the observation point. The opaque ring will block the out-of-phase light rays as long as its radius satisfies the equation

$$\pm\frac{\lambda}{r_1^2} = \frac{1}{Z'} + \frac{1}{Z}. \tag{10.24}$$

Equation (10.24) can be generalized to allow the radius of a ring to be calculated which will block a preselected zone, given the source and observation positions,

$$\pm\frac{n\lambda}{r_n^2} = \frac{1}{Z'} + \frac{1}{Z}.$$

The boundaries of the opaque zones used to block out-of-phase wavelets are seen to increase as the square root of the integers. An array of opaque rings constructed according to this prescription is called a *zone plate*; see Figure 10.23a.

By comparing Eq. (10.24) to Eq. (6.13) we see that a zone plate, with zone radii constructed according to the rule just derived, will perform like a lens with focal length

$$f = \pm\frac{r_1^2}{\lambda} \tag{10.25}$$

The zone plate's operation is based upon diffraction while a lens' operation is based upon the law of refraction. If the zone plate were a free-standing structure, it would be material property independent. You can build one for any wavelength; see Figure 10.23b. For this reason, the performance of a zone plate is different than the performance of a conventional lens.

The zone plate, shown in Figure 10.24, will act as both a positive and negative lens. What we originally called the source can now be labeled the object point, **O**, and what we called the observation point can now be labeled the image point, **I**. The light passing through the zone plate is diffracted into two paths labeled **C** and **D** in Figure 10.24. The light waves labeled **C** converge to a real image point **I**. For these waves the zone plate performs the role of a positive lens with a focal length given by the positive value of Eq. (10.25). The light waves labeled **D** in Figure 10.24 appear to originate from the virtual image point labeled **I**. For these waves the zone plate performs the role of a negative lens with a focal length given by the negative value of Eq. (10.24).

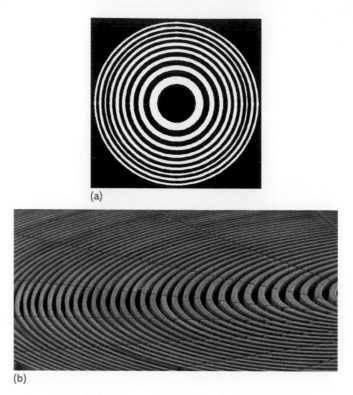

(a)

(b)

Figure 10.23 (a) Calculated amplitude zone plate with alternate opaque zones. (b) Phase zone plate for focusing X-rays produced using X-ray lithography [6].
Permission granted by Nature Communicaitons.

The zone plate will not have a single focus, as is the case for a refractive optical element, but rather will have multiple foci. As we move toward the zone plate, from the first focus, given by Eq. (10.24), the effective Fresnel zones will decrease in diameter. The zone plate will no longer obstruct out-of-phase Fresnel zones and the light intensity on the axis will decrease. However, additional maxima of the on-axis intensity will be observed at values of Z for which the first zone plate opening contains an odd number of zones. These positions can also be labeled as foci of the zone plate; however, the intensity at each of these foci will be less than the intensity at the primary focus.

Lord Rayleigh suggested that an improvement of the zone plate design would result if, instead of blocking every other zone, we shifted the phase of alternate zones by 180°. The resulting zone plate, called a *phase-reversal zone plate* (Figure 10.23b), would utilize more efficiently the incident light. R. W. Wood was the first to make such a zone plate. A zone plate in the microwave region of the spectrum can be constructed using polyethylene with an index of refraction of 2.25 at 33 GHz. A 9-mm-thick sheet of this plastic is about one wavelength thick. By cutting circular slots of radii given by Eq. (10.24) into a 9-mm sheet of polyethylene, an efficient, millimeter wave, Fresnel zone plate can be constructed. (Holography and X-ray lithography [7] provide a method of constructing the phase-reversal zone plates at shorter wavelengths.)

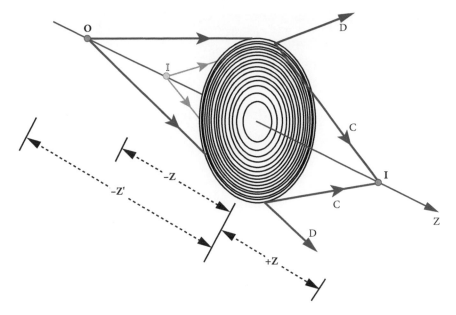

Figure 10.24 A zone plate acts as if it were both a positive and a negative lens.

The resolving power of a zone plate is a function of the number of zones contained in the plate. When the number of zones exceeds 200, the zone plate's resolution approaches that of a lens [8]. The number of zones has an inverse square relationship to the resolution.

$$N = r_N^2/\lambda f = \lambda f/4w^2.$$

10.3.3.1 Pinhole camera

A special case of the Fresnel zone plate is a simple aperture of radius r_1. Only the central zone would exist in this aperture and an image of an object will be generated in a plane whose position is determined by Eq. (10.24). An aperture designed using Eq. (10.24) produces images for a single object–image distance and a single wavelength but the performance of the device is tolerant of variations of these parameters. The single Fresnel Zone pinhole is very small compared to the other dimensions of the camera. If we use geometrical optics to analyze the camera's performance, we find that the pinhole's size prevents more than one ray from an object point from reaching the image plane. For this reason, we expect to observe little image blurring. Figure 10.25 contains an image recorded using such an aperture. The aperture generates acceptable but blurry images over a wide range of wavelengths and object distances.

The aperture containing a single Fresnel zone is called a *pinhole camera* and is quite easy to construct [9]. The camera used to produce Figure 10.25 was made of cardboard and had nominal dimensions of 35 × 35 × 45 mm. It was sized to use with 126 film cartridges. The pinhole was about 500 µm in diameter but the size is not critical. The pinhole can be produced in aluminum foil using a straight pin.

Figure 10.25 Photograph produced with a pinhole camera. It was a timed exposure and the wind moved the flowers.

From Figure 10.25, we can see several properties of a pinhole camera:

(1) *No linear distortion*: observe the tree trunks in all parts of the field.

(2) *Large depth of field*: near and far objects are in focus.

(3) *Long exposure times required*: the flowers moved in the breeze during exposure, resulting in a blurred image of the flowers.

The geometrical optics view of a pinhole camera is not adequate. The modulation transfer function approach discussed in the chapter on imaging is a more correct approach for the analysis of the imaging properties of the pinhole. When this approach is taken, one finds that the pinhole diameter given by Eq. (10.24) is too large [10]. The proper focal length, suggested by a detailed analysis, is

$$f = \frac{r_1^2}{3.8\lambda} \tag{10.26}$$

The pinhole constructed according to this formula covers about nine-tenths of the first Fresnel zone. The modification, Eq. (10.26), is a result of balancing the aberrations, due the neglect of wavefront curvature, against the resolving power, due to the aperture size.

10.3.3.2 Fermat's principle

The Fresnel zone construction provides physical insight into the interpretation given to Fermat's principle in Chapter 3. Fermat's principle states that if the optical path length of a light ray is varied in a neighborhood about the "true" path, there is no change in path length. By constructing a set of Fresnel zones about the optical path of a light ray, we can discover the size of the neighborhood.

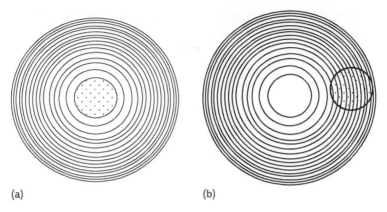

Figure 10.26 (a) A set of Fresnel zones have been constructed about the optical path taken by some light ray. If the optical path of the ray is varied over the crosshatched area shown in the figure, then the optical path length does not change. This crosshatched area is equal to the first Fresnel zone and is described as the neighborhood of the light ray. (b) The neighborhood defined in (a) is moved so that it surrounds an incorrect optical path for a light ray. We see that this region of space would contribute no wave amplitude at the observation point because of the destructive interference between the large number of partial zones contained in the neighborhood.

The rules for constructing a Fresnel zone require that all rays passing through a given Fresnel zone have the same optical path length. The true path will pass through the center of the first Fresnel zone constructed on the actual wavefront. A neighborhood must be the area of the first Fresnel zone, for it is over this area that the optical path length does not change. Figure 10.26a shows the neighborhood, about the true optical path, as a crosshatched region equal to the first Fresnel zone.

In Chapter 3 we stated that light waves did not travel over "wrong" paths because "The phase differences for those waves that travel over the 'wrong' paths are such that they destructively interfere." By moving the neighborhood defined in the previous paragraph to a region displaced from the "true" optical path, we can use the zone construction to see that this statement is correct. In Figure 10.26b the neighborhood is constructed about a ray that is improper according to Fermat's principle. We see that this region of space would contribute no energy at the observation point because of destructive interference between the large number of partial zones contained in the neighborhood.

10.4 **Problem Set 10**

(1) Use the table of Fresnel integrals to calculate the points $w = +1.5$, $w = -1.7$, and $w = -1.3$ in the diffraction pattern of a straight edge.

(2) A point source ($\lambda = 500$ nm) is placed 1 m from an aperture consisting of a 1-mm-radius hole, with an opaque disc 0.5 mm radius at its center. The observation point is 1 m away. What is the intensity with the aperture present, compared to the intensity without the aperture?

(3) A plane wave ($\lambda = 500$ nm) is normally incident on a 1-cm-diameter hole. How many Fresnel zones are visible when the aperture is viewed from a position on axis 0.5 m away?

(4) The innermost zone of a zone plate has a diameter of 0.4 mm. What is the focal length, and the first subsidiary focal length, when it is illuminated by a plane wave with $\lambda = 441.6$ nm?

(5)* A plane wave with a wavelength of $\lambda = 500$ nm is incident normal to a square hole, 4 mm on a side. The square hole is oriented with its sides parallel to the x- and y-axes. What is the intensity, relative to the incident intensity, 0.1 mm from the optical axis along the positive y-axis, 4 m from the hole?

(6) A plane monochromatic wave ($\lambda = 488$ nm) is incident normal to an opaque screen containing an aperture. The aperture is an annulus of radii 1 and 1.414 mm. What is the amplitude of the electric field on axis 2.05 m away from the screen in terms of the incident field, E_0?

(7) What is the longitudinal chromatic aberration of a pinhole camera with a pinhole diameter of 100 μm? Use C and F light in making the calculation.

(8) An argon laser of wavelength $\lambda = 488$ nm is focused to a point 50 cm in front of a pair of slits. The slits are each 1 mm wide and are spaced 0.5 mm apart. Use the Cornu spiral to generate a plot of the Fresnel diffraction pattern a distance 50 cm behind the slits. Plot a graph of the relative intensity versus w over the range $-2.0 \leq w \leq 2.0$.

(9) If the first zone of a zone plate has a diameter of 0.425 mm, what is the focal length when it is used in parallel light with a wavelength of $\lambda = 500$ nm?

(10) A 3-mm circular aperture is illuminated by $\lambda = 500$-nm plane waves. What are the locations of the first three maxima and minima along the optical axis?

REFERENCES

1. Abramowitz, M. and I. A. Stegun, *Handbook of Mathematical Functions*. New York: Dover, 1972.

2. Abramowitz, M. and I. A. Stegun, *Handbook of Mathematical Functions With Formulas, Graphs, and Mathematical Tables (AMS55)*. Available at: http://www.convertit.com/Go/ConvertIt/Reference/AMS55.ASP. ConvertIt.com, 2000.

3. Press, W. H., and S. A. Teukolsky, Fresnel integrals, cosine and sine integrals. *Computers in Physics* **6**: 670–2 (1992).

4. Jones, C. C., Diffraction pattern for a straight edge, 2000: Available from http://minerva.union.edu/jonesc/Photos%20Scientific.html.

5. Born, M., and E. Wolf, *Principles of Optics*, 7th edn. Cambridge, UK: Cambridge University Press, 1999.

6. Chang, C., and A. Sakdinawat, Ultra-high aspect ratio high resolution nanofabrication for hard X-ray diffractive optics. *Nature Communications* **5** (2014).

7. Vladimirsky, Y., et al., Demagnification in proximity x-ray lithography and extensibility to 25 nm by optimizing Fresnel diffraction. *J. Phys. D: Appl. Phys.* **32**(22): L114–18 (1999).

8. Stegliani, D. J., R. Mittra, and R. G. Semonin, Resolving power of a zone plate. *J. Opt. Soc. Am.* **57**: 610–13 (1967).

9. Pinhole camera. [cited June 2018] Available from https://en.wikipedia.org/wiki/Pinhole_camera.

10. Sayanagi, I., Pinhole imagery. *J. Opt Soc. Am.* **57**: 1091–9 (1967).

11 Imaging

11.1 Introduction

One of the most important applications of optics is the task of imaging. The theory of image formation developed independently from diffraction but as the capability to image improved, it became necessary to consider the limitations imposed upon imaging by diffraction. Most imaging tasks encountered in our daily experience involve incoherent images so we will examine incoherent illumination first.

11.1.1 Linear system

The approach we will take is based upon linear system theory [1]. In Chapter 1 we introduced Fourier theory that allowed us to describe the propagation of waves using a single cosine representation of a plane wave. The Fourier theory allowed the description of any desired waveform using a series of harmonics of the plane harmonic wave. This was based on the fact that solutions of the wave equation could be combined using the superposition principle to create a new solution to the wave equation. At that time, we did not explore the properties that supported this principle beyond description of wave propagation. Now we will want to expand our analysis to the theory of imaging using linear system theory. This approach to optics is called Fourier optics and was developed by Pierre-Michel Duffieux in 1941. Before discussing imaging, we need to outline the fundamental properties of a linear system.

A linear system is a mathematical model that can be defined in terms of an operator, $\mathcal{R}$, that converts an input f into an output g where in the consideration of optics, f and g are functions of time, t, or space, x. This can be written as

$$\mathcal{R}\{f(x)\} = g(x).$$

All linear systems must obey certain rules:

- *Homogeneity*: If a is a multiplicative constant and f is the input to the linear system, then

$$\mathcal{R}\{af(x)\} = a \cdot g(x).$$

As an example, if we double the input of a linear system, then the output must also double. If we make the object brighter, then the image will be brighter.

- *Additivity*: The output of a linear system produced when a sum of two functions is input is the sum of the individual outputs of the functions. We can combine this with homogeneity to create the principle of superposition

Modern Optics Simplified. B. D. Guenther. © B. D. Guenther 2020.
Published in 2020 by Oxford University Press. DOI: 10.1093/oso/9780198842859.001.0001

$$\mathcal{R}\{a \cdot f_1(x) + b \cdot f_2(x)\} = a \cdot g_1(x) + b \cdot g_2(x) \tag{11.1}$$

where a and b are constants with respect to x. If you and your friend join each other, then the image will contain both of you.

- *Shift Invariance*: We add an additional requirement to the linear system to create a shift-invariant linear system. If the system variable is space, then the system is space invariant and the location of the image of an object does not change appearance but only its location. If the system variable is time, then the system is time-invariant and the starting time does not change the output:

$$\mathcal{R}\Big\{f(x \pm x_0)\Big\} = g(x \pm x_0)$$

Linear, shift-invariant systems operate "independently" on each harmonic wave, and merely scale and shift them, which is why we can build a simple harmonic plane wave theory and expand it using Fourier theory. Also, this property allows us to use a convolution to describe the system performance.

11.1.1.1 Convolution

Linear theory allows us to replace the symbolic operator, R, by a convolution integral involving any two functions, $a(t)$ and $b(t)$. Mathematically, the convolution operation can be written

$$c(\tau) \doteq \int a(t) \bullet b(t - \tau)\, dt$$

We have written this in terms of time but space variables can also be used. We first saw the convolution integral when we used it to derive the array theorem in Chapter 9. To generate a feel for the effects of the convolution operation we select the set of two functions in Figure 11.1. The real function $a(\tau)$ is a sawtooth pulse and the real function $b(\tau)$ is an asymmetric rectangular pulse which we will label impulse response.

We overlay the two functions input (a) and the mirror image (time reversed) of the impulse (b) in Figure 11.1c, and calculate the area of overlap as we vary the displacement of b with respect to a. The displacement is given by τ and shown in Figure 11.1d for $\tau = 4$. The resulting correlation function is a plot of the area of overlap with respect to τ and is displayed in Figure 11.1e. Note that the effect of the impulse response is to smooth the input sawtooth shown in Figure 11.1a into the rounded version in Figure 11.1e. In optics the production of an out-of-focus photograph is generated by the convolution of a sharp image with a lens function. In the photographic community this is called *bokeh*.

Fourier theory provides another, sometimes simpler way of calculating the convolution called the convolution theorem that we will make use of later. Stated simply, convolution in the spatial domain transforms into multiplication in the spatial frequency domain:

$$\mathcal{F}\{a(t) \otimes b(t)\} = \mathcal{F}\left\{\int_{-\infty}^{\infty} a(t)b(\tau - t)\, dt\right\} = A(\omega)\, B(\omega). \tag{11.2}$$

11.1.1.2 Delta function

To properly generate the output of the linear system we must use the impulse response as the value of $b(t)$. The *impulse response* of the system is the output of the linear system in response

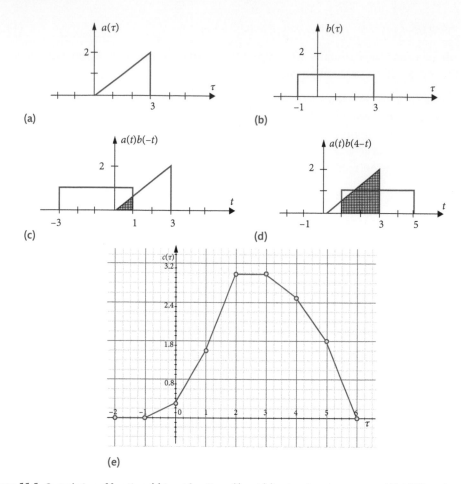

Figure 11.1 Convolution of functions (a) input function, $a(t)$ and (b) system impulse response, $b(t)$. (c) We make a mirror reflection of $b(t) \rightarrow b(-t)$ and overlay it on $a(t)$. We now slide $b(t)$ over $a(t)$ and integrate at each displacement to get area of overlap. In (d) we show b displaced by +4. (e) The correlation function is a smoothed version of the input, $a(t)$.

to a brief signal that we describe by the delta function. Paul Dirac (1902–84) introduced the delta function in his quantum textbook [2]. We introduced the delta function in Chapter 9, footnote 8. The precise mathematically definition of the delta function (also called a generalized function), based upon distribution theory, is given by the *sifting property* of the delta function:

$$\int_{-\infty}^{\infty} f(t)\delta\,(t - t_0)\,dt = f(t_0)\,. \tag{11.3}$$

A distribution is not an ordinary function but rather it is a method of assigning a number to a function. The assignment is expressed formally by an integral of the form of Eq. (11.3), where the delta function located at t_0 assigns the value $f(t_0)$ to the function $f(t)$. It should be emphasized that it is not the delta function symbol itself but rather it is the assignment operation that is defined.[1]

It is useful to note the Fourier transform of the delta function can be calculated by simply applying Eq. (11.3):

$$\mathcal{D}(\omega) = \int_{-\infty}^{\infty} \delta(t - t_0) e^{-i\omega t} dt = e^{-i\omega t_0}. \tag{11.4}$$

11.1.1.3 Impulse response of optical system

To measure the impulse response for an optical system we input, for example, light from a point source to get the spatial impulse response. Figure 11.2 displays the convolution operation visualized optically using a spatial impulse response. Here we have used a real impulse response of a microscope, an oblong-shaped diffuse light distribution labeled PSF (*point spread function*). The shape is probably due to astigmatism and maybe some coma. This experimentally obtained PSF does incorporate the wave properties of light unlike the spot diagrams used in most lens design programs. The linear system to be described is the microscope and the input function is light from two fluorescent objects (upper left of Figure 11.2). The convolution of the objects with the PSF is shown on the right side of the figure. The image is a blurred version of the objects degraded by the PSF of the microscope.

Using the linear system theory we briefly outlined, we will develop a model based on an archetype one-dimensional pinhole camera. We will assume the illumination is incoherent and

Figure 11.2 Convolution of PSF and Object to generate Image. Convolution illustrated: longitudinal (*XZ*) central slice of a 3D image acquired by a fluorescence microscope. The small object point in the upper right is still observable in the image.
By Default007 - Own work, Public Domain, https://commons.wikimedia.org/w/index.php?curid=877065.

---

[1] The delta function was introduced in a textbook by Dirac first published in the 1930s as a "convenient notation" and was used by physicists well before the mathematical foundations based on distribution theory was developed by Laurent Schwartz in the 1940s.

will measure the performance of the system using its impulse response or equivalently by the Fourier transform of the impulse response.

We will show that the incoherent imaging system is linear in intensity, while a coherent imaging system is linear in field amplitude. We will discover that the criterion for resolution must be applied with care to a coherent system. When properly applied, we will find that the resolution criterion requires a coherent optical system to have a larger aperture than an incoherent system to produce equivalent imaging capability.

The final topic we will discuss in coherent imaging is a brief review of holography.

11.2 **Incoherent Imaging**

An optical system utilizing incoherent illumination is linear in intensity. For spatially incoherent light, coherence theory states that all intensity pixels on a spatially incoherent source are uncorrelated (no interference is observable). This means that the total intensity is a sum of all the input intensities. To keep things simple, we will assume only one dimension is important, the x-direction. Because all pixel points on a spatially incoherent source are uncorrelated, we can represent a single-point object located at ξ' as a delta function[2]:

$$\left\langle \tilde{E}_0\left(\xi,t\right)\tilde{E}_0^*\left(\xi',t\right)\right\rangle = I_0\left(\xi\right)\delta\left(\xi-\xi'\right) \tag{11.5}$$

The image system we will model is shown in Figure 11.3; we will assume the source distribution consists of two incoherent point sources, A and B, equally spaced about the optical axis, a distance $2b$ apart. Because the incoherent system is linear, the source in Figure 11.3, containing A and B, is the sum of the independent intensity sources:

$$I(x) = I_0\left[\delta\left(x-b\right)+\delta\left(x+b\right)\right].$$

If we integrate $I(x)$ over x, the sifting property of the delta function, Eq. (11.3), will yield the two point sources at $x = \pm b$.

We will derive a performance metric for a simple imaging system using a one-dimensional pinhole with no refraction, Figure 11.3. The pinhole we will use is really a long, narrow aperture with a width $2a$ and the y-dimension simply ignored. Our objective is to determine how well the system resolves the two point objects.

Linear system theory states that an optical system with an amplitude impulse response, $\mathfrak{R}(x)$, and an input wave, $E_O(x,t)$, has a scalar wave output given by the correlation operation

$$\tilde{E}_I\left(x,t\right) = \int_{-\infty}^{\infty} \tilde{E}_O\left(\xi,t\right)\mathfrak{R}\left(x-\xi\right)d\xi,$$

where the subscript O and I represent object and image waves and ξ is the spatial dimension measured in the image plane.

---

[2] The correlation function is a delta function that is zero as soon as we move away from the source point. This is a mathematical way of saying that the neighbors of an image point have no correlation to the image point.

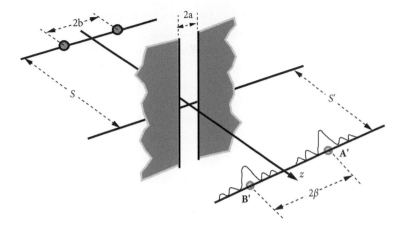

Figure 11.3 Geometry for imaging points A and B onto the positions A′ and B′ using an imaging pinhole imaging system with a rectangular aperture.

The time average of the square of the output wave's amplitude is the measured intensity distribution that forms our image. This intensity function is proportional to

$$\left\langle \tilde{E}_I(x,t)\,\tilde{E}_I^*(x,t) \right\rangle = \left\langle \int\limits_{-\infty}^{\infty} \tilde{E}_o(\xi,t)\,\Re(x-\xi)\,d\xi \right\rangle \left\langle \int\limits_{-\infty}^{\infty} \tilde{E}_o^*(\xi',t)\,\Re^*(x-\xi')\,d\xi' \right\rangle.$$

The spatial impulse response of the linear system is independent of time (shift-invariant property); thus, we may remove it, and its complex conjugate, from the time averages:

$$\left\langle \tilde{E}_I(x,t)\,\tilde{E}_I^*(x,t) \right\rangle = \int\limits_{-\infty}^{\infty}\int \left\langle \tilde{E}_o(\xi,t)\,\tilde{E}_o^*(\xi',t) \right\rangle \Re(x-\xi)\,\Re^*(x-\xi')\,d\xi\,d\xi'.$$

By using Eq. (11.3), the integration over ξ' can quickly be performed by applying the sifting property of the delta function. The result of the integration states that the output image of the optical system is the convolution of the input intensity with the square of the system's amplitude impulse response.[3]   ◾

11.2.1 One-dimensional imaging system

The amplitude impulse response for a one-dimensional slit was calculated in Eq. (9.33),

$$F(\omega_x) = \int_{-x_0}^{x_0} \cos\,\omega_x x\,dx = \frac{1}{\omega_x}[\sin\,\omega_x x]_{-x_0}^{x_0} = \frac{\sin\,\omega_x x_0}{\omega_x x_0},$$

where the spatial frequency is given by Eq (9.29)

$$\omega_x = -\frac{2\pi \sin\theta_x}{\lambda} = -\frac{2\pi\xi}{\lambda R_0}.$$

---

[3] Often called the point spread function.

We will use the spatial coordinate, ξ, rather than the spatial frequency, ω_x, in our discussions as the coordinate is directly related to image dimensions.

The square of the amplitude impulse response is physically equal to the Fraunhofer diffraction, intensity pattern of the system's aperture. Our system has no refraction so we can use the geometry of Figure 11.4 to determine the image B′ of the object B created by this pinhole camera, where the aperture is located at the origin of the coordinate system.

Using the variables displayed in Figures 11.3 and 11.4

$$x_0 = a \quad R_0 = S' \quad \xi = \beta$$

$$\mathcal{R}(\xi)\,\mathcal{R}^*(\xi) = s(\xi) = \frac{I_d(\xi)}{I_0}$$

$$\mathcal{R}(\xi)\,\mathcal{R}^*(\xi) = \frac{\sin^2\left(k\xi a \big/ S'\right)}{\left(k\xi a \big/ S'\right)^2} = \operatorname{sinc}^2\left(k\xi a \big/ S'\right). \tag{11.6}$$

This distribution is imaged onto the image points A′ and B′, whose locations at positions

$$\beta = \pm\frac{bS'}{S}$$

are found by using the geometry shown in Figure 11.4. In the image plane the images are given by geometry as

$$I(x') = I_0\left[\delta\left(x' - \beta\right) + \delta\left(x' + \beta\right)\right]. \tag{11.7}$$

The image intensity is obtained by convolving Eq. (11.3) with Eq. (11.7)

$$I(\xi) = \int_{-\infty}^{\infty} I_0\left[\delta\left(x' + \beta\right) + \delta\left(x' - \beta\right)\right]\sin^2\frac{ka}{S'}\left(x' - \xi\right)dx',$$

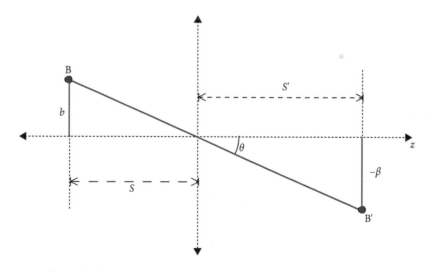

Figure 11.4 Geometric calculation of image **B′** of object **B** shown in Figure 11.3.

Using the sifting properties of the delta function we obtain

$$I(\xi) = I_0 \left[\operatorname{sinc}^2 \frac{ka}{S'} (\xi + \beta) + \operatorname{sinc}^2 \frac{ka}{S'} (\xi - \beta) \right],$$

The images are two sinc-squared functions located at $\pm\beta$.

We wish to discover the minimum value of b for which we can just resolve image points A′ and B′. One criterion that can be applied is Rayleigh's criterion:

11.2.1.1 Rayleigh's criterion

Two image points A′ and B′ in the image plane can be just resolved if the central diffraction maximum of one image falls on the first, diffraction pattern zero of the second image. See Figure 11.5.

The location of the central maximum and the first diffraction zero of the sinc function are obtained from

$$\frac{ka}{S'} (\xi - \beta) = 0 \qquad \frac{ka}{S'} (\xi + \beta) = \pi$$

Arranging the central maximum of image A′ to fall on the first minimum of image B′ results in

$$2\beta = \frac{\lambda S'}{2a}.$$

This result is equivalent to the result obtained in Chapter 9 using Gaussian wave theory for an optical system's minimum beam waist

$$w = \frac{2\lambda f}{D}.$$

From the geometric relationship between b and β in Figure 11.4, the minimum object separation that can be resolved in the image plane is given by

$$2b = \frac{\lambda S}{2a}. \tag{11.8}$$

The resolution criterion is usually stated in terms of the angular extent of the object

$$\sin \Psi \approx \tan \Psi = \frac{2b}{S};$$

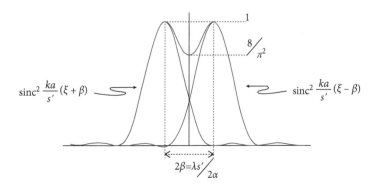

Figure 11.5 Minimum resolution for two slits.

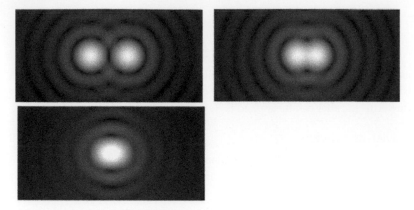

Figure 11.6 Resolving of two circular apertures. The resolution decreases from left to right with the middle image at Raleigh resolution limit.
By Spencer Bliven [Public domain], from Wikimedia Commons.

thus, the minimum angular extent of an object that can just be resolved by the optical system is

$$\sin \Psi \approx \frac{\lambda}{2a}.$$ (11.9)

While the math is a little harder, it is much easier experimentally to use a circular aperture of radius a to show the resolution limit. The resolution limit for two circular apertures is shown in Figure 11.6. For two circular apertures, the resolution criterion would be measured at the zero crossing of the Airy pattern (see Figure 9.32):

$$\sin \Psi \approx 1.22 \frac{\lambda}{2a}.$$ (11.10)

To develop a feeling for the magnitude of this criterion we will apply it to a common resolution problem. How far away can we resolve the separation of car headlights? We will assume they are spaced by 1.5 m, restrict the wavelength to 550 nm, and assume the pupil of your eye is dilated to 5 mm because it is dark and the headlights are on.[4] Applying the Rayleigh criterion

$$\theta_R = (1.22)/d$$

$$= \frac{(1.22)\left(550 \times 10^{-9}\right)}{5 \times 10^{-3}}$$

$$= 1.34 \times 10^{-4} \quad \text{rad}.$$

The distance from the eye to the car is

$$R = 1.5/\theta_R$$

$$= 1.5/1.34 \times 10^{-4}$$

$$= 11.2 \text{ km}.$$

[4] We assume with average illumination the pupil size of the eye is 4 mm, that is, 0.2 mrad.

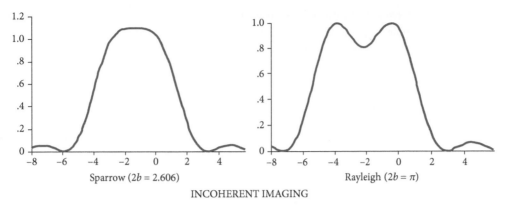

Figure 11.7 The Rayleigh and Sparrow criteria for a slit aperture and incoherent illumination.

11.2.1.2 Sparrow's criterion

The Rayleigh criterion is not a physical law but rather is a rule-of-thumb. For that reason, other criteria can be used; one of the most popular is *Sparrow's criterion* (Figure 11.7), which states the following:

> Two image points A′ and B′ in the image plane can be just resolved if the intensity is a constant as we move from image point A′ to point B′ in the image plane.

Mathematically, Sparrow's criterion is stated in terms of the second spatial derivative of the intensity (the definition of a stationary point):

$$\left[\frac{\partial^2 I(\xi)}{\partial \xi^2} \right]_{\xi=0} = 0. \tag{11.11}$$

For a circular aperture, the Sparrow criterion becomes

$$\sin \Psi = \frac{(2.976)\lambda}{2\pi a} = 0.95 \frac{\lambda}{2a}. \tag{11.12}$$

The Rayleigh and Sparrow criteria are too restrictive for general imagery and attempting to treat a general image as a continuous distribution of point sources is cumbersome, and often intractable, because of the convolution integral. Instead, another approach, based on linear system theory, is used. We move to spatial frequency space, via the Fourier transform, and describe the linear optical system in terms of its spatial frequency response. The imaging process is described in terms of the convolution integral. The most straightforward implementation of this would describe the imaging process by convolving the object with the impulse response of the optical system. It is usually much easier, however, to utilize the convolution theorem we first used in in Eq. (9.37),

$$\mathcal{F} \left\{ \int_{-\infty}^{\infty} f(t)g(\tau - t) \, dt \right\} = F(\omega) \, G(\omega), \tag{11.13}$$

and write the Fourier transform of the convolution as point-wise product[5] of the Fourier transform of the object function with the Fourier transform of the impulse response. It is useful to note that the theorem works the other way also

$$F(\omega) \otimes G(\omega) = \mathcal{F}\{f \cdot g\}. \tag{11.14}$$

11.2.1.3 Optical transfer function (OTF)

The Fourier transform of the impulse response of an optical system is called *the optical transfer function* (OTF). It draws its name from the analogy that can be formed between optical and electronic systems. In a linear electronic system, a sinusoidal input wave will result in a sinusoidal output wave, with an amplitude and phase modified by the transfer function of the electronic system. We will demonstrate, by using the one-dimensional model of Figure 11.3, that the OTF of an optical system similarly modifies the contrast of a simple sinusoidal object. It will become apparent that an imaging system can be characterized by the modification it produces on a distribution of single spatial frequency objects.

For the imaging system in Figure 11.3, the Fourier transform of the impulse response is from Eq. (11.4):

$$\tilde{S}(\omega_x) = \mathcal{F}\{s(x)\} = \int_{-\infty}^{\infty} \text{sinc}^2\left(\frac{kax}{S'}\right) e^{-i\omega_x x} dx. \tag{11.15}$$

Calculation of this integral is one way of obtaining the OTF. A second method is by calculating the autocorrelation of the pupil function[6] of the optical system. We will use this second technique but we should give some justification for our selection of the use of an autocorrelation approach.

In Chapter 4, the interference between two sources, a and b, was calculated to be

$$h(\tau) = \frac{AB}{2} \cos(\omega_0 \tau + \theta - \phi),$$

where phase difference $(\theta - \phi)$ generated the fringe pattern. The theoretical realization of the general result is calculated using cross-correlation integrals. In our discussion of imaging, our interest will be in the mathematical operation called autocorrelation. In Figure 11.8 we demonstrate how the autocorrelation of a simple rectangular pulse is calculated.

To calculate the correlation function we graphically slide one function across the second, measuring the overlapping area for each displacement, τ. The autocorrelation uses two identical functions for the calculation. The overlap of the function and its clone at τ is indicated in the figure by shading the overlap area. For $A(t)$ the area of overlap equals the area of the two pulses, $(A \cdot 2t_0 + A \cdot 2t_0)$, minus the area of each pulse not overlapped, $(A\tau + A\tau)$. The area is, thus,

$$4At_0 - 2A\tau.$$

---

[5] The point-wise product in a domain is obtained by multiplying the two functions at each value in the domain.

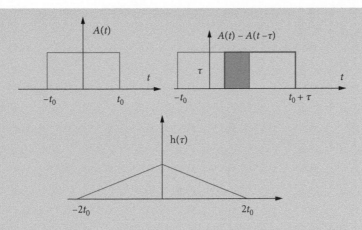

Figure 11.8 The calculation of the autocorrelation function of A(t), h(τ), can be generated if we simply slide one realization of A(t) over another copy and record the overlapping area, shown here as the shaded area.

We divide by the area of the pulses to normalize, yielding

$$h(\tau) = \begin{cases} 1 - \dfrac{|\tau|}{2t_0} & |\tau| \le 2t_0 \\ 0 & |\tau| > 2t_0 \end{cases}.$$

This autocorrelation looks very similar to the convolution operation. In fact if $a(x)$ is Hermitian,[7] the convolution and cross-correlation are identical:

$$a(x) \otimes b(x) = a(x) \oplus b(x).$$

---

[6] The Fourier transform of the PSF is the pupil function (aperture function) and describes the change in amplitude and phase of a wave transmitted through the aperture.

[7] The aperture function $a(x)$ is Hermitian whenever $a^*(x) = a(-x)$. In physically realizable optical systems $a(x)$ is real and even about the optical axis, satisfying both conditions for the aperture function to be Hermitian.

Having established the equality of the convolution and the cross-correlation, we can use the mathematical trick suggested by Eq. (11.2) to generate the OTF of our one-dimensional pinhole. The integrand of Eq. (11.15) is the product of two sinc functions,

$$\left[\sin\left(\frac{kax}{S'}\right)\right]\left[\sin\left(\frac{kax}{S'}\right)\right]^*.$$

It is not difficult to show that a sinc function is the Fourier transform of a rectangular function

$$\mathcal{F}\{\text{rect}(x)\} = \text{sinc}\left(\frac{kax}{S'}\right)$$

and the integral, Eq. (11.15), can now be viewed as the convolution of one rectangular function with itself. The rectangular function is equal to the aperture (pupil) function of a slit, the imaging system for our 1-d pinhole. Using our conclusion that the OTF of an imaging system is equal to the autocorrelation of the aperture function of the system we can calculate the OTF of the one-dimensional pinhole.

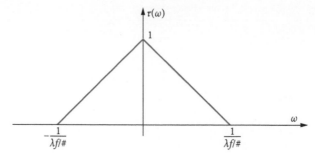

Figure 11.9 Incoherent transfer function for a rectangular aperture.

The autocorrelation of a rectangular function is a triangular function as we generated in Figure 11.8. If we plot $h(\tau)$ we obtain a triangle whose base is twice the width of the pinhole; this is the autocorrelation of the square aperture of width $2a$:

$$\tilde{S}(\omega_x) = \begin{cases} 2a\left(1 - \dfrac{S'}{2ak}\omega\right) & |\omega_x| \leq \dfrac{2ak}{S'} \\ 0 & \text{all other } \omega \end{cases}. \tag{11.16}$$

The normalized transfer function is defined as

$$\tau(\omega_x) = \frac{\tilde{S}(\omega_x)}{\tilde{S}(0)} = \begin{cases} 1 - \dfrac{S'}{2ak}\omega & \omega_x \leq \dfrac{2ak}{S'} \\ 0 & \text{all other } \omega \end{cases} \tag{11.17}$$

and is shown in Figure 11.9 where the cutoff is defined in Eq. (11.18).

11.2.1.4 2-d optical transfer function

The result obtained for the simple, one-dimensional model can be extended to two dimensions, allowing the optical transfer function for any imaging system to be generated by calculating the autocorrelation of the aperture function of the imaging system. The behavior of the transfer function shown in Figure 11.9 illustrates the general behavior of incoherent imaging systems; i.e., contrast is a decreasing function of spatial frequency, ω. We can obtain the properties of a typical imaging system by assuming a two-dimensional, circularly symmetric aperture. We replace S' by the focal length of the optical system and use the definition that the "f-number", $f/\#$, of the optical system as

$$f/\# = {f}/{2a}, \tag{11.18}$$

where a is now the radius of the aperture of the system.

To find the physical interpretation that should be associated with the OTF in Figure 11.9, we evaluate the response of the optical system, Figure 11.10, to a simple sinusoidal intensity distribution of width ℓ, located in plane O in Figure 11.10, whose production was illustrated in Figure 9.23:

$$I_0(x) = 1 + d\cos\omega_0 x. \tag{11.19}$$

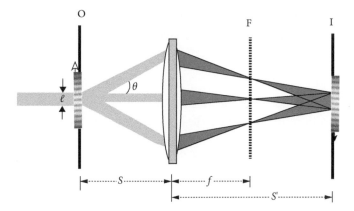

Figure 11.10 The object, Eq. (11.19), in plane O (a grating of width ℓ) is imaged by the lens in plane I. The spectrum of the object is produced in plane F. The impulse response of the lens is given by Eq. (11.15).

The image generation of this object is shown in Figure 11.10. In electronics, the constant d is called the modulation of the signal; in optics we define it as the *contrast* of the object, and is given by

$$C_0(\omega_0) = \frac{I_{\max} - I_{\min}}{I_{\max} + I_{\min}} = d. \tag{11.20}$$

We expect the contrast to exhibit a dependence on the spatial frequency and we emphasize this functional dependence in Eq. (11.20) by writing $C_O(\omega_0)$.

The intensity distribution of the image is

$$I_I(x) = \int_{-\infty}^{\infty} (1 + d\cos\omega_0\xi)\, s(x - \xi)\, d\xi = \mathcal{F}\{\mathcal{F}\{I_o(x)\}\mathcal{F}\{s(x)\}\}.,$$

where s is the impulse response of the imaging system. We can write the spatial frequency spectrum of the object as

$$\mathcal{F}\{I_O(x)\} = \delta(\omega) + \frac{d}{2}[\delta(\omega - \omega_0) + \delta(\omega + \omega_0)].$$

Because the object, Eq. (11.19), contains only one frequency, ω_0, the Fourier transform displays only two delta functions representing $\pm\omega_0$ and a delta function at zero frequency, shown on plane F in Figure 11.10. If we multiply this function by the transfer function shown in Figure 11.9 and perform the inverse transform, we obtain the image intensity, shown in plane I of Figure 11.10

$$I_I(x) = 1 + \tau(\omega_0)\, d\cos\omega_0 x.$$

The final image has its contrast reduced by

$$\tau(\omega_0) = 1 - \frac{S'\omega_0}{2ka}.$$

Frequencies in the object beyond the cutoff frequency $\left(\omega > \frac{1}{\lambda f/\#}\right)$ are not reproduced in the image.

11.2.1.5 Modulation transfer function (MTF)

We assume that the impulse response of the imaging system is $s(x)$ and the optical transfer function is $\mathbf{S}(\omega)$. In general, it is a complex function of the form

$$\tilde{S}(\omega_0) = |\tilde{S}(\omega_0)| \, e^{-i\varphi(\omega_0)},$$

where $|\tilde{S}(\omega_0)|$ is the modulus of the transfer function (affecting the contrast) and $\phi(\omega_0)$ is the phase (affecting the image position). The image intensity is then

$$I_I(x) = \int \left\{ \delta(\omega) + \frac{d}{2}[\delta(\omega - \omega_0) + \delta(\omega + \omega_0)] \right\} \tilde{S}(\omega) e^{-i\omega x} d\omega$$
$$= \tilde{S}(0) + \frac{d}{2} \left[\tilde{S}(\omega_0) e^{i\omega_0 x} + \tilde{S}(-\omega_0) e^{-i\omega_0 x} \right].$$

The impulse response of an incoherent system, $s(x)$, is always a real function because it equals the square of the amplitude impulse response. The Fourier transform of a real function is

$$\tilde{S}(\omega) = K[\cos\omega - i\sin\omega].$$

This function is Hermitian

$$\tilde{S}(-\omega) = \tilde{S}^*(\omega)$$

and

$$I_I(x) = \tilde{S}(0) + d \cdot \mathrm{Re}\left\{ \tilde{S}(\omega_0) e^{i\omega_0 x} \right\}$$
$$= \tilde{S}(0)\left[1 + d\frac{|\tilde{S}(\omega_0)|}{\tilde{S}(0)} \cos(\omega_0 x + \phi) \right]. \tag{11.21}$$

The use of the OTF to obtain an image of a simple sinusoidal function demonstrates that the image contrast is equal to the object contrast, modified by the modulus of the transfer function. Because d is called the modulation in electronics, the modifier of d is called *the modulation transfer function* (MTF) in optics. The image contrast of an optical system given by Eq. (11.21) is thus

$$C_I(\omega_0) = C_0(\omega_0)\frac{|\tilde{S}(\omega_0)|}{\tilde{S}(0)}$$
$$= C_0(\omega_0)\tau(\omega_0), \tag{11.22}$$

where $\tau(\omega_0)$ is the MTF. The units of the MTF are line pairs/mm where a line pair is a spatial cycle of light and dark.

This relationship suggests a method for measuring the MTF. The contrast in the image is measured using different spatial frequencies as object functions, as shown in Figure 11.11. Rather than collection data using individual targets as suggested in Figure 11.11, we might use specially designed targets like the spoke target[8] shown in Figure 11.12. In Figure 11.12a the test pattern of a spoke target is used to measure the modulation transfer function of an optical system [3]. The slanted edge technique can also be used [4]. In Figure 11.12a the optical system is diffraction limited and the MTF generated by scanning the spoke target is shown in 11.12b and is similar to Figure 11.9. From the MTF, the limiting resolution of the optical system used to generate Figure 11.12a is 2 μm.[9] Figure 11.11c is generated by a similar lens but with spherical

---

[8] The target is known as the Siemens Star. The Star and the Slanted Edge are standard methods for measuring the MTF.

[9] The lens was an $f/4$ operating at a wavelength of 500 nm.

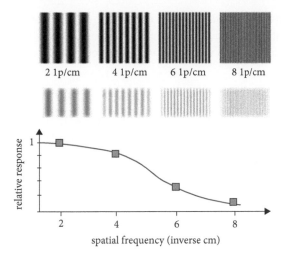

Figure 11.11 Determination of modulation transfer function.

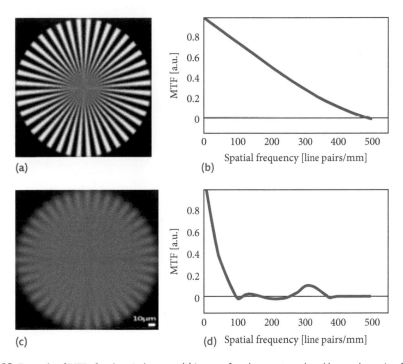

Figure 11.12 Example of MTF of real optical system. (a) Image of spoke target produced by an aberration free optical system. (b) 1-d OTF of the optical system. (c) Spoke target image produced by optical system with spherical aberration. (d) MTF of optical system with spherical aberration.

(Individual figures by Tom.vettenburg (Tom.vettenburg) [CC BY-SA 3.0 (https://creativecommons.org/licenses/by-sa/3.0)), from Wikimedia Commons.)

aberration. The MTF for this optical system is shown in 11.11d. The contrast goes to zero several times before finally reaching 500 lines/mm. One would be safe in assigning the resolution limit for this system to the point where the MTF first goes to 0, i.e., a resolution of 10 µm.

11.2.2 **Human vision**

In the early days of artificial intelligence and robotics it was though that the problem of vision was a simple one for undergraduates to attack as they developed their computational skills [5]. The problem was underestimated; simulating human vision continues to hold the position of a grand challenge and remains today an unfulfilled goal. Our vision system is not as simple to understand, as it is to use. A brief discussion of the parameters of the human imaging system will give us some understanding of the difficulty of trying to emulate our vision with devices we design.

Experimental measurements on the operation of the eye demonstrate that a great deal of signal processing takes place within the detector plane of the eye. Even with detector plane processing, about two-thirds of the brain is required for vision-related activities. The data needed by the brain come from the eye but the sensors and interconnects are actually an integral part of the brain. The field of view of each eye is 155° horizontal and 135° vertical. Within the brain the outputs of the two eyes are combined into a 220° image. The detectors within the eye are spread out over the retina (the back internal surface of the eye).

There are two classes of detectors:

1. *Rods*: 94% of the retinal detectors (about 150 million) are low-resolution, high-sensitivity detectors responding to a wavelength band around 496 nm. Their role is in motion and low light level detection (a rod will respond to one photon but conscious perception depends on the brain, resulting in a lowering of the quantum efficiency). The rods record no color information.

2. *Cones:* 6% of the detectors (about 7 million) are uniquely sensitive to one of three different color bands (419, 531, and 559 nm); these detectors do not have a sensitivity matching the rods but provide high-resolution three-color image data.

The distribution of detectors over the retina is not uniform; see Figure 11.13. The highest density of cones is found in the fovea where density is about 1.0×10^5/mm. The detector density in the fovea determines the maximum resolution of the eye. The fovea has a field of view of 2°; that is about twice the width of your thumbnail at arm's length. The average size of a cone in the fovea is about 2 µm, which results in a maximum human visual acuity of one arc-minute, equal to $1/60° = 0.016°$, or $\pi / [(180)(60)] = 2.9 \times 10^{-4}$ radians, or 290 µradians.[10] The fovea cone spacing is about 120 samples/degree.[11]

There are no rods in the fovea, which results in a "night blind spot" (about 5° to 10° wide) accompanied by an overall reduction in resolution in the entire nighttime image. The night

---

[10] For a young person at a reading distance of 25 cm, the aberration-free eye can resolve 0.04 mm on a page. The eye is not diffraction limited; therefore, designers use the resolution limit of 0.1 mm. Lower resolution is a bigger number.

[11] The maximum spatial frequency, according to the Nyquist theorem, is half of the sampling frequency or 60 cycles/degree.

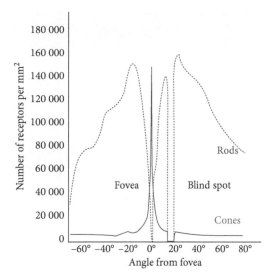

Figure 11.13 Cones are present at a low density throughout the retina, with a sharp peak in the population in the center of the *fovea*. Conversely, rods are present at high density throughout most of the retina, with a sharp decline in the fovea.

By Cmglee—Own work, CC BY-SA 3.0, https://commons.wikimedia.org/w/index.php?curid=29924570.

blind spot is observable when you stare directly at an object in low light levels. You can require the lost image by looking 10° away from the original object position.

There is also an "official" blind spot (scotoma) that occurs in both daylight and night because there are no sensors in the area where the optical nerve exits the eye. The brain processes the subimages so that the blind spot of each eye is filled in, as the brain generates the image, with an extrapolation of the surrounding image into the blind region. To produce an image at the maximum resolution over the eye's entire field of view requires that the eye be scanned. The scanning activity can be perceived by the illusion shown in Figure 11.14. The scanning also reduces the chance that the blind spots will be perceived.

Either one eye or both eyes can experience either voluntary or involuntary movement to bring an object in the field of view to focus on the fovea. This movement is required to allow the high-resolution fovea to explore the entire field of view of the eye. The motion also prevents a blurring of the image when the head is rotated.

The eye makes three types of movements. There are two movements that have a duration of about 1/100 second. The first is called *fixation,* where the eye maintains the focus of a single location in the field of view. A second movement is called *saccades*; it is associated with rapid movement of the focus between fixation points and cause an interruption in the vision process which keep us from seeing the motion in our own eyes. Finally, there is a rapid motion from ocular microtremors that have an average frequency of 84 Hz. These occur in conjunction with the other two motions and there is not agreement as to how the microtremors contribute to visual performance.[12] While there is no question that the eyes must move to allow all parts of the object's image to fall for a time on the fovea, the image must also move to prevent the image

---

[12] Microtremors provide a measure of neuronal activity, an involuntary response that provides a useful vital sign.

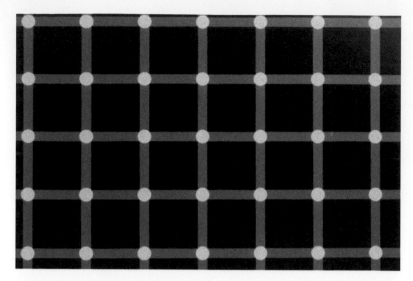

Figure 11.14 A scintillating grid illusion causes black dots to appear on the white dots at the intersections of the gray lines. If you focus on one of the white dots it will not show a black dot but remain white. This is a modified version of the Hermann grid. The fact that scanning eye movements are necessary to produce the scintillation effect sets it apart from the Hermann grid illusion. There is an optimum viewing distance for the illusion.
By User:Tó campos1 - Own work, Public Domain, https://commons.wikimedia.org/wiki/File:Grid_illusion.svg.

from fading away. The movement of the region of interest of the eye does not seem to follow a straightforward path to the creation of an understandable image. Figure 11.14 allows us to detect the motion of the eye through the scintillating grid illusion. It should be noted that the community does not agree on the origin of this illusion.

The eye acts as an adaptive optical system that changes lens shape to correct blur of objects as the eye moves about the field of view; it also changes its aperture to regulate light and control the depth of focus. This is all done by elements of the optical nerve at a place called the optic chiasm where the nerve fibers from each eye meet on the way to the brain. It is at this position that binocular vision and cooperative tracking of the two eyes take place. About half of the nerve fibers continue onto cerebral cortex where image interpretation takes place and onto the brain stem where adaptive optics integrate the subimages produced by scanning into a smooth image. There is also some visual processing occurring in the variety of interconnections within the retina, mapping the 10^8 detectors on the retina onto 10^6 nerve fibers in the optical nerve.

We do not understand the processes that take place in the brain to give us vision. To complicate the problem further, a baby is not born with the skills needed to see and the development of the needed skills is accomplished without any verbal communication skills. The visual performance is developed by the infant through observations and experimentation. The skills develop from birth to about 18 months and skill in eye–hand coordination and depth perception is not demonstrated until the child is two years old. Failure in proper development of the human imaging system can be treated if addressed early, but the development window is limited. For example, if for some reason one of the eyes fails to experience appropriate simulation, the brain will simply reject imagery from that eye and binocular vision can be permanently lost.

From the complexity of the hardware and the lack of understanding of the image processing taking place, it is easy to see why computer vision is taking so long to realize. An added complication is that we all do not have the same vision system. As an example, consider our ability to see color. The average person has cones sensitive to blue (419 nm), green (531 nm), and red (559nm) wavelengths. There is a population spread in the wavelength regions covered by each of the three bands, but the average of each band is the one used in the design of displays and cameras to match the general observer. This means that the vision systems we use limit our ability to detect departures from the average [6]. A small percentage of men and even smaller percentage of women lack sensitivity to one or more of the color bands. On the other response side there are a few women (2–3%) who are sensitive to four color bands. You can check your personal range of color response at this website [7].

The problem for many electronic imaging systems is that the human is the final arbitrator of the image quality and we do not understand the operations going on in human vision. It is difficult to include human perception and judgment in models that can predict quantitative or qualitative performance. The cost of collecting psychophysical measurements have led to the development of analytical models of human performance parameters that allow image quality assessment involving both the imaging system and the human. A typical test pattern used in these assessments is found in Figure 11.20. The end goal has still not been reached but there are books and reviews that can guide an assessment of the technology [8, 9].

11.3 Coherent Imaging

We learned in our discussion of coherence that a coherent imaging system is linear in field amplitude and the observed image intensity is calculated by first summing the amplitudes of the waves from the object and then squaring that summation. The impact of this will be seen as we discuss the theory developed by Abbe for imaging by a microscope.

In 1871, while developing a theory to guide the design of a microscope in collaboration with Karl Zeiss, Ernst Abbe (1840–1905) introduced a view of coherent imaging that has led to the modern approach by Duffieux called Fourier optics [10]. The key to the use of this theoretical construct is the fact that the light used is coherent. Abbe failed to publish the theoretical description of the theory before his death, which slowed its acceptance. Some five years after his death his lecture notes were finally used to publish a formal presentation of the theory. It was the first use of the Fourier integral as the theoretical basis for optics. Until the 1920s most comments about Abbe's theory discounted the importance of coherent imaging with the then-available light sources. It was not until H. H. Hopkins [11] demonstrated that the degree of coherence is determined by the numerical aperture of the illuminating condenser, relative to that of the microscope objective, that the theory's general applicability to microscopy was established.

Abbe's theory assumes the object can be decomposed into a number of elemental gratings; each grating diffracts light at an angle that is a function of the grating's period and orientation. The diffracted light beams are plane waves that are focused by an imaging lens to diffraction patterns of light in the lens' back focal plane. The limiting aperture of the optical system determines the individual shapes of the diffraction patterns, as we know from applying the

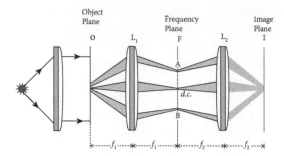

Figure 11.15 Two-lens imaging system. The first lens is a collimator that is to produce a plane wave that uniformly illuminates the object. In a microscope this would probably be Köhler illumination. The object is assumed to be a transparency containing the interference pattern given by Eq. (11.23). The object diffracts plane waves associated with the spatial frequency of the object and L_1 focuses the plane waves onto the frequency plane. A and B are images of the two pinholes used to produce the cosine interference pattern described by Eq. (11.22). The point labeled DC is the diffraction spot generated by the aperture in plane O containing Eq. (11.23) and is proportional to the light transmitted by the aperture.

array theorem. The diffraction patterns act as sources of waves that propagate from the focal plane to the image plane where they interfere to produce the image. This theory will be less confusing by the use of a set of optical components that are configured to allow us to separately study the focal plane and the image plane; see Figure 11.15.

Once Abbe's theory of imaging has been developed, a few applications to the field of signal processing will be discussed; all based on the premise of Abbe's theory that the Fourier transform of the object in the front focal plane is present in the back focal plane of an imaging lens. The applications to be discussed modify the Fourier transform of an object by:

(1) Changing its amplitude distribution,

(2) Changing its phase distribution, and

(3) Changing both the phase and the amplitude of the transform.

The objective of all three approaches is to modify the image produced by the optical system. The use of these three image modification techniques requires that the light is both temporally and spatially coherent. The demands on temporal coherence, made by the theory, are much less stringent than the demands on spatial coherence, as we will see in later discussions.

11.3.1 Abbe's theory of imaging

Rather than using a simple lens, as shown in Figure 11.10, to discuss Abbe's theory of imaging an extended object, we will use a "4f system"; see Figure 11.15.[13] The first element is an illumination subsystem that is not considered part of the 4f system. The design details of the illuminator depend on the light source but all source types have as their objective the production of uniform illumination. Rather than show a particular design we simply represent the ideal illuminator as a source and collimating lens that produces a uniform plane wave. To keep the system simple we use a transparency as the object. The object will have an amplitude transmission given by

---

[13] This generic system can be used to construct a microscope, telescope, or copy lens by modifying the focal lengths.

$$f(x,y) = \begin{cases} \dfrac{d}{2} + \dfrac{d}{2}\cos\omega_0 x & |x| \leq \dfrac{\ell}{2} \\[2mm] & |y| \leq \dfrac{\ell}{2} \\[2mm] 0 & \text{all other x and y} \end{cases}$$ (11.23)

The multiplicative factor, d, is known as the *amplitude modulation* of the grating. Because in optics we measure intensity rather than amplitude it is more useful to define object *contrast* from Eq. (11.20):

$$C = \frac{I_{max} - I_{min}}{I_{max} + I_{min}}.$$

For the amplitude object, Eq. (11.23), the object contrast is

$$C = \frac{2d}{1 + d^2} \approx 2d.$$

Note that the coherent contrast is different than the contrast we used for an intensity object. The extended object is located in the front focal plane of lens L_1 and the image of the object is located in the back focal plane of L_2. The image will undergo a magnification of $\mathcal{M} = f_2/f_1$. In Figure 11.15, the back focal plane of the lens, L_1, contains the Fourier transform of the object. This illustrates one of the two basic ideas contained in Abbe's theory. The light transmitted by the object is decomposed into a set of plane waves that are focused by a lens to sinc functions[14] in the back focal plane of the lens, L_1. We label the back focal plane of L_1 as the frequency plane. The location of each sinc function in the frequency plane is extracted from the definition of spatial frequency given in Chapter 9. It is a function of the angle the original diffracted plane wave makes with the optical axis as defined by the spatial frequencies (Eq. (9.24)):

$$\omega_x = -\frac{2\pi\xi}{\lambda f_1} \qquad \omega_y = -\frac{2\pi\eta}{\lambda f_1}$$ (11.24)

The second fundamental idea contained in Abbe's theory is that each amplitude function (a sinc function), in the back focal plane of the lens, acts as a source, producing a spherical wave. These spherical waves are transformed into plane waves by L_2. The plane waves will interfere at the image plane to produce an inverted image of the object.

A mathematical view of the process is that the lens L_1 produces a Fourier transform of the object in plane F, the back focal plane of L_1 called the frequency plane. The light distribution in the frequency plane acts as the source for lens L_2, which produces a Fourier transform of the frequency plane at the image plane, I. Thus, the field distribution in plane I is the Fourier transform of the Fourier transform of the object. This light distribution is the image of the object (of course the image is inverted so a minus sign will show up). We will assume that the two lenses, shown in Figure 11.15, have the same focal length, reducing the number of subscripts. The resulting the magnification of the system is unity, $\mathcal{M} = 1$.

The first question we want to address is what is the resolution limit of this coherent system. Answering this question is made easy by examining the simple sinusoidal grating, described

---

[14] We will assume the limiting aperture is the rectangular shape of the object given by Eq. (11.23).

by Eq. (11.23), in the object plane O. The source must have a spatial coherence length larger that the width and height defined by the sinusoidal grating. If the source meets the spatial coherence requirement, the amplitude distribution in plane F can be calculated using the Huygens–Fresnel integral; see Eq. (9.33) and the following,

$$\tilde{E}(x_0, y_0) = \frac{i\ell^2}{2\lambda f} e^{-2ikf} \operatorname{sinc}\left(\frac{\ell y_0}{\lambda f}\right)$$
$$\times \left\{ \sin\left(\frac{\ell x_0}{\lambda f}\right) + \frac{d}{2} \left[\operatorname{sinc} \frac{\ell}{\lambda f}\left(x_0 + \frac{\omega_0 f}{k}\right) + \operatorname{sinc} \frac{\ell}{\lambda f}\left(x_0 - \frac{\omega_0 f}{k}\right) \right] \right\}, \tag{11.25}$$

where f is the focal length of L_1. We have explicitly shown the second dimension but it can be ignored in our discussion. A schematic representation of the process described by Eq. (11.25) is shown in Figure 11.16 with the object at the left and its diffraction pattern located in plane F to the right of the lens. As predicted by Eq. (11.25), a light spot on the optical axis (often called the d.c.) and a spot right (A) and left (B) of the d.c. spot representing the frequency of the cosine object.

An added complication exists for coherent imaging systems. We not only have the intensity images of A and B, as we did in the incoherent case, but we now also have a light distribution due to interference between sources A and B (as required when the light is coherent). The intensity at some observation point, due to two point sources, was found in Chapter 4:

$$I_P = I_1 + I_2 + 2\sqrt{I_1 I_2}\, |\gamma_{12}|\cos\delta_{12}.$$

Calculating the square of the amplitude and ignoring the second dimension we can write the intensity distribution shown in Figure 11.16:

$$I(\xi) = 4a^2 I_0 \left\{ \operatorname{sinc}^2 \frac{ka}{S'}(\xi - \beta) + \operatorname{sinc}^2 \frac{ka}{S'}(\xi + \beta) + 2\gamma_{AB} \operatorname{sinc} \frac{ka}{S'}(\xi - \beta) \operatorname{sinc} \frac{ka}{S'}(\xi + \beta) \right\}. \tag{11.26}$$

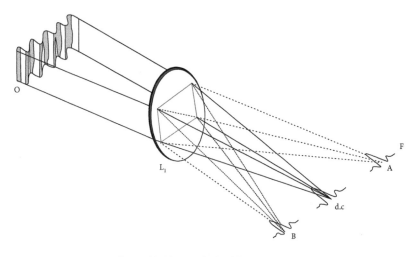

Figure 11.16 Fraunhofer diffraction pattern.

We now need to determine the resolution of the coherent imaging system in Figure 11.15 at plane I.

For complete coherence, the spatial coherence function is $\gamma_{AB} = 1$ and the image intensity is

$$I_I(\xi) = 4a^2 I_0 \left[\sin \frac{ka}{S'} (\xi - \beta) \sin \frac{ka}{S'} (\xi + \beta) \right]^2.$$

This implies that for coherent imaging

$$I_I(x) = \left[\int_{-\infty}^{\infty} E(\xi, t) \, \mathcal{R}(x - \xi) \, d\xi \right]^2. \tag{11.27}$$

The coherent system is linear in amplitude rather than intensity, as was the case for the incoherent system.

Rayleigh's criterion is inappropriate for coherent imaging. Figure 11.17 displays the calculated image of A and B separated by a distance equal to Rayleigh's criterion. The images, A′ and B′, are not resolved when objects A and B are in phase but A′ and B′ are easily resolved when objects A and B are out of phase.

Unlike Rayleigh's criterion, Sparrow's criterion can be used to obtain a minimum resolution for coherent imaging. We simply set the second derivative of Eq. (11.26) with respect to ξ equal to 0 to obtain the minimum angular resolution,

$$\sin \Psi = \frac{4.164}{ka} = 1.33 \frac{\lambda}{2a} \tag{11.28}$$

for a slit width of $2a$ and

$$\sin \Psi = \frac{4.60}{ka} = 1.46 \frac{\lambda}{2a} \tag{11.29}$$

for a circular aperture of radius a. Figure 11.18 shows the image of two, in-phase, coherent, point sources, separated by a distance equal to Sparrow's criterion.

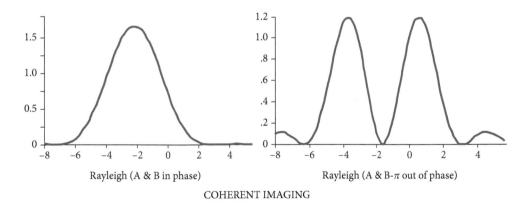

Rayleigh (A & B in phase)

Rayleigh (A & B-π out of phase)

COHERENT IMAGING

Figure 11.17 Application of the Rayleigh criterion with coherent sources A and B in Figure 11.1. The relative phase of the two sources determines whether resolution is possible, making this criterion useless for a coherent system.

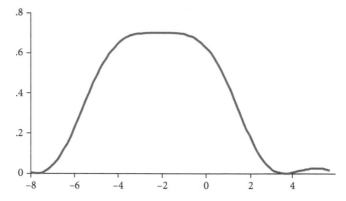

Figure 11.18 Resolution of two coherent sources in phase using Sparrow's criterion ($2b = 4.164$). If the sources are out of phase the discrimination between the two sources is better.

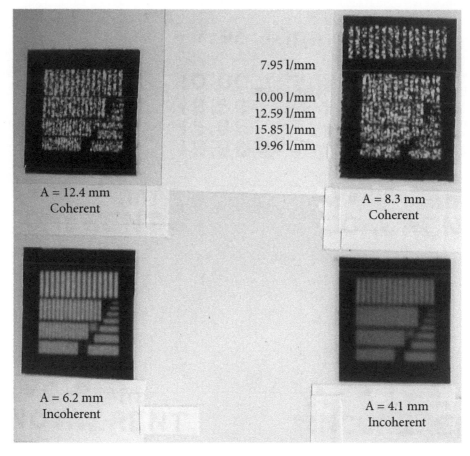

Figure 11.19 Images of a standard resolution chart containing rows of high contrast bars at different spatial frequencies. The top images were made using coherent light at apertures of 12 and 8 mm. The lower images were made with incoherent light at apertures of half the diameter used in the upper images, i.e., 6 and 4 mm. The smallest spatial frequency that can be detected in the coherent images is about the same as in the incoherent images. To produce the same resolution, the coherent system required an aperture twice the size required by the incoherent optical system. (Courtesy C. R. Christensen.)

Comparing Eqs. (11.28) and (11.29) to Eqs. (11.10) and (11.9) we see that the incoherent resolution for point objects is better than the coherent resolution. Figure 11.19 displays images of a standard resolution chart with the spatial frequencies associated with the various bars identified in the figure. Coherent images are made with two different apertures. Below each coherent image is shown an incoherent image with the same resolution. The incoherent image with a resolution equal to the coherent image resolution could be obtained with an aperture of about half the size of that used in the coherent image. These images of a standard resolution chart support the statement that the resolution of a coherent imaging system is less than that of an incoherent system.

Coherent light that travels different optical path lengths will interfere and produce an image with an intensity variation called speckle.[15] The intensity variations are due to interference effects for light traveling different optical paths that could be due to the atmospheric path and/or the surface roughness of the object. The speckle noise is often completely random and can be smoothed only by averaging statistically independent samples of the image, as is done in Figure 11.20.

Figure 11.20 shows the speckle noise effects on imaging using a test pattern consisting of a matrix of disks. The diameter of the disks in each row is identical but the diameters decrease by a factor of 2 as we move down the rows. The disks in each column have the same contrast but the contrast decreases from left to right. (Albert Rose developed this test pattern for the evaluation of photon noise in television systems [12].) The image at the lower right-hand portion of the figure, labeled $N = 1$, is immersed in fully developed speckle noise while the other images are created by recording multiple exposures (N denotes the number of exposures) of statistically independent images. The images contributing to the multiple exposed recordings differed only in their speckle statistics, and as N increases, the images become less coherent. The intensity averaging produced by the multiple exposures increases the signal-to-noise in the images, allowing small disks of lower contrast to be detected when N is large.

We introduced a transfer function for the incoherent imaging case. By analogy, we could define an amplitude transfer function for a coherent imaging system by applying the convolution theorem to

$$I_I(x) = \left[\int E(\xi, t) \mathcal{R}(x - \xi) \, d\xi \right]^2. \tag{11.30}$$

If we followed through the mathematical development, we would discover that the amplitude transfer function is the Fourier transform of the amplitude impulse response. The maximum spatial frequency that can be imaged by a coherent system is, therefore, one-half of the maximum spatial frequency imaged by the incoherent system. This result agrees with the images displayed in Figure 11.19. The contrast for a coherent imaging system does not decrease monotonically, as it does for an incoherent imaging system but instead remains equal to 1 up to the frequency cutoff. The coherent transfer function of a rectangular aperture is shown in Figure 11.21.

---

[15] The grainy intensity variation seen at the top of Figure 11.19 is called speckle noise and is due to interference between different wave fronts in coherent light. It is a random 2-dimensional distribution of fringes. In Figure 11.20 the effects of different amounts of speckle on resolution are shown.

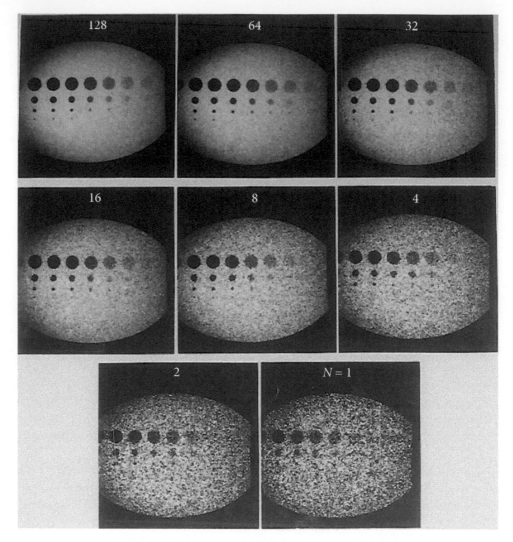

Figure 11.20 Images of a resolution chart developed by A. Rose have been produced using coherent light. The contrast of the chart elements decreases from left to right along each row and the size of the elements decreases from top to bottom along each column. The value of N shown on the images is equal to the number samples of the speckle field averaged to produce the image. The noise, most apparent in the images with low values of N, is called speckle and is a coherent noise that limits the imaging performance of a coherent optical system. The noise is due to interference between waves that have traveled over slightly different optical paths from the object to the image. The speckle can be reduced by statistically averaging a number of independent samples of the noise field, as was done in the multiple exposures shown here.

11.3.1.1 Coherency

The scale size of a diffraction pattern will change with the wavelength. If the temporal coherence of the source is small and a frequency spectrum of light is present, then the diffraction pattern formed by light on the blue end of the spectrum is smaller than the pattern formed by light on the red end, causing the sidebands for each wavelength to occur at different positions. Figure

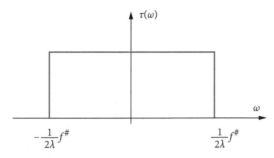

Figure 11.21 Coherent transfer function for a rectangular aperture.

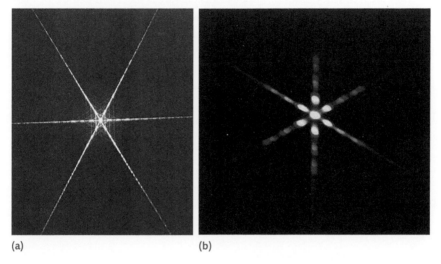

(a)   (b)

Figure 11.22 The Fraunhofer diffraction pattern produced by a nested set of triangles which form three sets of gratings oriented at 120° with respect to each other. (a) The illuminating light producing this diffraction pattern contained only a single wavelength. (b) The illuminating light was spatially coherent (small point source) white light. The spots are very broad due to the spread in wavelengths of the source. Note that the diffraction pattern of the red light is larger than the pattern for the blue light, as predicted by our equation.
(Nested triangles courtesy of N. George.)

11.22b shows the Fourier transform of a nested set of triangles, illuminated by a temporally incoherent light source. The spread in wavelengths present in the white light used to create Figure 11.22b causes the spots of light in the photograph to appear much larger than would be predicted by the size of the set of triangles. Figure 11.22a shows an identical Fraunhofer diffraction pattern from the nested set of triangles obtained using a single wavelength of light (488 nm in this case). Comparison of the spots of light in Figures 11.22a and 11.22b shows the effect of temporal coherence on the spots that make up the diffraction pattern.

We can establish the wavelength spread that would destroy the diffraction pattern observed in Figure 11.22b by assuming that the sinc functions for the two extremes in wavelength, λ_1 and λ_2, are separated by the width of the sinc function of λ_1. With this assumption, the allowable spread in wavelength is

$$\frac{\Delta\lambda}{\lambda} = \frac{\lambda_1 - \lambda_2}{\ell\omega_0} = \frac{4\pi^2}{\ell\omega_0}$$

$$= \frac{2\pi\,\Delta x}{\ell}. \tag{11.31}$$

This result states that as the resolution in the object increases (i.e., as the value of Δx decreases) or as the size of the object, ℓ, increases, the temporal coherence of the source must increase if the diffraction pattern is still to be observed.

11.3.1.2 Resolution

A logical question to ask is what would happen to the image of the object if we modified the light distribution in the frequency plane, F. Suppose an aperture is placed in plane F of Figure 11.15 with its symmetry axis aligned with the optical axis of the system. If the size of the aperture, denoted by D, were too small to allow the two sidebands of Eq. (11.25) to pass on to plane I, we would not see the sinusoidal grating at the image plane. Stated mathematically, if

$$D < \frac{2\omega_0 f}{k} + \frac{2\pi\lambda f}{\ell} = \frac{2f\lambda}{\Delta x} + \frac{2\pi\lambda f}{\ell},$$

then the optical system cannot resolve the grating. If D is less that this dimension, the light distribution in the image plane is a constant, equal to the Fourier transform of

$$\mathrm{sinc}\left(\frac{\ell y_0}{\lambda f}\right)\mathrm{sinc}\left(\frac{\ell x_0}{\lambda f}\right).$$

Any aperture less than the above dimension, but larger than a minimum dimension of

$$D = \frac{2\pi\lambda f}{\ell},$$

yields an image that is a uniform light distribution over a rectangle the size of the grating. If a small aperture equal to the minimum dimension could be moved up or down in the frequency plane so that it allowed either of the sidebands to pass, we would find minimal change in the appearance of the image plane. Because each of the three components is a simple sinc function, the light distribution in plane I, when any one of the three components is allowed to pass, is the image of the rectangular aperture with uniform illumination, i.e., a plane wave with limited extent. If the zero order is passed through, the plane wave will propagate parallel to the z-axis, whereas if a sideband is passed through, the plane wave will propagate at an angle with respect to the z-axis.

By neglecting the effect of a finite-sized grating, the minimum diameter required to resolve a grating with resolution Δx can be simplified to

$$D = \frac{2f\lambda}{\Delta x}. \tag{11.32}$$

This result should be compared to the diameter of an optical system needed to produce a Gaussian beam waist of w, calculated in Chapter 9,

$$D = \frac{2\lambda f}{w}.$$

The minimum resolution of an optical system is associated with the minimum spot size that the optical system can produce. This is the Rayleigh criterion we introduced in Figure 11.5 for a pinhole camera.

11.3.1.3 Oblique illumination

We now set the size of the aperture in plane F equal to the value

$$D = \frac{f\lambda}{\Delta x} + \frac{2\pi\lambda f}{\ell}$$

and move its symmetry axis above the optical axis, so that the zero order and one sideband are allowed to pass through to the image plane. This configuration reduces the number of diffraction patterns in the frequency plane to two. The two diffraction patterns, sinc functions in the example we are considering, act as point sources. The transition from the frequency plane to the image plane is equivalent to Young's experiment.

In the image plane, light from the two sources will interfere to produce a sinusoidal interference pattern. From Eq. (4.61) the period of this interference pattern is

$$T = \frac{\lambda D}{h},$$

yielding a spatial frequency of the sinusoidal interference pattern equal to

$$\omega = \frac{2\pi}{T} = \frac{kh}{D}.$$

In the example we are considering, the distance to the observation plane is given by $D = f$ and the point source separation is

$$h = \frac{\omega_0 f}{k};$$

thus, the spatial frequency of the interference pattern is ω_0, the frequency of the input object.

The peak intensities of the two sources in plane F are

$$I_1 = \left(\frac{\ell^2}{2\lambda f}\right)^2 \qquad I_2 = \left(\frac{\ell^2}{2\lambda f}\right)^2 \frac{d^2}{4}.$$

From Eq. (4.12) we can write the intensity distribution in the image plane, I, as

$$I = \left(\frac{\ell^2}{2lf}\right)\left[\left(1 + \frac{d^2}{4}\right) + d\cos\omega_0 x\right]. \tag{11.33}$$

From our definition of the contrast (or modulation) of the object above, $C_0 = d$, the contrast of the image is given by

$$C_I = \frac{d}{1 + \left(\frac{d}{2}\right)^2}.$$

The image of the object retains the object's spatial frequency content, but at a reduced contrast.

If we neglect the term associated with the size of the grating, we find that the aperture diameter required to resolve the grating is half the size of the aperture needed to pass both sidebands. (This is equivalent to single-sideband transmission in communication theory.) If the noise associated with the object is small enough to allow detection of the image at the reduced contrast, then the aperture (or equivalently, the spatial bandwidth) of the imaging system can be reduced by one-half with no loss in resolution.

Instead of displacing the aperture in the frequency plane above the optical axis, we could illuminate the object with a plane wave incident from below the optical axis at an angle, γ (see Figure 11.23):

$$\gamma = \frac{x_0}{f} = \frac{\omega_0}{2k} = \frac{\lambda}{2\Delta x}.$$

The off-axis illumination causes the diffraction pattern in the frequency plane to move upward, half the distance between the d.c. and first-order sideband:

$$x_0 = \frac{\omega_0 f}{2k}.$$

For the off-axis illumination shown in Figure 11.23, an aperture of size

$$D_\theta = \frac{f\lambda}{\Delta x},$$

centered upon the optical axis, will pass the d.c. component and the first-order sideband. We therefore find that by illuminating the object at an angle with respect to the optical axis, the imaging system is able to resolve the object grating with an aperture only one-half the size needed to resolve the same grating with illumination incident normal to the grating.

A microscope can be designed using the concept shown in Figure 11.23 and is said to operate with oblique illumination. Figure 11.24 show images of the surface of a leaf using three different types of illumination. Under each image is displayed the aperture used to illuminate the object. If the images were made in the system from Figure 11.15, the apertures would be found in the frequency plane.

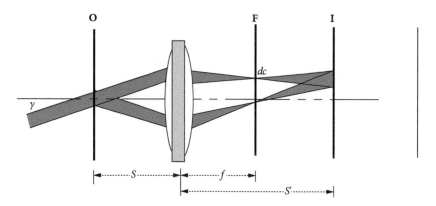

Figure 11.23 Off-axis illumination allows an aperture to provide the same resolution as an aperture twice its size operating with illumination parallel to the optical axis.

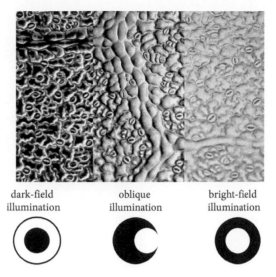

dark-field   oblique   bright-field
illumination   illumination   illumination

Figure 11.24 Comparison of images of a cast of the surface structure of a leaf showing both nerves and stomata. On the left, the dark-field image highlights the edged of the images associated with higher spatial frequencies. In the center, oblique illumination allows resolution of finer detail. On the right is bright field illumination that produces low contrast images. The bottom row of images shows the dark region in the aperture used in the Frequency plane to control spatial frequencies used to create the images in the top row. Graphic courtesy of Wim van Egmond (Copyright: © Wim van Egmond http://www.micropolitan.org)

11.3.1.4 Frequency filtering

A more complicated object than the sinusoidal one we have considered would contain many more frequencies but the analysis, just completed, would still apply, as demonstrated in Figure 11.24. An obstruction in the frequency plane, shown at the bottom of Figure 11.24, would prevent the lower frequencies from contributing to the image, resulting in a display of only the edges of the object. This so-called *dark-field illumination* microscopy technique uses a circular obstruction[16] to prevent the d.c. component from propagating past the frequency plane. The image generated, shown in Figure 24, is an edge-enhanced image. This technique permits the observation of objects that are transparent but scatter a small amount of light. The scattered light is at high spatial frequencies and misses the d.c. stop; the majority of the light, which is unscattered, is removed from the image by the obstruction.

An obstruction in the frequency plane can also restrict the maximum frequency that could pass through the imaging system, limiting the resolution of the resulting image, as shown on the right of Figure 11.24. Another example of low-pass spatial filtering of an image is shown in Figure 11.25.

It is possible to use the diffraction patterns, displayed in plane F, produced by different areas of a complex scene to rapidly sort aerial images, as is illustrated by the images contained in Figure 11.26. Of particular interest is the diffraction patterns produced by surface waves on

---

[16] The circular obstruction is sometimes called a *d.c. stop* because it is used to remove the d.c. component of the diffraction field.

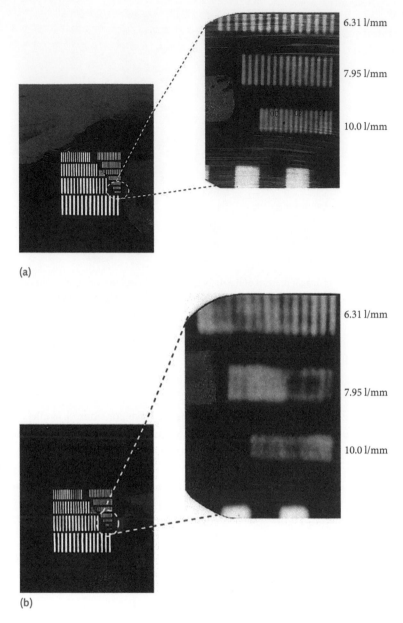

6.31 l/mm

7.95 l/mm

10.0 l/mm

(a)

6.31 l/mm

7.95 l/mm

10.0 l/mm

(b)

Figure 11.25 (a) A resolution chart imaged by an optical system with a maximum spatial frequency of 63 l/mm. (b) Optical low-pass filtering performed by introducing a circular aperture in the frequency plane of an imaging system similar to that shown in Figure 11.24. The resolution of the optical system has been reduced by about a factor of 10. Now we can just resolve 6.3 l/mm.

the ocean. From the orientation and scale of this diffraction pattern the ocean wave pattern can be analyzed.

A frequent application of amplitude spatial filtering is the use of a simple circular aperture to smooth the output intensity of a laser, for example, the illumination in Figure 11.15. The fundamental (TEM$_{00}$) mode of a laser, when a conventional Fabry–Perot resonator serves as the laser's feedback mechanism, is a Gaussian wave with amplitude

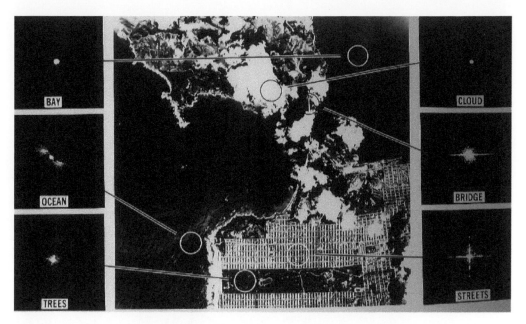

Figure 11.26 An aerial photograph illuminated by a laser is surrounded by Fraunhofer diffraction patterns taken of small areas in the photograph. The symmetry and extent of the diffraction pattern can be used to recognize geographical classes in the photograph. For example, the diffraction pattern of the ocean can be used to analyze the wave structure and direction on the ocean surface.

(Courtesy of Robert Leighty, U.S. Army Engineering Topographic Laboratory.)

$$E = Ae^{-\frac{x^2+y^2}{w^2}},$$

where w is the Gaussian beam width that was defined in Chapter 9; see Eq. (9.77b). This is an ideal representation of the beam. The actual laser beam departs from this simple theoretical distribution due to imperfections in the laser. We can treat the departures from the ideal beam by adding a correction, $E_r = E + \Delta E$, to the laser's light field. The fluctuations described by ΔE will create spatial frequencies at much higher values than a perfect Gaussian would contain. To construct a spatial filter that will remove the noise in the optical beam, we pass the laser beam through a microscope objective and place in the back focal plane of that lens, of focal length f, an aperture with a diameter of

$$D_{pinhole} = \frac{2.44\,(\lambda \cdot f)}{\rho_{beam}}$$

$$f = {160}/{\mathcal{M}},$$

(11.34)

where $\mathcal{M}$. is the magnification of the microscope objective,[17] ρ is the beam diameter, and f is the objective focal length. The fluctuations at high spatial frequencies will be eliminated by the pinhole and pass only an ideal beam.

---

[17] The magnification of the microscope objective (given as the power as, for example, 10×) depends on its focal length, f, and the tube length (distance between objective and eyepiece focal planes). For the DIN standard the tube length is 160 mm. There is also a JIS standard that is 170 mm and several microscope manufactures use their own length.

If we filtered a Gaussian beam by the filter, we would no longer have a Gaussian beam because we cut off the Gaussian outer edges. It is more useful to try to overfill the aperture of the microscope objective with a nearly plane wave. At the pinhole the microscope objective would generate a light distribution shown in Figure 9.32, the Airy pattern with a diameter given by Eq. (11.34). Conveniently the edge of the filter would fall in the dark region of the Airy pattern.

A typical experimental arrangement used to produce a plane wave is shown in Figure 11.27. A laser beam strikes a microscope objective from the left. The wave is focused down to a spot by the microscope objective and a pinhole is positioned so that it allows only the Airy disk to pass onto a second lens. As is shown schematically in Figure 11.27, the original beam's wavefront departs from a plane wave but the pinhole removes the higher spatial frequencies contained in the initial beam to produce a final beam with a plane wavefront. Figure 11.28 shows an intensity distribution over the beam wavefront before (on the left) and after spatial filtering. The focal length of the second lens is usually much larger than the microscope objective, in order to magnify the diameter of the beam.

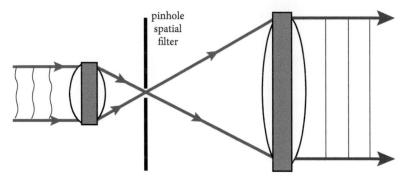

Figure 11.27 Spatial filter to produce a plane wave. A laser beam strikes a microscope objective from the left. The beam's wavefront departs from a plane wave, as shown by squiggles in the wavefronts in the figure. The wave is focused down to a spot by the microscope objective and a pinhole is positioned to allow only the d.c. component to pass on to a second lens. Often the focal length of the second lens is selected to produce a larger diameter beam of light.

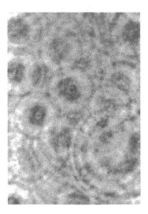 

Figure 11.28 A laser beam before spatial filtering is shown on the left and after spatial filtering on the right.

11.3.2 **Apodization**

The amplitude distribution of a wave, after passing through an aperture, has a central maximum but also has a number of secondary maxima, called sidelobes in radio frequency antenna design. It is desirable when operating an imaging system near its diffraction limiting resolution to reduce the sidelobes. A process known as apodization[18] can do this.

We have just discussed modifying a beam by removing spatial frequencies with an opaque screen in the spatial frequency plane. By modifying the diffracting aperture's edge, the secondary maxima, contained in the Fraunhofer diffraction pattern, can also be modified. A simple example will make this application of spatial filtering clear.

Assume the aperture of the optical system is a simple slit and ignore the y dependency:

$$f(x) = \begin{cases} 1 & -\frac{b}{2} \leq x \leq \frac{b}{2} \\ 0 & \text{all other } x \end{cases}.$$

The Fraunhofer diffraction pattern produced by this aperture is equal to the Fourier transform of the aperture:

$$F(\omega_x) = \int_{-\frac{b}{2}}^{\frac{b}{2}} e^{-i\omega_x x} dx$$

$$= \frac{b \sin \frac{\omega_x b}{2}}{\frac{\omega_x b}{2}}.$$

This result is identical to that obtained in Chapter 9; see Eq. (9.33). We now apodize the aperture by changing its transmission function to a cosine function

$$f(x) = \begin{cases} \cos \frac{\pi x}{b} & -\frac{b}{2} \leq x \leq \frac{b}{2} \\ 0 & \text{all other } x \end{cases}.$$

The Fraunhofer diffraction pattern produced by this aperture is

$$F_a(\omega_x) = \int_{-\frac{b}{2}}^{\frac{b}{2}} \cos \frac{\pi x}{b} e^{-i\omega_x x} dx$$

$$= \cos \frac{\omega_x b}{2} \left[\frac{1}{\omega_x + \frac{\pi}{b}} - \frac{1}{\omega_x - \frac{\pi}{b}} \right].$$

[18] Apodization, derived from the Greek word meaning foot, has as its objective the removal of the secondary diffraction maxima produced by an aperture with sharp boundaries.

The apodization smoothens the aperture's edge from a discontinuous change in transmission to a continuous change. The result of apodization is a diffraction amplitude given by

$$F_a(\omega_x) = \frac{2\pi}{b}\left[\frac{\cos\dfrac{\omega_x b}{2}}{\left(\dfrac{\pi}{b}\right)^2 - \omega_x}\right].$$

Comparing the amplitudes of the diffraction patterns for these two slit functions, we find that the amplitude on the optical axis is given by $E_a(0) = 2b/\pi$ and $E(0) = b$. Apodization has reduced the intensity, on the optical axis of the apodized aperture, to 41% of the intensity of the slit aperture.

If the widths of the central peaks of the diffraction patterns are compared, we find that the apodized central maximum is broader. The first minimum in the diffraction from a slit occurs when $\omega_x = 2\pi/b$. The first minimum in the apodized aperture's diffraction pattern occurs when $\omega_x = 3\pi/b$. The central spot in the diffraction pattern from the apodized aperture is wider than the central spot produced by the slit.

From the analysis so far, the advantage provided by apodization is not clear. The advantage provided by apodization becomes apparent when the intensity in the sidelobes of the diffraction patterns is compared. The first secondary maximum, in the diffraction pattern for the apodized aperture, occurs when $\omega_x b/2 = 2\pi$. The amplitude of the diffraction wave at this secondary maximum is $2b/15\pi$. The ratio of the secondary maximum intensity to the on-axis intensity for the apodized pattern is

$$\left[\frac{F_a\left(\dfrac{4\pi}{b}\right)}{F_a(0)}\right]^2 = \left[\frac{\dfrac{2b}{15\pi}}{\dfrac{2b}{\pi}}\right] = 0.0044.$$

When we calculate the same intensity ratio, for the diffraction pattern from a rectangular aperture, we obtain 0.047.

The advantage of apodization is now clear. The relative intensity of the first secondary maximum in the apodized aperture's diffraction pattern is a factor of 10 smaller than that found in the diffraction pattern of the rectangular aperture. A comparison of the two diffraction patterns is shown in Figure 11.29. The improvement in sidelobe intensity has come at the expense of on-axis intensity and central beam width. In photography apodization is used to produce the aesthetic quality of blur about the subject of the photograph called bokeh.

11.3.2.1 Phase and amplitude filtering

Apodization is also used in telescope optics to improve the dynamic range of the image. Coupled with phase filtering, this can be used to aid in imaging planets orbiting a star. Figure 11.30[19] was taken using a small, 1.5-m portion of the Palomar Observatory's Hale Telescope,

---

[19] The star in Figure 11.30 is called HR8799 and the planets are called HR8799 b, c, and d. The three planets are thought to be similar to Jupiter but in orbit to their host star at roughly 24, 38, and 68 times the distance between our Earth and Sun where Jupiter is 5 times the Earth–Sun distance.

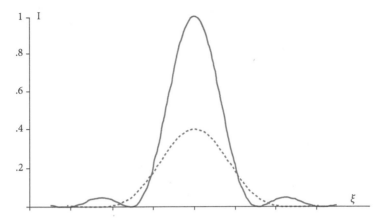

Figure 11.29 The solid curve is the diffraction pattern for a slit and the dotted curve is the diffraction pattern for a slit apodized by a cosine function.

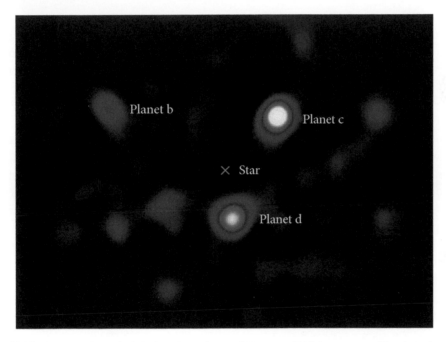

Figure 11.30 This image shows the light from three planets orbiting a star 120 light-years away. The planets' star is located at the spot marked with an "X."
By NASA/JPL-Caltech/Palomar Observatory—http://www.nasa.gov/topics/ universe/ features/ exoplanet20100414-a. html. Public Domain: https://commons. wikimedia. org/w/ index. php? curid=10016784.

north of San Diego, California. This is the first time a picture of planets beyond our solar system has been captured using a telescope with a modest-sized mirror—previous images were taken using larger telescopes.

Typical optical designs used to generate Figure 11.30 [13, 14] can be arranged to fit the discussion system in Figure 11.31. The apodization of the entrance pupil and an amplitude filter

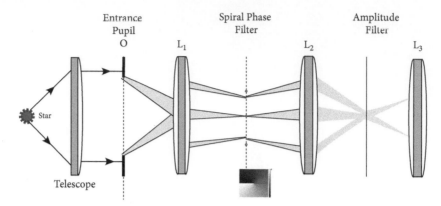

Figure 11.31 A general design of an optical vortex coronagraph. The entrance pupil and amplitude filter are designed to maximize contrast and minimize the star's intensity using apodizing techniques. The phase filter, one of a variety of types employed, is used to create destructive interference of the on-axis star.

designed to block light falling in the outer rim of the star image decrease the star's contribution to the image. The phase filter introduces destructive interference of the star's emission. The system is designed for three different wavelength regions to allow near white light performance.

11.3.2.2 Phase filtering

A second phase filtering technique, called the *Schlieren method* [15], replaces the circular obstruction of the dark-ground technique shown in Figure 11.24 with a knife edge that interrupts all spatial frequencies on one side of the d.c. component. This technique produces an intensity distribution proportional to the phase change produced by the object. The Schlieren method is used in aerodynamics to measure density changes associated with airflow (Figure 11.32).

A special case of the Schlieren method, *the Foucault knife-edge test*, is used to evaluate lens and mirrors. This technique brings a knife edge into the d.c. spot in the back focal plane of an optical component under test (see Figure 11.33). Aberrations in the optical system will cause the light from different portions of the optical element to be focused at different positions along the optical axis. Interrupting light at different positions along the optical axis can identify the portion of the lens producing these rays.

11.3.2.3 Phase contrast

An approach to microscopy that grew directly out of the Abbe theory was developed by *Fritz Zernike* (1888–1966) to generate images of microscopic organisms. Zernike received the Nobel Prize for this invention, called the *phase contrast microscope*. This microscope is now used extensively in the study of biological specimens.

Rather than removing the d.c. spot in the frequency plane, as is done in the dark-field microscope of Figure 11.24, Zernike shifted the phase of this portion of the spatial frequency spectrum by $\pi/2$ (a quarter-wave) relative to the phase of the rest of the spatial spectrum. The technique allows the observation of transparent objects that absorb very little light, by

Figure 11.32 A Schlieren image of supersonic air flow around a model of an X-15 aircraft. By NASA (Great Images in NASA Description) (public domain), via Wikimedia Commons.

converting phase changes introduced by the transparent object into intensity changes. The conversion from spatial phase variations to equivalent variations in the intensity is accomplished by placing a transparent plate in the frequency plane of the optical system with a quarter-wave coating placed over the low-frequency portion of the spatial spectrum.

To understand the Zernike phase contrast microscope, assume that the object absorbs no light but rather has a transmission function that is a spatial modulation of only the phase

$$f(x) = e^{i\phi(x)}.$$

We will also assume that the index of refraction is very nearly that of its surroundings. For a living organism in water this approximation is very good and allows the exponent to be approximated by the first two terms of its power series

$$f(x) \approx 1 + \phi(x).$$

The Fourier transform of this transmission function is of the form

$$E(\omega_x) = \int\limits_{-\infty}^{\infty} [1 + i\phi(x)] e^{-i\omega_x x} dx = \mathcal{F}\{1\} + \mathcal{F}\{\phi(x)\} = A(\omega_x) + i\Phi(\omega_x).$$

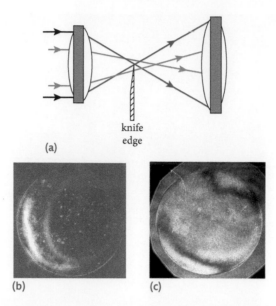

(a)

knife
edge

(b)   (c)

Figure 11.33 (a) Foucault knife-edge test. In the example shown, a knife edge is brought into the focus of the marginal rays of a lens with spherical aberration. Some of the paraxial rays are not interrupted, resulting in the observation of light over the lower portion of the inner segment of the lens image. (b) A knife-edge test of the mirror whose interferogram is shown in (c).

The functions $A(\omega)$ and $\Phi(\omega)$ have a relative phase difference of 90°, as indicated by the multiplicative factor i.[20] If the relative phase of the two functions can be retarded or advanced, we will demonstrate that it is possible to produce an image of the phase object.

We construct a spatial filter that modifies only the phase of the light, by the deposition of a quarter-wave dielectric layer on an optically flat plate. By using a flat plate of transparent material, which is made thicker near the optical axis, we can shift the phase of the low spatial frequencies by more than the high spatial frequencies. This phase plate is placed in the F plane, of Figure 11.15, which represents a microscope when the proper values for f_1 and f_2 are selected.

$A(\omega)$ is the diffraction pattern of the entire aperture and because there is no amplitude modulation in the aperture, it is non-zero only near $\omega = 0$; i.e., $A(\omega)$ contains only low spatial frequencies. The phase plate will only modify the phase of $A(\omega)$ because of the limited extent of the dielectric layer. The diffraction pattern, modified by the phase plate, is

$$E'(\omega_x) = A(\omega_x) e^{\pm i \frac{\pi}{2}} + i\Phi(\omega_x) = \pm i\left[A(\omega_x) \pm \Phi(\omega_x)\right].$$

(The positive and negative signs allow the use of a quarter-wave layer in the d.c. region of the spectrum (a positive phase plate) or a quarter-wave layer in the high-frequency region of the spatial spectrum (a negative phase plate) to obtain the relative phase shift.)

---

[20] $e^{\pm i(\pi/2)} = \pm i.$

The image at plane I, of Figure 11.15, is the inverse transform of the phase-modified signal. The inverse transform is given by $f' = \pm i(1 \pm \phi)$ and has an intensity distribution (neglecting ϕ^2) given by $1 \pm \phi(x)$. To the degree that the approximation of small spatial phase variation is correct, the phase filter provides an image whose variation in intensity is directly proportional to the variation in phase in the object.

The actual optical design is slightly more complicated than implied by the above discussion (see Figure 11.34). To increase the amount of available light and to increase the depth of field, which will eliminate the confusion of out-of-focus images, a source in the form of an annular ring is used. With an annular source, the phase plate must have a quarter-wave layer also in the form of a ring; see Figure 11.34b.

11.3.2.4 Phase and amplitude

The Fourier transform of a real object is a function of both amplitude and phase. Normally, we only operate on one of those functions. It is possible to record both the amplitude and phase of the Fourier transform by using holographic principles that we will discuss next. A. Vander Lugt was the first to demonstrate the creation of a complex Fourier transform by adding a reference beam to the frequency plane of Figure 11.15 [16]. Interference between the reference beam and the Fourier transform creates an intensity pattern that contains both amplitude and phase information. A complex Fourier transform of a car is shown in Figure 11.35. Because of limitations in the dynamic range of the recording medium, only a range of bandwidths can be recorded. Shown in Figure 11.35 are recordings made for two different spatial frequency bandwidths. One of the filters shown in Figure 11.35 can be used to identify the position of the car in a dynamic scene to allow tracking of the car in the scene. The very low frequencies are not recorded. That is an advantage because in most applications low-frequency information is not useful.

11.4 **Holography**

There is a special coherent imaging technique called holography. It allows us to use both the amplitude and phase of a coherent optical wave to make a 3-dimensional image of an object. The basic idea is rather simple. We start off with a beam from a monochromatic source. We split the beam into two parts. One we call the reference and the other illuminates the object we want to record, creating a scattered beam called the signal beam (Figure 11.36). We recombine the reference beam with the signal beam. Interference fringes are produced and recorded onto what will become the hologram.

Here is a crude math model with all the complexity hidden in a generalized notation. The signal beam and the reference beam consist of two waves with amplitudes and phases that vary in space:

Signal   Reference

$$a = a_0\left(x,y\right)e^{-i\phi(x,y)} \quad A = A_0\left(x,y\right)e^{-i\Psi(x,y)}.$$

The interference pattern that we record is the intensity recording given by

$$I(x,y) = AA^* + aa^* + A^*a + Aa^*$$
$$= \mid A(x,y)^2 \mid + \left|a(x,y)\right|^2 + 2A_0a_0\cos\left[\Psi(x,y) - \phi(x,y)\right].$$

(11.35)

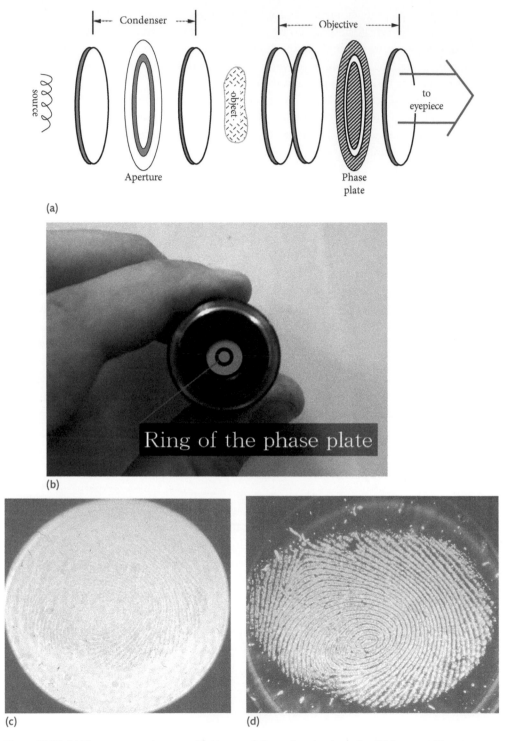

(a)

(b)

(c)

(d)

Figure 11.34 (a) Phase contrast microscope. The aperture is focused on the phase plate (b). Low spatial frequencies are given a quarter-wave phase shift relative to higher spatial frequencies. (c) A conventional image of a fingerprint produced by skin oils. (d) The fingerprint of (c) viewed through a phase contrast microscope.

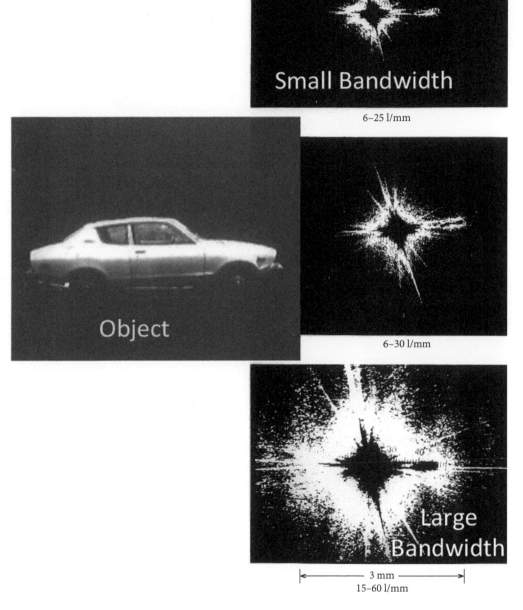

Small Bandwidth

6–25 l/mm

6–30 l/mm

Object

Large Bandwidth

|← 3 mm →|
15–60 l/mm

Figure 11.35 The recording of a Fourier transform of the image of a car. The recording of the transform is band limited. The functioning frequencies are those that are bright. The actual bandwidth of each transform is shown as l/mm.

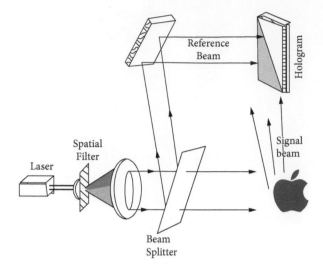

Figure 11.36 Schematic of the experimental creation of a holographic recording.

The phase difference between the two waves is recorded as a modulation of the average intensity of the two added waves.

Figure 11.37 is a picture of the recording process.

11.4.1 **Recording**

We assume a linear recording.[21] The average energy needed for the recording is

$$E_0 \approx \left(|A|^2 + |a|^2 \right) \tau_e,$$

where τ_e is the exposure time for the recording. After exposure and developing the transmission of the recording is the transmission coefficient

$$t_H(x, y) = t_0 + \beta \left(A^* a + A a^* \right)$$
$$\beta = \beta' \tau_e$$

The factor t_0 is the transmission averaged over the recording material and β' is the energy contained in the interference term. We can use Young's two-slit theory from Chapter 9, Eq. (9.46) to model the recording process (Figure 11.38). The two sources form spherical waves that the lens turns into two plane waves.

The angle the plane waves make with the optical axis is given by

$$\theta_1 = -{x_1}/{f} \quad \theta_2 = {x_2}/{f}.$$

---

[21] Linearity assumption is not important. Non-linear terms are simply refracted away from the signal beam at large angles.

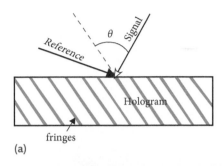

(a)

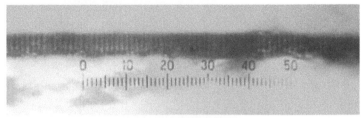

(b)

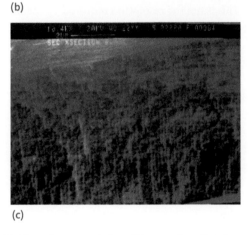

(c)

Figure 11.37 (a) Schematic representation of the recording of interference fringes for hologram. The fringes bisect the angle 2θ between the reference and signal beams. (b) An absorption hologram made at $2\theta < 10°$. The bright lines are layers of silver in a conventional developed film emulsion representing bright fringes. (c) A phase recording in a developed photopolymer at $2\theta > 30°$. The bright fringes become voids in the photopolymer after development. The use of an offset beam invented by E. N. Leith and J. Upatnicks [17] made the hologram a practical imaging method.

The transmission through each slit is A and a, connecting our discussion to Eq. (11.35). The intensity of the interference pattern in the back focal plane of the lens is

$$I_D = I_0 \left(\frac{\sin \alpha}{\alpha} \right)^2 \cos^2 \beta_h$$
$$= \frac{I_0}{2} (1 + \cos 2\beta_h).$$

The frequency of the fringes in the interference pattern is

$$\beta_h = \frac{k}{2} (x_1 + x_2) \frac{\xi}{f} = \frac{\pi h \xi}{\lambda f},$$

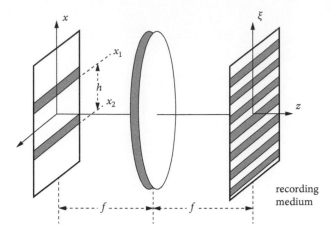

Figure 11.38 Simple hologram recording.

where h is the fringe spacing. The recording medium must be able to provide the resolution needed to record this fringe pattern. This is not the resolution required by the image but rather a much higher resolution required by the interference pattern that carries the image.

11.4.2 **Reconstruction**

We take whatever steps are needed to turn the light distribution into a recording and then illuminate it with a reconstruction (or playback) wave $B(x,y)$. The wave transmitted by the recording will have amplitude[22]

$$B(x,y)\, t_h(x,y) = t_0 B + \beta A^* B a + \beta A B a^*.$$

If we restrict the playback beam to be a reproduction of the reference beam, then $B = A$. We can propagate that same beam in the opposite direction to give $B = A^*$. The beam transmitted by the recording will be one of the two values

$$\beta |A|^2 a^*(x,y) \quad \beta |A|^2 a(x,y).$$

In the recording and playback we have shown the ability to recall an optical wave, both its amplitude and phase, from a stored memory (Figure 11.39).

We will try a different geometry to make a hologram (Figure 11.40). We start with a single point source at B. We will call the wave traveling to the recording point P over the distance t the

---

[22] We need to generate a physical interpretation of A or A^*. The two functions are

$$A = |A|\, e^{i(\omega t - kz)} \quad A^* = |A|\, e^{-i(\omega t - kz)},$$

where we assume the wave is restricted to traveling in the z-direction. If A moves to the right, then A^* travels in the reverse direction, to the left, and backward in time ($-t$). If we store a hologram of a wave traveling at an angle θ to the z-direction, then the two playback waves will have the temporal behavior of the playback beam but the spatial direction will be

$$A \propto e^{-ikx\sin\theta} \quad A^* \propto e^{-ik\sin(-\theta)}.$$

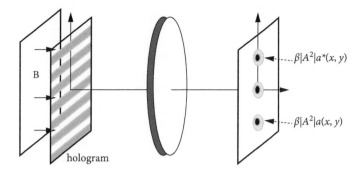

Figure 11.39 Playback of simple hologram. This displays the advantage for having the reference and signal beam at an angle. Unwanted beams occur at different angles such as the undiffracted beam t_0B on axis. Nonlinear recorded signals are not shown but would occur at angles beyond the angles for a and a^*.

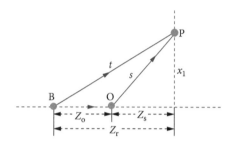

Figure 11.40 Simple model for recording holographic image.

reference beam. A second beam from B travels over a distance Z_0 is scattered by O and travels the distance s to the point P (Figure 11.40). This is the signal beam.

The phase difference between the signal and reference beam is

$$\delta = kt - k(Z_0 + s)$$
$$= k(Z_s - Z_r + t - s).$$

Using the binomial expansion we can write the phase difference as

$$\delta \approx k \left(Z_s - Z_r + Z_r + \frac{x_1^2}{2Z_r} Z_s - \frac{x_1^2}{2Z_s} \right)$$
$$\approx \frac{kx_1^2}{2} \left(\frac{1}{Z_r} - \frac{1}{Z_s} \right)$$

The recording behaves like a zone plate with a focal length, f:

$$\frac{1}{f} = \frac{1}{Z_r} - \frac{1}{Z_s}.$$

If the recorded fringes are square waves, then m is an integer identifying one of a number of foci. If we record a sinusoidal fringe pattern, then $m = 1$:

$$\frac{1}{f_\pm} = m\frac{\lambda_2}{\lambda_1}\left(\frac{1}{Z_r} - \frac{1}{Z_s}\right).$$

This equation allows for the playback wavelength to be different than the recording wavelength.

Figure 11.41 is a picture of the recording. We do not have to record the entire interference pattern. If we only record the shaded rectangle, then the recording will perform as shown in Figure 11.42. The two images correspond to the signal wave and its conjugate. The fact that the reference and signal beams were at an angle allows the separation of the two images.

A detailed study of the hologram recording process can lead to the use of more complicated recording geometries. Figure 11.43 is a display hologram made by using the real image of a hologram as the object of a second holgrahic recording. This proceedure reduced the coherency requirements on the playback illumination.

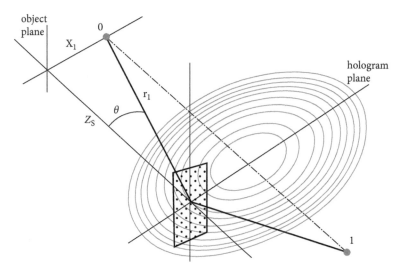

Figure 11.41 Fresnel zone picture of holography. The hologram recording is the rectangle filled with dots.

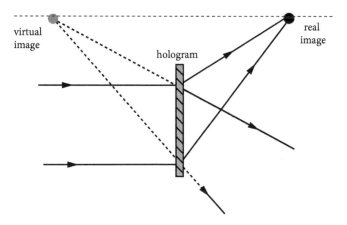

Figure 11.42 Real and virtual images produced by hologram.

Figure 11.43 Display hologram generated by using the real image of a hologram as the object in the recording of a reflective hologram. Sunlight was used as the playback illumination.

11.5 **Problem Set 11**

(1) What diameter telescope aperture is needed to resolve a double star with a separation of 10^8 km if the star is located 10 light-years from the Earth? (Assume $\lambda = 500$ nm.)

(2) A collimated beam ($\lambda = 600$ nm) is incident normally onto a 1.2-cm-diameter, 50-cm focal length converging lens. What is the size of the Airy disk?

(3) From Rayleigh's criterion, what would be the smallest angular separation between two equally bright stars that could be resolved by the 200-inch Hale telescope? Assume $\lambda = 550$ nm.

(4) The eye's maximum sensitivity is at 550 nm. If the eye's aperture is 2.2 mm and the distance from the aperture to the retina is 20 mm, then what is the diameter of the formed Airy disk?

(5) What is the Airy disk diameter in terms of the $f/\#$?

(6) What is the resolution of a 50-mm, $f/2$ lens that is diffraction limited?

(7) Which of the following systems are linear?

    1. $g(x) = e^{\pi} f(x)$

    2. $g(x) = f(x) + 1$

    3. $g(x) = x f(x)$

    4. $g(x) = \left[f(x) \right]^2$

(8) A system is time invariant if its output depends only on relative time of the input, not absolute time. To test if this quality exists for a system, delay the input by t_0. If the output shifts by the same amount, the system is time invariant:

- i.e., system output $f(t) \rightarrow g(t)$
- time invariant system response $f(t - t_o) \rightarrow g(t - t_o)$.

Is $f(t) \rightarrow f(at) \rightarrow g(t)$ (a pulse compressor) time invariant?

(9) Graphically calculate the convolution of the rectangular pulses: $\prod(x)$ with $\prod\left(^x/_2\right)$:

$$g(x) = \prod(x) * \prod\left(^x/_2\right) = \int \prod\left(^{x'}/_2\right) * \prod\left(-[x' - x]\right) dx'$$

We define the rectangular pulse as

$$\prod(x) = \begin{cases} 1 & |x| < 1 \\ 0 & \text{all other } x. \end{cases}$$

REFERENCES

1. Reynolds, G. O., et al., *New Physical Optics Notebook: Tutorials in Fourier Optics*, 1st edn. Bellingham, WA: SPIE-International Society for Optical Engineers, 1989.

2. Dirac, P. A. M., *The Principles of Quantum Mechanics*, 4th edn. International Series of Monographs on Physics. Oxford: Clarendon Press, 1958.

3. Loebich, C., et al., Digital camera resolution measurement using sinusoidal Siemens stars, in R. A. Martin, J. M. DiCarlo, and N. Sampat (Eds), *IS&T, SPIE Electronic Imaging Conference, San Jose, CA*, pp. 1–11. Bellingham, WA: SPIE, 2007.

4. Burns, P. D., Slanted-edge MTF for digital camera and scanner analysis, in *Proc. PICS Conf. 2000*, pp. 135–8. Springfield, VA: Society for Imaging Science and Technology, 2000.

5. Papert, S. A., The summer vision project, in *AIM-100.ps*. Artificial Intelligence Lab Publications, 1966.

6. Jordan, G., S. Raphael, and H. L. Sueur, *Tetrachromacy Project*, 2018. Available from: https://research.ncl.ac.uk/tetrachromacy/thescience/.

7. Derval, D., *Derval Color Test*, 2015. Available from: https://www.linkedin.com/pulse/25-people-have-4th-cone-see-colors-p-prof-diana-derval.

8. Keelan, B. W., *Handbook of Image Quality Characterization and Prediction*. New York: Marcel Dekker, 2002.

9. Sonawane, S., and A. M. Deshpande, Image quality assessment techniques: an overview. *International Journal of Engineering Research & Technology (IJERT)* **3**(4): 2013–17 (2014).

10. Duffleux, P. M., *The Fourier Integral and its Applications to Optics*, 2nd edn. New York: Wiley, 1983.

11. Hopkins, H. H., On the diffraction theory of optical images. *Proceedings of the Royal Society of London, Series A* **217**: 408–32 (1953).

12. Rose, A., *Vision: Human and Electronic*. Technology & Engineering. New York: Plenum Press, 1973.

13. Zimmerman, N. T., et al., Shaped pupil Lyot coronagraphs: high-contrast solutions for restricted focal planes. *J. Astron. Telesc. Instrum. Syst.* **2**(1) (2016).

14. Liu, C., et al., Design and experimental test of an optical vortex coronagraph. *Research in Astronomy and Astrophysics* **17**(6) (2017).

15. Settles, G. S., *Schlieren and Shadowgraph Techniques: Visualizing Phenomena in Transparent Media*, 2nd edn. Berlin: Springer-Verlag, 2001.

16. VanderLugt, A., Signal detection by complex spatial filtering. *IEEE Transactions on Information Theory* **10**: 139–45 (1964).

17. Leith, E. N., and J. Upatnieks, Wavefront reconstruction with continuous-tone objects. *J. Opt. Soc. Am.* **53**: 1377–81 (1963).

Index